# Groundwater Assessment and Modelling

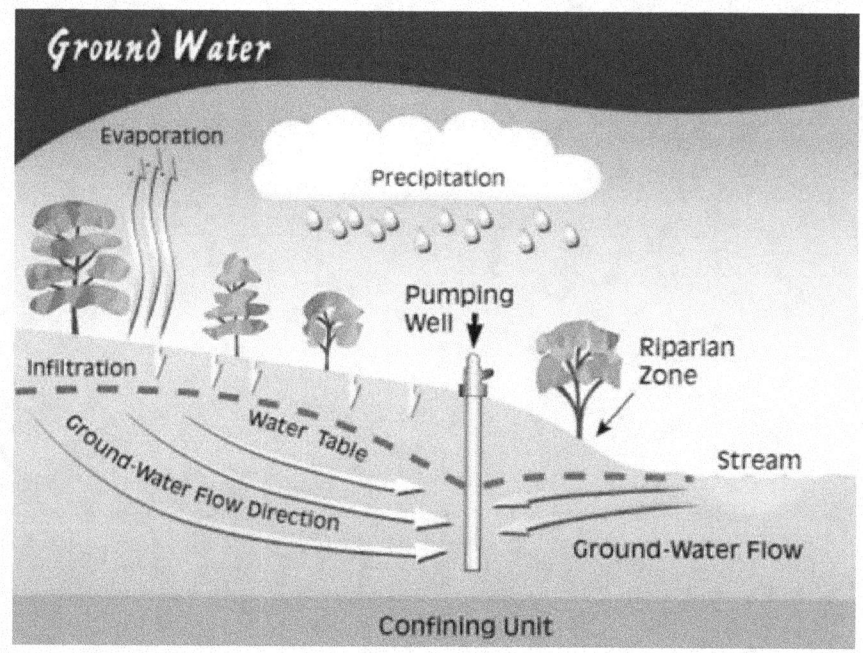

## C. P. Kumar

# Groundwater Assessment and Modelling

C. P. Kumar

# Preface

Groundwater development has shown phenomenal progress in our country during past few decades. There has been a vast improvement in the perception, outlook and significance of groundwater resource. Groundwater is a dynamic system. It is dynamic in the sense that the state of any hydrological system is changing with time, and in the sense that we are continually developing new scientific techniques to evaluate these systems.

The total annual replenishable groundwater resource of India is around 431 BCM. Inspite of the national scenario on the availability of groundwater being favourable, there are many areas in the country facing scarcity of water. This is because of the unplanned groundwater development resulting in fall of water levels, failure of wells, and salinity ingress in coastal areas. The development and over-exploitation of groundwater resources in certain parts of the country have raised the concern and need for judicious and scientific resource management and conservation.

A complexity of factors - hydrogeological, hydrological and climatological, control the groundwater occurrence and movement. The precise assessment of recharge and discharge is rather difficult, as no techniques are currently available for their direct measurements. Hence, the methods employed for groundwater resource estimation are all indirect. Groundwater being a dynamic and replenishable resource is generally estimated based on the component of annual recharge, which could be subjected to development by means of suitable groundwater structures.

Mathematical models are tools, which are frequently used in studying groundwater systems. In general, mathematical models are used to simulate (or to predict) the groundwater flow. Predictive simulations must be viewed as estimates, dependent upon the quality and uncertainty of the input data. Model conceptualization is the process in which data describing field conditions are assembled in a systematic way to describe groundwater flow processes at a site. The model conceptualization aids in determining the modelling approach and which model software to use.

Taking the base from my lecture notes delivered in various training courses during last 26 years and further editing and additions, I have developed this book titled *"Groundwater Assessment and Modelling"*. The book is intended to provide a comprehensive treatise related to assessment and modelling of groundwater. It includes chapters on assessment of groundwater potential, groundwater data requirement and analysis, basic concepts and guidelines for groundwater modelling, groundwater modelling software, modelling of unsaturated flow, modelling of sea water intrusion, and impact of climate change on groundwater resources.

I hope this book will be quite useful for undergraduate and postgraduate students (water resources engineering), field engineers and researchers working in the area of assessment, development and management of groundwater resources. Any comments or suggestions to improve the contents of book are welcome.

C. P. Kumar

# Contents

# 1-Water Resources of India

## Introduction

Water resources of a country constitute one of its vital assets. India receives annual precipitation of about 4000 km$^3$. The rainfall in India shows very high spatial and temporal variability and paradox of the situation is that Mousinram near Cherrapunji, which receives the highest rainfall in the world, also suffers from a shortage of water during the non-rainy season, almost every year. The total average annual flow per year for the Indian rivers is estimated as 1869 km$^3$. The total annual replenishable groundwater resources are assessed as 431 km$^3$. The annual utilizable surface water and groundwater resources of India are estimated as 690 km$^3$ and 396 km$^3$ per year respectively.

The surface water and groundwater resources of the country play a major role in agriculture, hydropower generation, livestock production, industrial activities, forestry, fisheries, navigation, recreational activities, etc. According to National Water Policy (2002) in the planning and operation of systems, water allocation priorities should be broadly as: (i) drinking water, (ii) irrigation, (iii) hydropower, (iv) ecology, (v) agro-industries and non-agricultural industries, and (vi) navigation.

Traditionally, India has been an agriculture-based economy. Hence, development of irrigation to increase agricultural production for making the country self-sustained and for poverty alleviation has been of crucial importance for the planners. Accordingly, the irrigation sector is assigned a very high priority in the 5-year plans. Giant schemes like the Bhakra Nangal, Hirakud, Damodar Valley, Nagarjunasagar, Rajasthan Canal project, etc. were taken up to increase irrigation potential and maximize agricultural production. Long-term planning has to account for the growth of population.

With rapid growing population and improving living standards, the pressure on our water resources is increasing and per capita availability of water resources is reducing day by day. Due to spatial and temporal variability in precipitation, the country faces the problem of flood and drought syndrome. Over-exploitation of groundwater is leading to reduction of low flows in the rivers, declining of the groundwater resources, and salt water intrusion in aquifers of the coastal areas. Over canal-irrigation in some of the command areas has resulted in waterlogging and salinity problem. The quality of surface water and groundwater resources is also deteriorating because of increasing pollutant loads from point and non-point sources. The climate change is expected to affect precipitation and water availability.

## Rainfall and River System

Most of the rainfall in India takes place under the influence of South West monsoon between June to September except in Tamil Nadu where it is under the influence of North-East monsoon during October and November. The rainfall in India shows great variations, unequal seasonal distribution, still more unequal geographical distribution and the frequent departures from the normal. It generally exceeds 1000 mm in areas to the East of Longitude 78 degree. It extends to 2500 mm along almost the entire West Coast and Western Ghats and over most of

Assam and Sub-Himalayan West Bengal. The Peninsula has large areas of rainfall less than 600 mm with pockets of even 500 mm.

India is a land of many rivers and mountains. Its geographical area of about 329 MHa is criss-crossed by a large number of small and big rivers, some of them figuring amongst the mighty rivers of the world. The rivers and mountains have a greater significance in the history of Indian cultural development, religious and spiritual life. It may not be an exaggeration to say that the rivers are the heart and soul of Indian life. India is a union of States with a federal set up. Politically, the country is divided into 28 States and 7 Union Territories. A major part of India's population is rural and agriculturally oriented for whom the rivers are the source of their prosperity.

India is blessed with many rivers. Twelve of them are classified as major rivers whose total catchment area is 252.8 million hectare (MHa). Of the major rivers, the Ganga - Brahmaputra Meghana system is the biggest with catchment area of about 110 MHa which is more than 43 percent of the catchment area of all the major rivers in the country. The other major rivers with catchment area more than 10 MHa are Indus (32.1 MHa), Godavari (31.3 MHa), Krishna (25.9 MHa) and Mahanadi (14.2 MHa). The catchment area of medium rivers is about 25 MHa and Subernarekha with 1.9 MHa catchment area is the largest river among the medium rivers in the country.

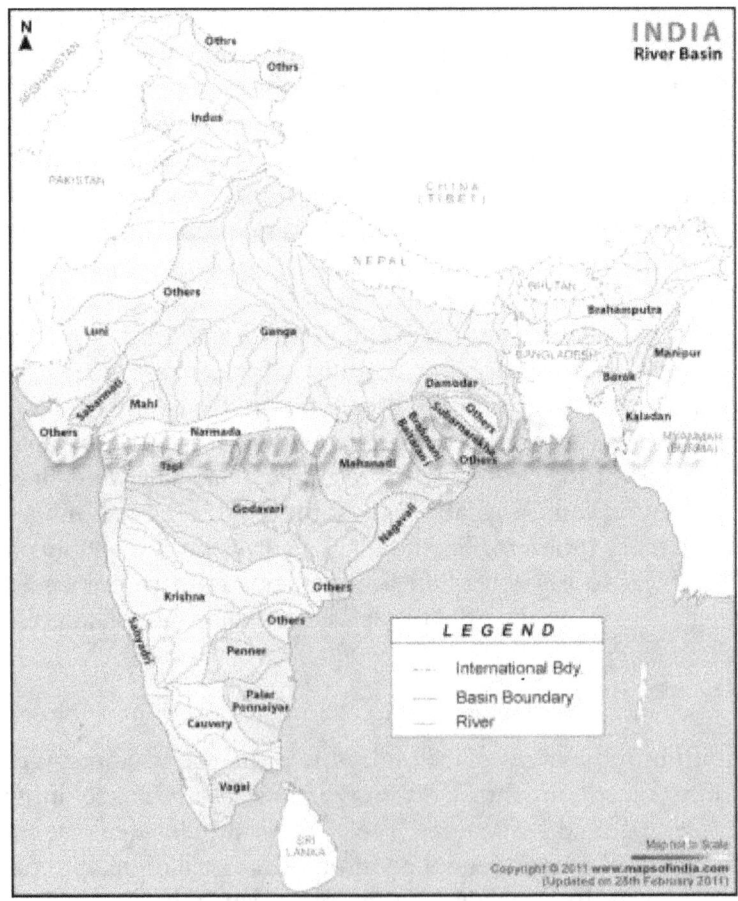

Average annual water availability in various river basins is given in the following table.

| S. No. | Name of the River Basin | Average annual availability (Cubic Km/Year) |
|---|---|---|
| 1. | Indus (up to Border) | 73.31 |
| 2. | a) Ganga | 525.02 |
| | b) Brahmaputra ,Barak & Others | 585.60 |
| 3. | Godavari | 110.54 |
| 4. | Krishna | 78.12 |
| 5. | Cauvery | 21.36 |
| 6. | Pennar | 6.32 |
| 7. | East Flowing Rivers Between Mahanadi & Pennar | 22.52 |
| 8. | East Flowing Rivers Between Pennar and Kanyakumari | 16.46 |
| 9. | Mahanadi | 66.88 |
| 10. | Brahmani & Baitarni | 28.48 |
| 11. | Subernarekha | 12.37 |
| 12. | Sabarmati | 3.81 |
| 13. | Mahi | 11.02 |
| 14. | West Flowing Rivers of Kutch, Sabarmati including Luni | 15.10 |
| 15. | Narmada | 45.64 |
| 16. | Tapi | 14.88 |
| 17. | West Flowing Rivers from Tapi to Tadri | 87.41 |
| 18. | West Flowing Rivers from Tadri to Kanyakumari | 113.53 |
| 19. | Area of Inland drainage in Rajasthan desert | NEGLIGIBLE |
| 20. | Minor River Basins draining into Bangladesh & Burma | 31.00 |
| | **Total** | **1869.35** |

## Man's Influence on Hydrological Cycle

The hydrological cycle is being modified quantitatively and qualitatively in most of the river basins of our country as a result of the developmental activities such as construction of dams and reservoirs, land use change, irrigation, etc. Such human activities affecting the hydrological regime can be classified into four major groups: (i) activities which affect river runoff by diverting water from rivers, lakes, and reservoirs or by groundwater extraction, (ii) activities modifying the river channels, e.g. construction of reservoirs and ponds, levees and river training, channel dredging, etc. (iii) activities due to which runoff and other water balance components are modified due to impacts of basin surface e.g. agricultural practices, drainage of swamps, afforestation or deforestation, urbanization, etc. and (iv) activities which may induce climate changes at regional or global scale, e.g. modifying the composition of atmosphere by increasing the 'greenhouse' gases or by increased evaporation caused by large scale water projects. For understanding these effects appropriately, hydrological modelling approaches have to be adopted.

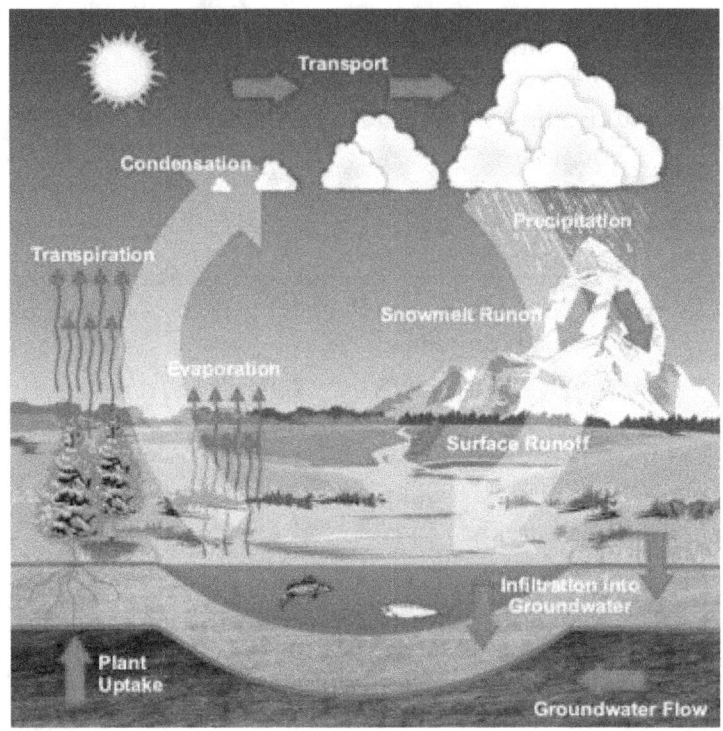

**Water Resources Management in India**

In view of the existing status of water resources and increasing demands of water for meeting the requirements of the rapidly growing population of the country as well as the problems that are likely to arise in future, a holistic, well planned long-term strategy is needed for sustainable water resources management in India. The water resources management practices may be based on increasing the water supply and managing the water demand under the stressed water availability conditions. Data monitoring, processing, storage, retrieval and dissemination constitute the very important aspects of the water resources management. These data may be utilized not only for management but also for the planning and design of the water resources structures. In addition to these, now-a-days decision support systems are being developed for providing the necessary inputs to the decision makers for water resources management. Also, knowledge sharing, people's participation, mass communication and capacity building are essential for effective water resources management. Some important aspects of such strategies are described as follows.

**Drought and its Management**

The Irrigation Commission, 1972 has identified 67 drought prone districts comprising of 326 Talukas located in 8 States having an area of 49.73 MHa. Subsequently, the National Commission on Agriculture, 1976 identified few more drought prone areas with a slightly different criteria. The erstwhile Drought Area Study and Investigation Organisation of

C.W.C. setup in 1978 started with 99 districts after considering the list of districts identified by the Irrigation Commission and also by the National Commission on Agriculture for carrying out further studies. For the studies, Central Water Commission adopted the same criteria as followed by the Irrigation Commission, 1972 i.e. drought is a situation occurring in an area:

- When the annual rainfall is less than 75% of the normal in 20% of the years examined.
- Less than 30% of the cultivated area is irrigated.

Central Water Commission adopted a smaller unit viz. Talukas for drought identification studies instead of districts and therefore, number of drought affected Talukas were identified as 315 out of a total of 725 Talukas in 99 districts. Accordingly, out of 108 MHa area of 99 districts, only 51.12 MHa spread over 74 districts have been considered as drought districts. Thus, in comparison to total geographical area of the country (329 MHa), about 1/6th is drought prone.

Irrigation has proved to be the most effective drought proofing mechanism and single biggest factor in bringing about a large measure of stability in agricultural production. The total geographical area of the drought districts is 108 MHa out of which 81 MHa is culturable (75%), gross sown area is 61.9 MHa (57.4%) and the gross irrigated area is 14.3 MHa. About 23.23% of the total cropped area is irrigated in the drought districts as against an all India average of 30.15%.

The planning and management of the effects of drought appear to have a low priority due to associated randomness and uncertainty in defining the start and end of droughts. Further, most of the drought planning and management schemes are generally launched after persisting drought conditions. The traditional system of drought monitoring and estimating losses by crop cutting needs replacement with real time remote sensing, GIS, GPS and modelling techniques for ensuring transparency and quick response. Scope of losses may be extended to groundwater depletion, damage to perennial trees, plantations, orchards and depletion in fertility of livestock. Food, fodder, agricultural inputs and water banks may be established in vulnerable areas instead of their storage in surplus regions to avoid transport bottlenecks during the drought. Robust and rainfall independent off-farm livelihood opportunities may be targeted in the drought mitigation strategy. Conjunctive use of surface water and groundwater, aquifer recharge and watershed management with community participation is another important policy paradigm shift to be internalized fully. The following approaches can be adopted to minimize the effect of drought.

- ➢ Larger thrust for watershed development under Drought Prone Area Program.

- ➢ Dry land farming and water resource development. Instead of capital intensive engineering works for soil and moisture conservation, encourage simple and low cost structures which can be completed in a short time with the help of local skills.

- ➢ Dovetail crop production activities into the watershed project along with soil conservation activities.

- ➢ Take up large scale dry land farming demonstrations.

- ➢ Limited irrational and over exploitation of groundwater resources.

- ➢ Undertake research on the efficacy and economics of sprinkler and drip irrigation systems.

- ➢ Construction of suitable water harvesting structures for the purpose of conservation and optimal use of surface water and recharge of underground aquifers.

- ➢ Renovation and restoration of old tanks/farm ponds in the villages.

- ➢ Afforestation and pasture development.

- ➢ Animal husbandry and fodder development.

- ➢ People's participation in drought proofing.

**Floods and its Management**

Floods occur in almost all rivers basins of the country. Heavy rainfall, inadequate capacity of rivers to carry the high flood discharge, inadequate drainage to carry away the rainwater quickly to streams/rivers are the main causes of floods. Ice jams or land slides blocking streams; typhoons and cyclones also cause floods. Excessive rainfall combined with inadequate carrying capacity of streams resulting in over spilling of banks is the cause for flooding in majority of cases.

Among all natural disasters, floods are the most frequent in India. Floods in the eastern part of India, viz. Orissa, West Bengal, Bihar and Andhra Pradesh in the recent past, are striking examples. According to the information published by different government agencies, the tangible and intangible losses due to floods in India are increasing at alarming rate.

Rashtriya Barh Ayog (RBA) constituted by the Government of India in 1976 carried out an extensive analysis to estimate the flood-affected area in the country. RBA in its report in 1980 has assessed the area liable to floods as 40 million hectares. It was determined by summing up the maximum area affected by floods in any one year in each state during the period from 1953 to 1978 for which data was analyzed by the Ayog. This sum has been corrected for the area that was provided with protection at that time and for the protected area that got affected due to failure of protection works during the period under analysis to arrive at the total area liable to floods in the country as per break-up given below.

| State | Area liable to Floods (Million Ha.) |
|---|---|
| 1. Andhra Pradesh | 1.39 |
| 2. Assam | 3.15 |
| 3. Bihar | 4.26 |
| 4. Gujarat | 1.39 |
| 5. Haryana | 2.35 |
| 6. Himachal Pradesh | 0.23 |
| 7. Jammu & Kashmir | 0.08 |
| 8. Karnataka | 0.02 |
| 9. Kerala | 0.87 |
| 10. Madhya Pradesh | 0.26 |
| 11. Maharashtra | 0.23 |
| 12. Manipur | 0.08 |
| 13. Meghalaya | 0.02 |
| 14. Orissa | 1.40 |
| 15. Punjab | 3.70 |
| 16. Rajasthan | 3.26 |
| 17. Tamil Nadu | 0.45 |
| 18. Tripura | 0.33 |
| 19. Uttar Pradesh | 7.336 |
| 20. West Bengal | 2.65 |
| 21. Delhi | 0.05 |
| 22. Pondichery | 0.01 |
| **Total** | **33.516** |

Floods being a natural phenomena, total elimination or control of floods is neither practically possible nor economically viable. Hence, flood management aims at providing a reasonable degree of protection against flood damage at economic costs.

In India, systematic planning for flood management commenced with the Five Year Plans, particularly with the launching of National Program of Flood Management in 1954. During the last few decades, different methods of flood protection, structural as well as non-structural, have been adopted in different states depending upon the nature of the problem and local conditions. Structural measures include storage reservoirs, flood embankments, drainage channels, anti-erosion works, channel improvement works, detention basins etc. and non-structural measures include flood forecasting, flood plain zoning, flood proofing, disaster preparedness etc.

Flood forecasting has been recognized as one of the most important, reliable and cost-effective non-structural measures for flood management. Recognizing the crucial role it can play, Central Water Commission, Ministry of Water Resources has set up a network of forecasting stations covering all important flood prone inter-state rivers. The forecasts issued by these stations are used to alert the public and to enable the administrative and engineering agencies of the States/UT's to take appropriate measures. Central Water Commission started flood-forecasting services in 1958 with the setting up of its first forecasting station on Yamuna at Delhi Railway Bridge.

The National Flood Control Program was launched in 1954 for the first time, in the country. Since then, sizeable progress has been made in the flood protection measures. Nearly one third of the flood prone area had been afforded reasonable protection by 1976. During the period of two decades, considerable experience had also been gained in planning, implementation and performance of the flood protection and control measures. Advancement in technology had taken place not only in India but also in the world over. It, therefore, became necessary to conduct an in depth study of the country's approach and programs of flood control measures and formulation of a flood control policy. It required detailed study of various problems concerned with flood control measures and aspects like soil conservation and afforestation. The Government of India, therefore, decided to set up the Rashtriya Barh Ayog (National Flood Commission) in 1976 to evolve a coordinated, integrated and scientific approach to the flood control problems in the country and to draw out a national plan fixing priorities for implementation in the future.

Though the RBA report was submitted in 1980 and accepted by Government, not much progress has been made in the implementation of its recommendations. As such, Ministry of Water Resources set up an Expert Committee in 2001 to review the implementation of recommendations of Rashtriya Barh Ayog (RBA). The Committee which was to identify the bottlenecks faced by the State Governments in its implementation and to examine and suggest measures for implementation of RBA recommendations for effective flood management in the country, has since submitted its report. The report of the committee is under examination in the Ministry of Water Resources.

**Groundwater Resources**

The behavior of groundwater in the Indian sub-continent is highly complicated due to the occurrence of diversified geological formations with considerable lithological and chronological variations, complex tectonic framework, climatological dissimilarities and various hydro-chemical conditions. Studies carried out over the years have revealed that aquifer groups in alluvial / soft rocks even transcend the surface basin boundaries. Broadly, two groups of rock formations have been identified depending on characteristically different hydraulics of groundwater, viz. porous formations and fissured formations.

The groundwater resources of the country have been estimated for freshwater based on the guidelines and recommendations of the GEC-1997. The total annual replenishable groundwater resources of the country have been estimated as 431 billion cubic meter (BCM) or km$^3$. Keeping 35 BCM for natural discharge, the net annual groundwater availability for the entire country is 396 BCM. The annual groundwater draft is 243 BCM out of which 221 BCM is for irrigation use and 22 BCM is for domestic and industrial use.

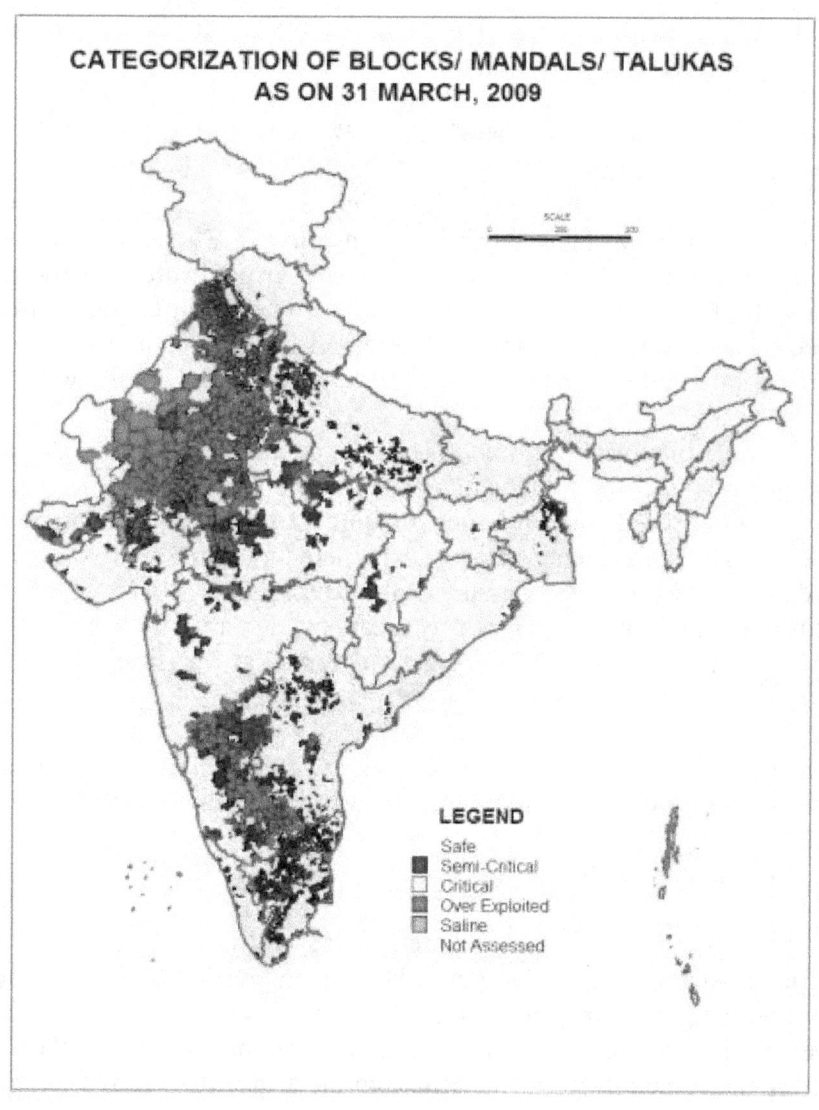

9

The stage of groundwater development in the country is 61%. The development of groundwater in different areas of the country has not been uniform. Highly intensive development of groundwater in certain areas in the country has resulted in over exploitation leading to decline in the levels of groundwater and sea water intrusion in coastal areas. Out of 5842 numbers of assessment administrative units (Blocks/Taluks/Mandals/Watershed), 802 units are "overexploited", 169 units are "critical", 523 units are "semi-critical", 4277 units are "safe" and 71 units are "saline".

## Groundwater Management

Groundwater resources occur in dynamic state and hence subjected to periodic changes. In order to augment the depleting groundwater resources, it is essential that the surplus monsoon runoff that flows into the sea, is conserved and recharged to augment groundwater resources. Groundwater storage that could be feasible has been estimated as 214 billion cubic meters (BCM) of which 160 BCM is considered retrievable. Central Groundwater Board has prepared a conceptual plan for artificial recharge to groundwater for the country. Out of total geographical area of 3,28,7263 sq. km. of the country, an area of 4,48,760 sq. km. has been identified suitable for artificial recharge. The total quantity of surplus monsoon runoff to be recharged works out as 36.4 BCM.

The Central Groundwater Board has prepared a Manual and subsequently a Guide on Artificial Recharge to Groundwater which provides guidelines on investigation techniques for selection of sites, planning & design of artificial recharge structures, economic evaluation and monitoring of recharge facility. These are of immense use to States/ U.T.s in planning and implementation of recharge schemes for augmentation of groundwater in various parts of the country. Groundwater Management Studies (GWMS) are essential to update the scenario of groundwater occurrence, availability and utilization in terms of quality and quantity with reference to the previous studies. The objectives of the GWMS are as follows:

1.  To depict the groundwater regime in terms of quantity and quality as on the date.
2.  Ascertaining the factors influencing the groundwater scenario.
3.  Identification of problems and issues pertaining to groundwater and provide suitable object oriented management strategy for implementation.
4.  To update the existing database on groundwater regime.
5.  To demarcate the groundwater worthy areas.
6.  To recommend suitable follow up action/ remedial measures/ administrative and technical measures for the specific problems.

Priority is given to "Over-Exploited" and critical areas, hard rock area, coastal area, drought prone area, naturally contaminated area, urban area, water logged area etc.

## Conjunctive Use of Surface Water and Groundwater

Large canal infrastructure network for providing irrigation has been the prime goal of the Government of India, since the first five-year plan, which continued up to seventh five-year plan. In some of the irrigation project commands such as Sarda Sahayak in UP, Gandak in Bihar, Chambal in Rajasthan, Nagarjuna Sagar in Andhra Pradesh, Ghataprabha and Malaprabha in Karnataka etc., problems of waterlogging are being faced. The main reason for excessive use of surface water as compared to groundwater is its much lower price for irrigation as compared to the cost incurred in using groundwater.

Waterlogging problems could be overcome if conjunctive use of surface water and groundwater is made. Groundwater utilization for irrigation in waterlogged areas can help to lower the groundwater table and reclaim the affected soil. Over-exploitation of groundwater in areas like Mehsana in Gujarat; parts of Meeurt and Varanasi districts in Uttar Pradesh, Coimbatore in Tamil Nadu and Karnal district in Haryana etc. have resulted in mining of groundwater. Many research workers have focused the causes of waterlogging. Several groundwater flow modelling studies have focused on assessing the waterlogged areas and measures to control problems of waterlogging and salinization. It is desirable that the irrigation needs for fulfilling crop water requirements should be satisfied by judicious utilization of available canal water in conjunction with groundwater so as to keep the water table within the acceptable range. Thus, the optimal conjunctive use of the region's surface and groundwater resources would help in minimizing the problems of waterlogging and groundwater mining.

**Water Conservation and Rainwater Harvesting**

Water conservation implies improving the availability of water through augmentation by means of storage of water in surface reservoirs, tanks, soil and groundwater zone. It emphasizes the need to modify the space and time availability of water to meet the demands. This concept also highlights the need for judicious use of water. There is a great potential for better conservation and management of water resources in its various uses. On the demand side, a variety of economic, administrative and community-based measures can help conserve water. Also, it is necessary to control the growth of population since large population is putting massive stress on all natural resources.

Since agriculture accounts for about 69% of all water withdrawn, the greatest potential for conservation lies in increasing irrigation efficiencies. Just a 10% improvement in irrigation efficiency could conserve enough water to double the amount available for drinking. In India, sprinkler irrigation is being adopted in Haryana, Rajasthan, Uttar Pradesh, Karnataka, Gujarat and Maharashtra. The use of sprinkler irrigation saves about 56% of water for the winter crops of bajra and jowar, while for cotton, the saving is about 30% as compared to the traditional gravity irrigation. An important supplement to conservation is to minimize the wastage of water. In urban water supply, for example, almost 30% of the water is wasted due to leakages, carelessness, etc. while most metro cities face deficit in supply of water. It is, therefore, imperative to prevent wastage. In industries also, there is a scope for economy in the use of water. Prices of water for all uses should be fixed, keeping in mind its economic value, control of wastage, and the ability of users to pay. As water is becoming scarcer, pricing will be an important factor in avoiding wastage and ensuring optimal use.

Rainwater harvesting is the process to capture and store rainfall for its efficient utilization and conservation to control its runoff, evaporation and seepage. Some of the benefits of rainwater harvesting are:

> ➢ It increases water availability
> ➢ It checks the declining water table

> It is environmentally friendly
> It improves the quality of groundwater through dilution, mainly of fluoride, nitrate, and salinity, and
> It prevents soil erosion and flooding, especially in the urban areas.

Even in ancient days, people were familiar with the methods of conservation of rainwater and had practised them with success. Different methods of rainwater harvesting were developed to suit the geographical and meteorological conditions of the region in various parts of the country. Traditional rainwater harvesting, which is still prevalent in rural areas, is done by using surface storage bodies like lakes, ponds, irrigation tanks, temple tanks, etc. For example, Kul (diversion channels) irrigation system which carries water from glaciers to villages is practised in the Spiti area of Himachal Pradesh. In the arid regions of Rajasthan, rainwater harvesting structures locally known as Kund (a covered underground tank), are constructed near the house or a village to tackle drinking water problem. In Meghalaya, Bamboo rainwater harvesting for tapping of stream and spring water through bamboo pipes to irrigate plantations is widely prevalent. The system is so perfected that about 18–20 litres of water entering the bamboo pipe system per minute is transported over several hundred meters.

There is a need to recharge aquifers and conserve rainwater through water harvesting structures. In urban areas, rainwater will have to be harvested using rooftops and open spaces. Harvesting rainwater not only reduces the possibility of flooding, but also decreases the community's dependence on groundwater for domestic uses. Apart from bridging the demand–supply gap, recharging improves the quality of groundwater, raises the water table in wells/bore-wells and prevents flooding and choking of drains. One can also save energy to pump groundwater as water table rises. These days rainwater harvesting is being taken up on a massive scale in many states in India. Substantial benefits of rainwater harvesting exist in urban areas as water demand has already outstripped supply in most of the cities.

## Water Quality Conservation

India is endowed with diverse geological formations from oldest achaeans to recent alluviums and characterized by varying climatic conditions in different parts of the country. The natural chemical content of groundwater is influenced by depth of the soils and sub-surface geological formations through which groundwater remains in contact. In general, greater part of the country, groundwater is of good quality and suitable for drinking, agricultural or industrial purposes. Groundwater in shallow aquifers is generally suitable for use for different purposes and is mainly of Calcium Bicarbonate and mixed type. However, other types of water are also available including Sodium-Chloride water. The quality in deeper aquifers also varies from place to place and is generally found suitable for common uses. There is salinity problem in the coastal tracts and high incidence of fluoride, arsenic, iron & heavy metals etc. in isolated pockets have also been reported. The main groundwater quality problems in India are as follows.

## Salinity

Salinity in groundwater can be of broadly categorized into two types, i.e. inland salinity and coastal salinity

### Inland Salinity

Inland salinity in groundwater is prevalent mainly in the arid and semi-arid regions of Rajasthan, Haryana, Punjab and Gujarat, Uttar Pradesh, Delhi, Andhra Pradesh, Maharashtra, Karnataka and Tamil Nadu. There are several places in Rajasthan and southern Haryana where EC values of groundwater is quite high making water non-potable. In some areas of Rajasthan and Gujarat, groundwater salinity is so high that the well water is directly used for salt manufacturing by solar evaporation. Inland salinity is also caused due to practice of surface water irrigation without consideration of groundwater status. The gradual rise of groundwater levels with time has resulted in water logging and heavy evaporation in semi-arid regions lead to salinity problem in command areas.

### Coastal Salinity

The Indian subcontinent has a dynamic coast line of about 7500 km length. It stretches from Rann of Kutch in Gujarat to Konkan and Malabar coast to Kanyakumari in the south to northwards along the Coromandel coast to Sunderbans in West Bengal. The western coast is characterized by wide continental shelf and is marked by backwaters and mud flats while the eastern coast has a narrow continental shelf and is characterized by deltaic and estuarine land forms. Groundwater in coastal areas occurs under unconfined to confined conditions in a wide range of unconsolidated and consolidated formations. Normally, saline water bodies owe their origin to entrapped sea water (connate water), sea water ingress, leachates from navigation canals constructed along the coast, leachates from salt pans etc. In India, salinity problems have been observed in a number of places in most of the coastal states of the country. Problem of salinity ingress has been conspicuously noticed in Minjur area of Tamil Nadu and Mangrol - Chorwad - Porbander belt along the Saurashtra coast. Under Hydrology

Project (Phase-II), National Institute of Hydrology, Roorkee has completed a purpose driven study related to seawater intrusion in Minsar river basin in Porbandar district of Gujarat.

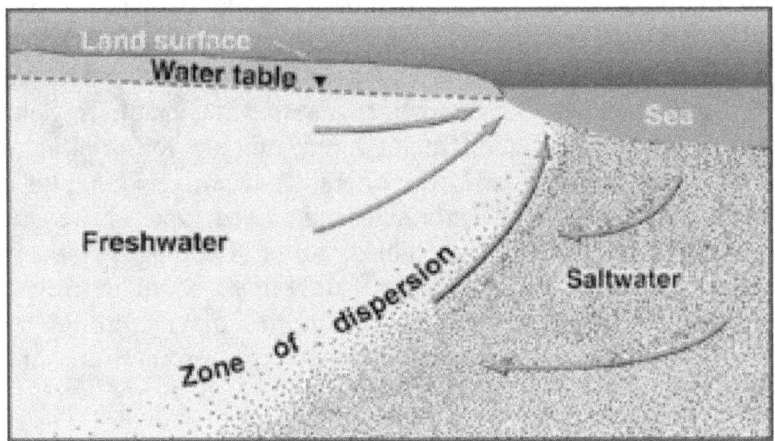

## Fluoride

85% of rural population of the country uses groundwater for drinking and domestic purposes. High concentration of fluoride in groundwater beyond the permissible limit of 1.5 mg/l poses the health problem. The occurrences of fluoride beyond permissible limit (> 1.5 mg/l) has been observed based on the chemical analysis of water samples collected from the groundwater observation wells.

## Arsenic

The occurrence of arsenic in groundwater was first reported in 1980 in West Bengal. In West Bengal, 79 blocks in 8 districts have arsenic beyond the permissible limit of 0.05 mg/l. The most affected districts are on the eastern side of Bhagirathi river in the districts of Malda, Murshidabad, Nadia, North 24 Parganas and South 24 Parganas and western side of the districts of Howrah, Hugli and Bardhman. The occurrence of arsenic in groundwater is mainly in the intermediate aquifers upto the depth of 100 m. The deeper aquifers are free from arsenic contamination. Apart from West Bengal, arsenic contamination in groundwater has been found in the states of Bihar, Chhattisgarh, Uttar Pradesh and Assam. Arsenic in groundwater has been reported in 15 districts In Bihar, 9 districts in U.P. and one district each in Chhatisgarh and Assam states. The occurrence of arsenic in the states of Bihar, West Bengal and Uttar Pradesh is in alluvium formation but in the state of Chhattisgarh, it is in the volcanic rocks exclusively confined to N-S trending Dongargarh-Kotri ancient rift zone. It has also been reported in Dhemaji district of Assam.

## Iron

High concentration of iron (>1.0 mg/l) in groundwater has been observed in more than 1.1 lakh habitations in the country. Groundwater contaminated by iron has been reported from the states of Andhra Pradesh, Assam, Bihar, Chhattisgarh, Goa, Gujarat, Haryana, J&K, Jharkhand, Karnataka, Kerala, Madhya Pradesh, Maharashtra, Manipur, Meghalaya, Orissa, Punjab, Rajasthan, Tamil Nadu, Tripura, Uttar Pradesh, West Bengal and UT of Andaman & Nicobar.

## Nitrate

Nitrate is a very common constituent in the groundwater, especially in shallow aquifers. The source is mainly from anthropogenic activities. High concentration of nitrate in water beyond the permissible limit of 45 mg/l causes health problems. High nitrate concentration in groundwater in India has been found in almost all hydrogeological formations.

Implementation of water pollution prevention strategies and restoration of ecological systems are integral components of all development plans. To preserve our water and environment, we need to make systematic changes in the way we grow our food, manufacture the goods, and dispose off the waste. In India, agriculture is the biggest user and polluter of water. If pollution by agriculture is reduced, it would improve water quality and would also eliminate cost incurred for treatment of diseases. Like all other inputs, there is an optimal quantity of fertilizer for given conditions and excess application does not improve the crop yield. Pricing of fertilizers and pesticides as well as appropriate legislation to regulate their use will also go a long way in stopping indiscriminate use. Industries need to carefully treat their waste discharges. Manufacturers may reduce water pollution by reusing materials and chemicals and switching over to less toxic alternatives. Industrial symbiosis, in which the unusable wastes from one product/firm become the input for another, is an attractive solution. Also, there is a need to encourage reductions or replacement of toxic chemicals, possibly through fiscal measures.

## Urban Water Supply and Sanitation

The urban water supply and sanitation sector in the country is suffering from inadequate levels of service, an increasing demand-supply gap, poor sanitary conditions and deteriorating financial and technical performance. Supply of water is highly erratic and unreliable. Transmission and distribution networks are old and poorly maintained, and generally of a poor quality. Consequently physical losses are typically high, ranging from 25 to over 50 per cent. Low pressures and intermittent supplies allow back siphoning, which results in contamination of water in the distribution network. Water is typically available for only 2-8 hours a day in most Indian cities. The situation is even worse in summer when water is available only for a few minutes, sometimes not at all.

Sanitation and water management should be looked at simultaneously. Too often attention is focused on drinking water supply, leaving sanitation and wastewater treatment for later. However, for every 100 litres of water going into a house about 90 litres will have to leave the plot again. Water supply is an institutional process and an institutional framework for effective water supply and sanitation has to comply with the functions of policy, regulation and sector organization, management of quality, infrastructure and on-site sanitation.

It is desirable to think of water supply, sanitation and wastewater in an integrated way. Urban centres in India are facing an ironical situation today. On one hand, there is the acute water scarcity and on the other, the streets are often flooded during the monsoons. This has led to serious problems with quality and quantity of groundwater. This is despite the fact that all these cities receive good rainfall. However, this rainfall occurs during short spells of high intensity. Because of such short duration of heavy rain, most of the rain falling on the surface tends to flow away rapidly leaving very little for recharge of groundwater. As water shortage increases, alternative sources of water supply are gaining importance. These include sewage recycle, rainwater harvesting, etc.

It should be made mandatory for each industry to install water management solutions to recycle its waste water for reuse. Major step in this front is through the development of industrial effluent recycle solution which integrates physio-chemical, biological and membrane separation processes for optimum water recovery. They achieve water management through water recycle and source reduction, and waste management through product recovery and waste minimization.

## People Participation and Capacity Building

For making the people of various sections of the society aware about the different issues of water resources management, a participatory approach may be adopted. Mass communication programs may be launched using the modern communication means for educating the people about water conservation and efficient utilization of water. Capacity building should be perceived as the process whereby a community equips itself to become an active and well-informed partner in decision making. The process of capacity building must be aimed at both increasing access to water resources and changing the power relationships between the stakeholders. Capacity building is not only limited to officials and technicians but must also include the general awareness of the local population regarding their responsibilities in sustainable management of the water resources. Policy decisions in any water resources project should be directed to improve knowledge, attitude and practices about the linkages between health and hygiene, provide higher water supply service levels and to improve environment through safe disposal of human waste. Sustainable management of water requires decentralized decisions by giving authority, responsibility and financial support to communities to manage their natural resources and thereby protect the environment.

## References

1.      Central Ground Water Board, Dynamic Ground Water Resources of India (as on 31 March 2009). Ministry of Water Resources, Government of India, November 2011, 225 p.

2.      Deorah, Amrit. Water Problem in India and How to Solve it. http://www.ezilon.com/articles/articles/1766/1/Water-Problem-in-India-and-How-to-Solve-it , Published 06/12/2006.

3.      Kumar, Rakesh, R. D. Singh, K. D. Sharma (2005). Water Resources of India. Current Science, Vol. 89, No. 5, 10 September 2005, pp. 794-811.

4.      Ministry of Water Resources (Government of India) Website, http://mowr.gov.in/

5.      National Water Policy, Ministry of Water Resources, New Delhi, 2002.

# 2-Water Balance Analysis

## Introduction

Water balance techniques, one of the main subjects in hydrology, are a means of solution of important theoretical and practical hydrological problems. On the basis of the water balance approach, it is possible to make a quantitative evaluation of water resources and their change under the influence of people's activities. The study of the water balance structure of lakes, river basins, and ground-water basins forms a basis for the hydrological substantiation of projects for the rational use, control and redistribution of water resources in time and space. Conducting water balance estimation provides you with a comprehensive understanding of the water flow system and water resources in your area.

In the natural environment, water is almost constantly in motion and is able to change state from liquid to a solid or a vapour under appropriate conditions. Conservation of mass requires that, within a specific area over a specific period of time, water inflows are equal to water outflows, plus or minus any change of storage within the area of interest. Put more simply, the water entering an area has to leave the area or be stored within the area.

A water balance analysis can be used to assess the current status and trends in water resource availability in an area over a specific period of time. Water balance estimates are often presented as being precise. In fact, there is always uncertainly, arising from inadequate data capture networks, measurement errors and the complex spatial and temporal heterogeneity that characterizes hydrological processes. Consequently, uncertainty analysis is an important part of water balance estimation as is quality control of information before used. When the data sources are imprecise, it is often possible to omit components that do not affect changes. For example, it is possible to omit storage from an annual water balance if year-on-year storage changes (such as reservoirs) are negligible.

This article starts with a description of the different elements of water balance equations and goes on to discuss the water balance of the unsaturated zone, the surface water balance, the groundwater balance, and the integrated water balance.

## Equations for Water Balances

The water balance is defined by the general hydrologic equation, which is basically a statement of the law of conservation of mass as applied to the hydrologic cycle. In its simplest form, this equation reads

Inflow = Outflow + Change in Storage                                    ...(1)

Water balance equations can be assessed for any area and for any period of time. It is worth noting that the word 'area' is commonly used in the professional jargon to mean 'volume', i.e. a certain part of a three-dimensional flow domain. The process of 'making an overall water balance for a certain area' thus implies that an evaluation is necessary of all inflow, outflow, and water storage components of the flow domain as bounded by the land surface, by the impermeable base of the underlying groundwater reservoir, and by the imaginary vertical planes of the area's boundaries.

The water balance method has four characteristic features. They are:

➢ A water balance can be assessed for any subsystem of the hydrologic cycle, for any size of area, and for any period of time;
➢ A water balance can serve to check whether all flow and storage components involved have been considered quantitatively;
➢ A water balance can serve to calculate the one unknown of the balance equation, provided that the other components are known with sufficient accuracy;
➢ A water balance can be regarded as a model of the complete hydrologic process under study, which means it can be used to predict what effect the changes imposed on certain components will have on the other components of the system or subsystem.

**Components of Water Balances**

**Time** - Water balances are often assessed for an average year. But waterlogging and salinity problems are not of the same duration or frequency throughout the world. In some regions, they are permanent (e.g. in marshy areas, which are topographic depressions with a permanently high water table caused by a combination of surface and subsurface inflow). In others, they are temporary (e.g. in areas of incidentally high rainfall or in irrigation areas that receive large quantities of surface water only during the irrigation season). In both cases, the water table rises to an unacceptable level because the natural drainage of the area cannot cope with the excessive recharge of the groundwater reservoir.

If the water table remains high for long periods, crop yields will diminish. In areas where waterlogging occurs, it is necessary to assess water balances not only for an average year, but also for specific years and even for specific seasons (e.g. the growing season, the irrigation season, or, in irrigation areas in arid and semi-arid climates, the period of leaching the soil to prevent salinization).

**Flow Domain** - Let us say that we want to make a water balance study of a certain surface area.
We can choose from two types of flow domains. They are:

- Flow domains comprising physical entities (e.g. river catchments and groundwater basins);

- Flow domains comprising only parts of physical entities (e.g. irrigation schemes and areas with shallow water tables). Let us assess the water balance of the river catchment shown in Figure 1. Suppose that field work has shown that the boundary of the catchment coincides with the groundwater divide. The divide can be regarded as an impermeable boundary because no groundwater flows across it.

Let us assume that the area lies in a humid climate, where changes in water storage usually follow an annual cycle. By choosing dates that are one year apart when we decide the beginning and the end of the period for which the water balance is to be assessed, we can usually ignore the change in water storage. Rainfall is measured at several meteorological stations in the catchment. The runoff from the area is measured at the outlet. Because impermeable bedrock comes close to the land surface at the outlet, preventing groundwater outflow, all water from the catchment area leaves the area as stream flow.

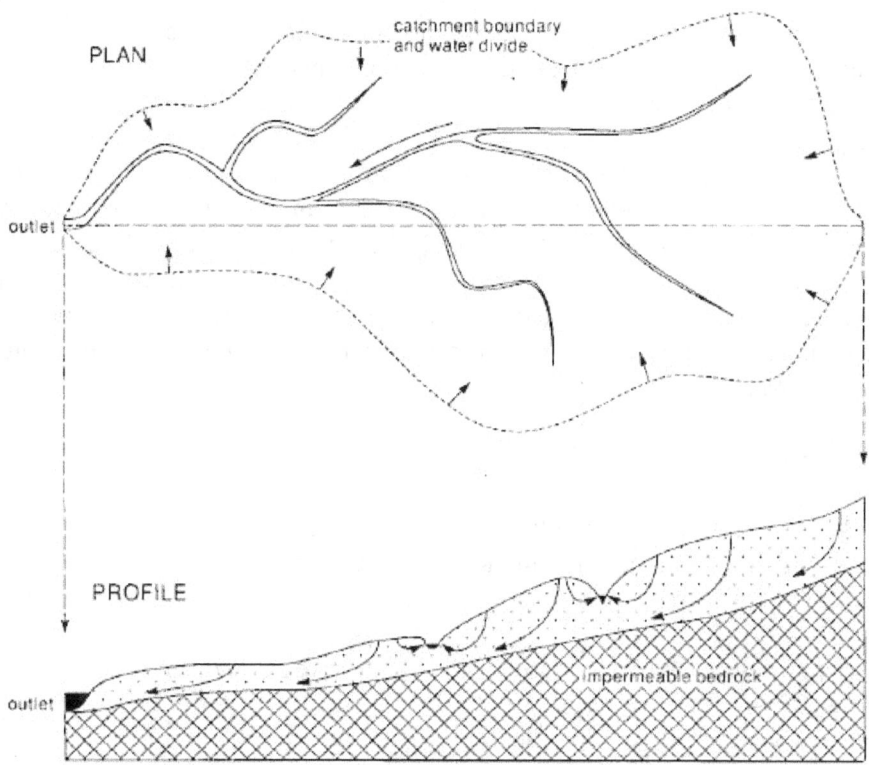

Figure 1: River catchment with a single outlet in bedrock

The overall water balance equation for the area then reads

Rainfall - River Outflow = Evapotranspiration                               ...(2)

From this equation, we can solve the unknown evapotranspiration. Dalton (1802) was among the first to use a catchment water-balance method to correlate the measured rainfall and streamflow data with the estimated evaporation data for England and Wales, as reported by Dooge (1984). It should be noted that Equation (2) is a strongly simplified version of the general water balance equation.

Artificially determined areas such as irrigation areas and areas in need of drainage usually cover only part of a river catchment or groundwater basin. Therefore, it is necessary to account for surface and subsurface inflow and outflow across the vertical planes of the boundaries of these areas. If we determine all their inflow, outflow, and water storage components, we can assess the overall water balance. This is how water balance studies for subsurface drainage are usually done.

In overall water balances, we consider the flow domain vertically - from the soil surface to the impermeable base of the groundwater reservoir. The impermeable base may consist of massive hard rock or of a clay layer whose permeability for vertical flow is so low that it can be regarded as impermeable. Three reservoirs occur in this flow domain: at the surface itself, in the zone between the surface and the water table, and in the zone between the water table and the impermeable base. Because the reservoirs are hydraulically connected, it is often necessary to assess partial water balances for each of them in order to specify the drainable surplus. These water balances are referred to here as the surface water balance, the water balance of the unsaturated zone, and the groundwater balance. We shall discuss them in more detail in the sections that follow.

It is important to note that, at certain depths, there can be clay layers that behave more like aquitards than like aquicludes. The occurrence of these aquitards implies the presence of one or more confined aquifers underneath. In principle, then, it is possible to consider either a multiple aquifer system as a whole or the shallow aquifer alone. In water balance studies for subsurface drainage, it is common to consider only the shallow aquifer. This approach makes it necessary to consider the possible interaction between the deeper, confined water and the shallow, unconfined water.

**Water Balance of the Unsaturated Zone**

For any drainage study, it is absolutely essential to understand the water regime in the unsaturated zone, which extends from the land surface to the water table. It is in this zone that favourable conditions for crop growth must be created.

Some components of a water-balance study of the unsaturated zone are:

- Determine the soil-water storage;
- Assess the soil-water balance and define the relation between it, the water balance of the underlying saturated zone (zone below the water table), and the hydro-meteorological factors;
- Assess the infiltration, evaporation and evapotranspiration, seepage and percolation, and groundwater movement.

Clearly, for large areas, the time and money necessary to conduct such a study would be prohibitive. It would be better to do the research on balance plots or in a pilot area whose soil and hydrology are representative of conditions in the surrounding area.

The unsaturated zone consists of pores that are filled partially with water and partially with air. It can be referred to sometimes as the aeration zone or the vadose zone (a term derived from the Latin word vadosus, meaning 'shallow'). The name 'unsaturated' can be misleading because there are portions of the zone that may actually be saturated even though the pressure of the water is below atmospheric pressure. Examples of saturated portions are the capillary fringe above the water table, the rain-saturated topsoil, and the saturated layers of clay that hold water more tightly than the underlying coarser sediments.

Figure 2 shows the three subzones of the unsaturated zone: the soil-water zone, the intermediate vadose zone, and the capillary fringe. The soil-water zone extends from the surface down through the major root zone of crops and vegetation. It is not saturated except for the times when the land surface receives water from precipitation or (in irrigated areas) for irrigation. Its thickness varies with soil type and with the types of crops and vegetation, ranging from less than one metre to several metres.

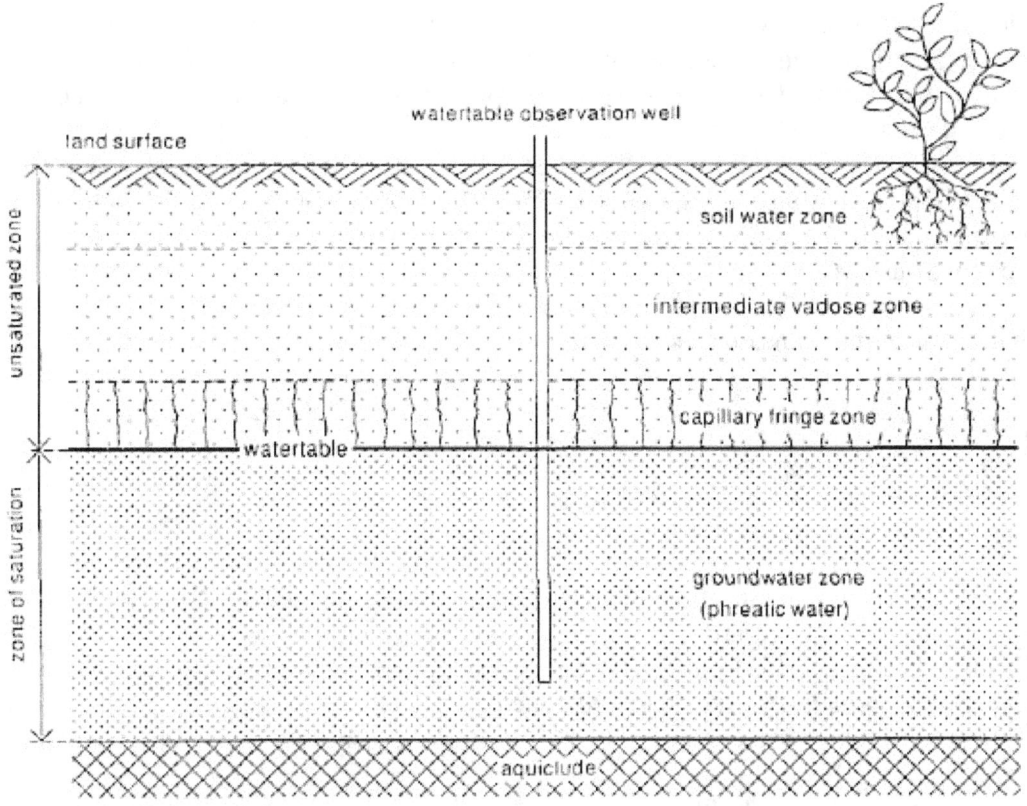

Figure 2: Three Subzones of the Unsaturated Zone

The intermediate vadose zone extends from the lower boundary of the soil-water zone to the upper limit of the capillary fringe. Its thickness varies from zero in areas with a shallow water table to many tens of metres in areas with a deep water table. Any excess water from rain or irrigation that has escaped evapotranspiration in the overlying soil-water zone, passes the intermediate vadose zone as piston flow and eventually reaches the water table, i.e. the zone of saturation. In areas with a shallow water table, the boundary zones merge and we can ignore the processes occurring in the intermediate vadose zone.

The capillary fringe extends from the water table up to the maximum height of capillary rise, which varies with soil texture. The thickness of the capillary fringe varies inversely with the grain size of the soil. Just above the water table, almost all the pores contain water. At a somewhat higher level, only the smaller, connected pores contain water. Higher still, only the smallest of these contain water. Note that the water pressure in the capillary fringe is less than atmospheric, which means that water from this zone will not flow into a well, drain, or open borehole.

In areas with a shallow water table, the capillary fringe may extend into the root zone of the crops and vegetation. A vertical flux from the saturated zone may then develop and move up into the unsaturated zone, from where it is removed by evapotranspiration. The rate of capillary rise, and the subsequent evaporation at the surface, decrease as the depth of the water table increases.

Infiltrating rain and irrigation water increase the soil-water content and can cause the water table to rise. The time required for the infiltrating water to reach the water table increases in proportion to the depth of the water table. Clearly then, if we want to assess the water balance of the unsaturated zone, we must consider all waters that infiltrate into it due to precipitation, irrigation, and seepage. We must know not only the maximum water-holding capacity of the soil, but also the amount of moisture stored in the zone, the actual rate of evapotranspiration of the crops, the percolation to the groundwater, and the rate of capillary rise from the groundwater.

The water balance of the unsaturated zone reads

$$I - E + G - R = \frac{\Delta W_u}{\Delta t} \qquad \qquad ...(3)$$

where,

I       = rate of infiltration into the unsaturated zone (mm/d)
E      = rate of evapotranspiration from the unsaturated zone (mm/d)
G      = rate of capillary rise from the saturated zone (mm/d)
R      = rate of percolation to the saturated zone (mm/d)
$\Delta W_u$ = change in soil water storage in the unsaturated zone during computation interval (mm)
$\Delta t$      = computation interval of time (d)

The common assumption is that the flow direction in the zone is mainly vertical, so no lateral flow components occur in the water balance.

In Figure 3, a rise in the water table $\Delta h$ (due to downward flow from, say, infiltrating rainwater) is depicted during the time interval $\Delta t$. Conversely, during a period of drought, we can expect a decline in the water table due to upward flow from capillary rise and to subsequent evapotranspiration by the crops and natural vegetation. In both situations, it should be certain that the position of the water table at the beginning and the end of the time interval is what accounts for the change in the volume of the unsaturated zone and for the inherent change in soil-water storage.

Note that in areas with deep water tables, the component G will disappear from the water balance equation of the unsaturated zone. Most of the components of Equation (3) cannot be measured in the field. Some components can be assessed only from combinations of other, partial water balances.

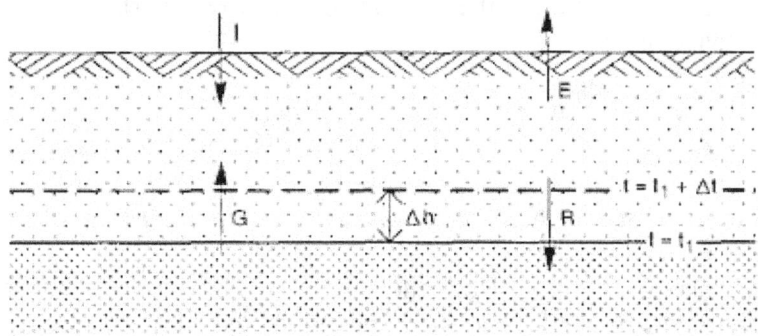

Figure 3: Water Balance Components of the Unsaturated Zone

**Water Balance at the Land Surface**

Because the rate of infiltration (I) in Equation (3) is the recharge into the unsaturated zone, its value is related to the inflow and outflow components of the surface water balance. These components are:

- Water that reaches the land surface from precipitation;
- Water that enters the water balance area by lateral surface inflow and leaves it by lateral surface outflow;
- Water that evaporates from the land surface.

The difference between the components is due to changes in surface water storage. Infiltration in the unsaturated zone can therefore be expressed by the following equation

$$I = P - E_0 + 1000 \frac{Q_{si} - Q_{so}}{A} - \frac{\Delta W_s}{\Delta t} \qquad \qquad \text{...(4)}$$

where,

| | |
|---|---|
| P | = precipitation for the time interval $\Delta t$ (mm) |
| $E_0$ | = evaporation from the land surface (mm/d) |
| $Q_{si}$ | = lateral inflow of surface water into the water balance area (A) (m$^3$/d) |
| $Q_{so}$ | = lateral outflow of surface water from the water balance area (A) (m$^3$/d) |
| A | = water balance area (m$^2$) |
| $\Delta W_s$ | = change in surface water storage (mm) |

Note that we can make a water balance analysis very much simpler by selecting a suitable area and a suitable time period. For example, if we choose a rainy period, we shall not have to consider evaporation. Conversely, if we choose a dry period, precipitation can be eliminated. We can select an area without any inflow and outflow of surface water, or we can select a time period that has the same surface water storage at the beginning and end.

In irrigated areas, the major input and output of a water balance are usually determined by two artificial components, namely the application of water for irrigation and (in arid zones) for leaching the soil, and the removal of excess irrigation water (surface drainage) and excess groundwater (subsurface drainage).

Figure 4 shows the components of the surface water balance in an area of basin irrigation. On the left, an irrigation canal delivers surface water to an irrigation basin ($Q_{ib}$). A portion of this water is lost through evaporation to the atmosphere ($E_{ob}$). Another portion infiltrates at the surface of the basin ($I_b$), increasing the soil-water content in the unsaturated zone. Any surface water that is not lost through either evaporation or infiltration is discharged downslope by a surface drain ($Q_{ob}$). Both the irrigation canals and the surface drains lose water through evaporation ($E_{oc} + E_{od}$) to the atmosphere and through seepage to the zone of aeration ($I_c + I_d$).

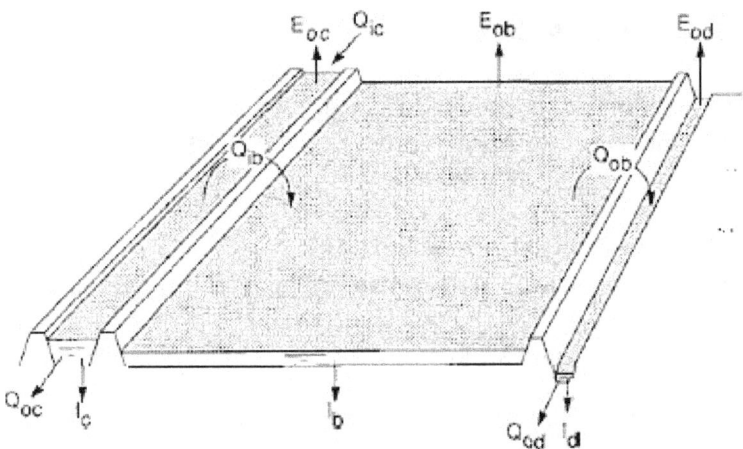

Figure 4: Surface Water Balance Components for a Basin-Irrigated Area

We can still describe the surface water balance in this area with Equation (4) if we substitute ($Q_{ic} + Q_{id}$) for $Q_{si}$, and ($Q_{oc} + Q_{od}$) for $Q_{so}$. Note that infiltration (I) now comprises the combined effect of the infiltration of rainfall, the infiltration of irrigation water at the fields, and the seepage losses of the irrigation canals and the surface drains to the unsaturated zone. Note also that E, comprises not only evaporation of rainwater that did not infiltrate into the soil, but also evaporation of water in canals, basins, and drains.

In other areas, irrigation is practised with borders, furrows, sprinklers, and (where the terrain is sloping) contour line ditches. In principle, we use the same components to make surface water balances for these areas as we did for basin irrigation areas.

**Groundwater Balance**

The water balance for the saturated zone, also called the groundwater balance, can generally be expressed as follows (see Figure 5)

$$R - G + 1000 \frac{Q_{gi} - Q_{go}}{A} = \mu \frac{\Delta h}{\Delta t} \qquad \qquad \ldots (5)$$

where,

$Q_{gi} = Q_{gih} + Q_{giv}$ = total rate of groundwater inflow into the shallow unconfined aquifer $(m^3/d)$

$Q_{go} = Q_{goh} + Q_{gov}$ = total rate of groundwater outflow from the shallow unconfined aquifer $(m^3/d)$

$Q_{gih}$ = rate of horizontal groundwater inflow into the shallow unconfined aquifer $(m^3/d)$

$Q_{goh}$ = rate of horizontal groundwater outflow from the shallow unconfined aquifer $(m^3/d)$

$Q_{giv}$ = rate of vertical groundwater inflow from the deep confined aquifer into the shallow unconfined aquifer $(m^3/d)$

$Q_{gov}$ = rate of vertical groundwater outflow from the shallow unconfined aquifer into the deep confined aquifer $(m^3/d)$

$\mu$ = specific yield, as a fraction of the volume of soil (-)

$\Delta h$ = rise or fall of the water table during the computation interval (mm)

and the other symbols as defined earlier.

When the layer beneath the shallow unconfined aquifer is impermeable, the rates of vertical groundwater inflow and outflow equal zero, the total groundwater inflow equals the horizontal groundwater inflow, and the total groundwater outflow equals the horizontal groundwater outflow.

The base of an unconfined aquifer is, in reality, seldom impermeable; it is the first clay layer struck at some depth during borehole drilling. In sandy areas, groundwater underlying the 'impermeable' base is confined. In discharge areas of the groundwater system, the aquifer receives confined water from beneath, and the quantity of inflow per computation interval of time must be included in the water balance. The total groundwater inflow is then equal to the sum of horizontal and vertical inflow.

In irrigation areas, the water table in the unconfined aquifer can be appreciably higher than the piezometric surface in the deep aquifer. The resulting downward seepage from the shallow aquifer to the deep aquifer, over the time interval $\Delta t$, must then be included in the water balance. The total groundwater outflow then equals the sum of horizontal and vertical outflow. This flow constitutes what is called the 'natural drainage' of the area. In areas with an operational field drainage system, the drain discharge should be a separate component of the water balance.

We can determine the horizontal groundwater inflow and outflow through the boundaries of the area by using water table contour maps, which show the direction of groundwater flow and the hydraulic gradient, and by considering transmissivity at the boundary. We can determine upward and downward seepage through an underlying semi-confined layer by considering vertical gradients and the shallow aquifer's hydraulic resistance. And we can calculate the change in storage by using groundwater hydrographs and the specific yield or drainable pore space of the shallow aquifer.

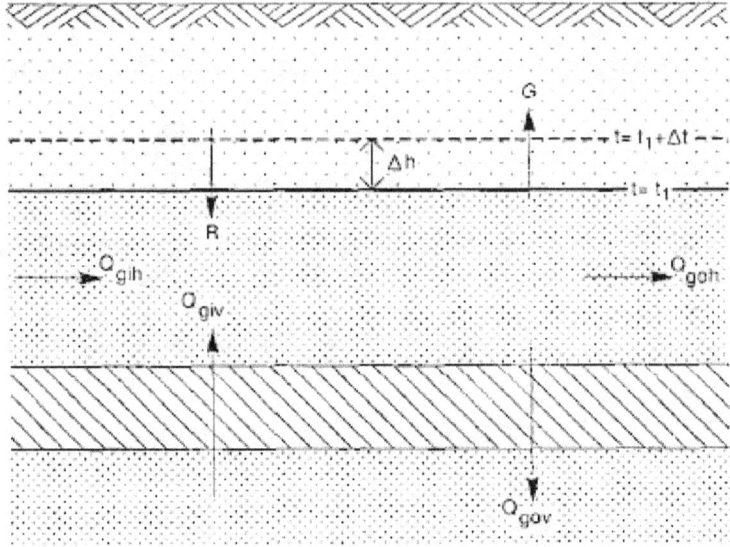

Figure 5: Groundwater Balance Components of the Shallow
Aquifer of a Multiple Aquifer System

To get the data necessary for these direct calculations of horizontal and vertical groundwater flow, and of the actual amount of water going into or out of storage, we must install deep and shallow piezometers and conduct aquifer tests.

In some areas with limited surface water resources, groundwater is used both for human consumption and for irrigation. When this occurs, the rate of groundwater abstraction must be accounted for in the water balance. If pumped wells provide irrigation water, we must keep track of the amount of return flow, i.e. the portion of the total groundwater abstraction that returns to the deeper layers and so recharges the groundwater reservoir. Return flow must also be accounted for in the water balance.

According to Equation (5), we can calculate the value of the net percolation as $R^* = R - G$. In areas with deep water tables, there is no upward flux by capillary rise, and so the actual percolation equals the calculated net percolation. In areas with shallow water tables, it is possible to determine only the net percolation.

**Integrated Water Balances**

The partial water balances that we discussed in the three previous sections are often combined to form integrated water balances. For example, by combining Equations (3) and (4), we get the water balance of the topsoil.

$$P - E_0 - E + G - R + 1000\frac{Q_{si} - Q_{so}}{A} = \frac{\Delta W_s + \Delta W_u}{\Delta t} \qquad \qquad ...(6)$$

To assess the net percolation $R^* = R - G$, we can use Equation (6). We can also assess this value from the groundwater balance (Equation 5). And, if sufficient data are available, we can use both of these methods and then compare the net percolation values obtained. If the values do not agree, the degree of discrepancy can indicate how unreliable the obtained data are and whether or not there is a need for further observation and verification.

Another possibility is to integrate the water balance of the unsaturated zone with that of the saturated zone. Combining Equations (3) and (5), we get the water balance of the aquifer system.

$$I - E + 1000\frac{Q_{gi} - Q_{go}}{A} = \frac{\Delta W_u}{\Delta t} + \mu\frac{\Delta h}{\Delta t} \qquad \qquad \text{...(7)}$$

We can assess the infiltration from Equation (7), provided we can calculate the total groundwater inflow and outflow, the change in storage, and the actual evapotranspiration rate of the crops. We can also assess the infiltration from the surface water balance (Equation 4). And, if sufficient data are available, we can follow the same procedure we followed above.

Finally, let us integrate all three of the water balances described in the previous sections. This overall water balance reads

$$P - E_0 - E + 1000\frac{Q_{si} - Q_{so}}{A} + 1000\frac{Q_{gi} - Q_{go}}{A} = \frac{\Delta W_u}{\Delta t} + \frac{\Delta W_s}{\Delta t} + \mu\frac{\Delta h}{\Delta t} \qquad \qquad \text{...(8)}$$

Equation (8) shows that the vertical flows I, R, and G (all important linking factors between the partial water balances) disappear in the overall water balance. Nevertheless, these linking factors determine to a great extent whether there are drainage problems or not.

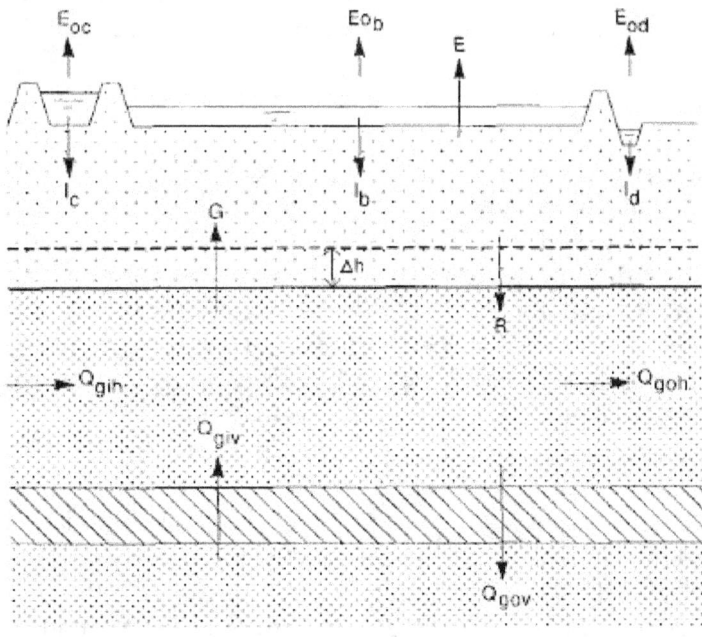

Figure 6: Overall Water Balance Components for a Basin-Irrigated Area

For an example, let us look at Figure 6, which shows all the terms of the overall water balance for an area with basin irrigation. The overall water balance can be described by Equation (8) if we make the following substitutions:

$$E_0 = E_{oc} + E_{ob} + E_{od}$$
$$Q_{si} = Q_{ic} + Q_{id}$$
$$Q_{so} = Q_{oc} + Q_{od}$$
$$Q_{gi} = Q_{gih} + Q_{giv}$$
$$Q_{go} = Q_{goh} + Q_{gov}$$

When water balances are assessed for a hydrologic year, changes in storage in the various partial water balances can often be ignored or reduced to zero if the partial balances are based on long-term average conditions. In Equations (3) to (8), the sum of the various inflow components then equals the sum of the various outflow components. Drainage designers use this concept of steady state frequently.

**Practical Applications**

Water balance analyses are particularly useful in land reclamation and drainage projects because they provide insight into:

- The sources of local groundwater flow, i.e. the difference between the outflow and inflow of groundwater in the study area;

- The portions of rainfall and irrigation water that infiltrate at the land surface, evaporate from the surface or from the unsaturated zone, or leave the surface as overland flow;

- The quantity of, and the monthly or annual changes in, groundwater flow;

- The quantity of groundwater that must be drained artificially to maintain the water table at a suitable depth, i.e. the drainable surplus.

We can ask ourselves if the knowledge we gain by investigating an experimental plot or a tract of several hectares is valid for the rest of the area under study (e.g. a large river basin or a delta plain). The answer is 'no' because throughout the river basin there are spatial variations in rainfall, evaporation, land use, soil, and hydrogeological conditions. The best way to obtain the water balance for the greater area is to divide the basin into hydrogeological sub-areas. The division should be based on water table contour maps of the aquifer system and on well hydrographs. Variations in the spacing of the water table contours reflect differences in the lithology of the aquifer and thus in the transmissivity. An analysis of the available well hydrographs makes it possible to select those wells whose water level fluctuations are similar. With this information, we can distinguish the hydrogeological sub-areas whose water tables react similarly to the processes of groundwater recharge and discharge. If we assess monthly water balances for these sub-areas, and then add up all the corresponding components of the balances, we obtain the annual water balance for the entire basin. These basin discretizations require sufficient and accurate data on the water table fluctuations throughout the basin, on the specific yield of the unsaturated zone, on the thickness and hydraulic conductivity of the saturated zone, and on soil water content. They also require a network of stream-gauging stations to supply accurate data on the surface water inflow and outflow of the sub-areas.

A river catchment usually consists of a network of natural drainage channels that eventually join the main river. Each of these channels drains a certain area (sub-catchment). Because, over the catchment, there are variations in rainfall distribution, soil and hydrogeological conditions, and land use and vegetation, the sub-catchments react differently during a hydrologic event (e.g. a rainstorm). The peak flows may show not only a marked difference, but also a time lag.

The river runoff depends not only on the hydrology of the catchment, but also on the distribution and intensity of the rainfall, and on the evapotranspiration and storage capacity of the unsaturated zone. When considering the soil water content, we must be aware that different soil types have different water retention curves. These curves characterize the ability of the soil to retain water during gravity drainage or drying. We must also be aware of the difference in the relation between capillary flux and soil water content when a soil is absorbing water through infiltration and when it is losing water through drainage. One reason for this difference (hysteresis) is the entrapment of air in the soil during the wetting period, which means that the pores of the soil do not completely fill with water even when the capillary flux is zero.

Finally, we must realize that the depth of the water table plays a crucial role in evaluating the discharge from an area. If the groundwater is deep, say many metres below the ground surface, we only have to consider the groundwater inflow and outflow for a specified area. In areas with shallow water tables, however, say less than 3 m below the ground surface, there is considerable groundwater outflow due to capillary rise, i.e. an upward flux from the water table, which can bring the groundwater to the land surface, where it is then lost to evaporation (if the soil is barren) or to evapotranspiration (if the soil has a vegetative cover). The loss of groundwater to the root zone of crops can also be great, especially in the dry season, when the soil water storage is partly depleted. But the size of these losses depends on the soil type. Very fine sandy loam, for example, has a very high capacity for capillary rise into the rootzone.

We must also realize that the time interval we choose for a water balance study can affect the result of our calculations. As an example, let us say that there was a heavy rainstorm just before the beginning of our chosen time interval. Although the storm will not appear in the water balance, its effects will - on soil water content, groundwater recharge, and groundwater flow. Under these circumstances, it is advisable to shift the time interval so that it begins before the rainstorm and ends a few days after the rain has ceased. If we can not do all our measurements and observations on the same day, then we should at least do them on consecutive days.

Meteorological conditions vary throughout the year and from one year to another. Accordingly, there are wet, dry, and normal years. Because of these variations, we must investigate how much the hydrologic conditions in the year under study deviate from those in a normal or average year. A study of this kind requires long-term records of precipitation, evapotranspiration, and river discharge.

# References

1.  Dalton, J. (1802). Experiments and observations to determine whether the quantity of rainfall and dew is equal to the quantity of water carried off by the rivers and raised by evaporation, with an enquiry into the origin of springs. Mem. Proc. Lit. Phil. Soc., Manchester 5, part 2, pp. 346-372.

2.  de Ridder, N.A. and J. Boonstra (1994). Analysis of Water Balances. In. Drainage Principles and Applications (Editor-in-Chief: H. P. Ritzema), ILRI Publication 16, The Netherlands, pp. 601-633.

3.  Dooge, J.C.I. (1984). The Water of the Earth. Hydrol. Sci. J., 29, 2, 6, pp. 149-176.

# 3-Overview of Groundwater Hydrology Aspects

## Introduction

Ground water is a distinguished component of the hydrologic cycle. Surface water storage and groundwater withdrawal are traditional engineering approaches which will continue to be followed in future. The uncertainty about the occurrence, distribution and quality aspect of the groundwater and the energy requirement for its withdrawal impose restriction on exploitation of groundwater. In spite of its uncertainty, groundwater has some obvious advantages. These advantages are: groundwater is much protected from pollution; it requires little treatment before it use; it is available almost everywhere; it can be developed with little gestation period and can be supplied at a fairly steady rate. It does not require any distribution system and its interference with land resources is minimum. In hilly region, the groundwater emerging as springs can serve as a viable source for supply of drinking water. In a canal command area, use of groundwater controls waterlogging problems. The role of groundwater is conspicuous during period of drought.

## Groundwater Situation in India

India is a vast country having diversified geological setting. Variations exhibited by the rock formations, ranging in age from the Archaean crystallines to the recent alluvia, are as great as the hydrometeorological conditions. Variations in the land forms are not insignificant. It varies from sea level at the coasts to the lofty peaks of the snow clad Himalayan mountains attaining staggering altitude upto about 9000 metres. Variations in the nature and composition of rock types, the geological structures, geomorphological set up and hydrometeorological conditions have correspondingly given rise to widely varying groundwater situation in different parts of the country.

In the high relief areas of the northern and north-eastern regions occupied by the Himalayan ranges, the various conspicuous hill ranges of Rajasthan, the Central and Southern Indian regions, the presence of very steep slope conditions and geological structures offer extremely high runoff and thus very little scope for rain water to find favourable conditions of storage and circulation as groundwater. The large alluvial tract extending over 2000 km in length from Punjab in the West, to Assam in the East often referred to as Sindhu-Ganga-Brahmaputra Plain, is perhaps, the most potential and important region from the view point of groundwater resources.

Almost the entire Central and Southern India is occupied by a variety of hard rocks with hard sediments (including carbonate rocks) in the intertectonic and major river basins. Rugged topography, hard and compact nature of the rock formations, the geologic structures and meteorological conditions have yielded an environment which allows groundwater storage in the weathered residium and its circulation in the underlying fracture systems. The hard rock terrains, river valleys and abandoned channels, wherever having adequate thickness of porous material, act as potential areas for groundwater storage and development. It is observed that sustained yield from fracture system down to 200 m is possible in certain hard rock areas of Peninsular India.

There is a wide variation in the chemical quality of groundwater in the country, reflecting the diverse geo-hydrology, hydrometeorology, topographic and drainage conditions and artificially imposed conditions, such as surface water irrigation. The coastal and deltaic tracts, particularly in the East Coast, are covered with vast and extensive alluvial sediments. Though these tracts are productive in terms of water yield yet the overall groundwater regime in coastal areas suffer from salinity hazards. Groundwater development in coastal areas should be regulated such that contamination of the fresh groundwater body with sea water is avoided. Salinity in groundwater also exist in inland areas, apart from coastal areas, in parts of Punjab, Haryana, Uttar Pradesh, Rajasthan and Gujarat. This is generally confined to arid and semi-arid regions. In some of the canal command areas where there is progressive rise of water table, it brings about waterlogging conditions resulting in increase of salinity in the top soils and rendering cultivable land barren. Conjunctive use of surface water and groundwater should be practiced as an effective method to check the waterlogging conditions in such areas.

**Assessment of Groundwater Resources**

Groundwater regime is a dynamic system in which water is absorbed at the land surface and eventually recycled back to the surface. The groundwater movement occurs through the porous unconsolidated sediments and through interconnected openings in the rocks that mantle the earth. The occurrence and movement of groundwater depend on the geohydrological characteristics of the subsurface formations. These natural deposits vary greatly in their lithology, texture and structure and differ in respect of their hydrological characteristics. The framework in which groundwater occurs is as varied as those rocks and as intricate as their deformation, which has progressed through geologic time. The possible combinations of variety and intricacy are virtually infinite. It has been, therefore, experienced that groundwater investigations at a given site almost always exhibit a certain uniqueness.

For estimating regional groundwater potential, it has been recommended that the general manner in which the regime functions must be identified. The potential for recharge to the groundwater regime in an area depends on the amount and pattern of annual precipitation in relation to the potential evaporation in the area and to the occurrence of any surface or subsurface inflow from adjacent areas. Most of this potential recharge is commonly intercepted by the soil veneer and eventually return to the atmosphere through process of evapotranspiration or dissipate through surface runoff. The amount that actually contributes to groundwater recharge varies seasonally and from year to year. It is generally difficult to quantify the recharge to groundwater from various sources. Similarly, groundwater discharge may be difficult to quantify because of temporal variations, especially if it occurs at a number of scattered locations, either at the land surface in the form of springs, gaining streams, lakes, ponds, marshes or growths of phreatophytes, or at depth through permeable formations.

For quantifying the groundwater resource available, two different concepts based on the existing hydrogeological situations are used:

1. Quantity concept - for unconfined (water table) aquifers.
2. Rate concept  - for confined aquifers.

*(a) Unconfined (water table) aquifers*

Groundwater resources can be classified as Static and Dynamic. The Static resource can be defined as the amount of groundwater available in the permeable portion of the aquifer below the zone of water level fluctuation. The Dynamic resource can be defined as the amount of groundwater available in the zone of water level fluctuation. The usable groundwater resource is essentially a dynamic resource which is recharged annually or periodically by rainfall, irrigation return flows, canal seepage, influent seepage etc. The most important component of recharge to the aquifer is the direct infiltration of rain water, which varies according to the climate, topography, soil and subsurface geological characteristics. A part of applied irrigation water both from groundwater and surface water resources, reach groundwater depending on the efficiency of irrigation system and soil characteristics. Influent streams also recharge the groundwater body depending on the drainage density, width of streams and on the texture of river bed material. Other sources of recharge are
percolation from canal systems, reservoirs, tanks etc.

*(b) Confined aquifers*

For the confined aquifers which are hydrogeologically separate from shallow water table aquifers, the groundwater assessment is done by rate concept. The groundwater available in a confined aquifer equals the rate of flow of groundwater through this aquifer. The rate of groundwater flow available for development in a confined aquifer in the area can be estimated by using Darcy's law.

**Groundwater Modelling**

Groundwater modelling is a tool that can help analyse many groundwater problems. It begins with a conceptual understanding of the physical problem. The next step in modelling is translating the physical system into mathematical terms. In general, the final results are the familiar groundwater flow equation and transport equations. These equations, however, are often simplified, using site-specific assumptions, to form a variety of equation subsets. An understanding of these equations and their associated boundary and initial conditions is necessary before a modelling problem can be formulated.

Mathematical groundwater models essentially comprise (i) appropriate differential equations governing the groundwater behaviour in a domain of interest and (ii) an algorithm to solve the equations - analytically or numerically. The analytical solutions are generally based upon many restrictive assumptions e.g. homogeneity; isotropy; time and (or) space independent withdrawals, recharge, boundary conditions; regular geometry of the domain. On the other hand the numerical solutions can follow the real world conditions quite closely and in fact, subject to the data availability, the only assumptions which need to be made, are the validity of Darcy's law and the continuity equation. However, the data inadequacy mostly does necessitate many assumptions even in numerical solutions. Nevertheless such solutions can generally provide far more realistic solutions than what can be provided by the analytical solutions.

# Modelling of Flow in Unsaturated Zone

Soil moisture conditions of a basin control the rainfall-runoff phenomena. It has been shown that performance of rainfall-runoff models is sensitive to the methods used to specify the effective precipitation and the antecedent moisture conditions. Study of soil moisture movement on a continuous basis considering both the storm and interstorm periods will not only determine antecedent moisture conditions but also would enable determination of recharge due to several rainfall events and evaporation losses from soil moisture storage.

Flow of water in the unsaturated zone is a complex phenomenon involving transfers of water, air and vapour through dynamic flow pathways under the influence of hydraulic, temperature, density and osmotic gradients in a compressible porous medium. In hydrological modelling, the physical representation of water flow in unsaturated zone has been made using Richards equation which combines Darcy's law with the equation of continuity. Richards equation is based on the assumptions that the fluid is incompressible and the flow takes place under isothermal condition. Both simulation and laboratory experiments have shown that under certain conditions, the effects of air, vapour and thermal fluxes may be important in the unsaturated zone. Theoretical reasoning would therefore suggest that the use of Richards equation as a simulator of the unsaturated zone should be expected to be inaccurate in some circumstances. Limiting conditions for the use of Richards equation for simulating processes in unsaturated zone in the field have not, however, been established and it is usually assumed, with some justification, that the error resulting from ignoring the multi-phase nature of the flow process may be negligible, relative to the problems of estimating the single-phase flow parameters for field situations. The complexity in physical representation of unsaturated fluid flow would further manifest in greater requirements for parameter and input data. These would include characteristic curves for the transport of air and vapour, thermal capacity and diffusivity, densities of water and air, and upper boundary conditions for air pressures, humidity and temperature in addition to water fluxes.

Another important aspect of soil moisture study which needs to be undertaken is soil moisture forecast. The soil moisture forecast will be useful for water management. A forecast of soil moisture is the prior estimate of the future state of soil water in the zone of aeration. The forecast variables include soil moisture status in the root zone depth, time of occurrence of permanent wilting condition, continuing period of permanent wilting condition, period for which the root zone depth remains in saturated state, time at which root zone attains field capacity, depth fraction of root zone depth that remains at less than field capacity. Roles of a short term forecast i.e. a forecast of future value of an element of the regime for a period ending upto two days from the issue of the forecast and a medium term or extended forecast for a period ending between two and ten days, in water management need to be investigated.

The controlling factors of soil moisture may be classified under two main groups viz. climatic factors and soil factors. Climatic factors include precipitation data containing rainfall intensity, storm duration, interstorm period, temperature of soil surface, relative humidity, radiation, evaporation and evapotranspiration. The soil factors include soil matric potential and water content relationship, hydraulic conductivity and water content relationship of the soil, saturated hydraulic conductivity and effective medium porosity. Besides these factors, the information about depth to water table is also required.

The continuous variation of soil moisture with time and depth can be known by solving the Richards equation. The presentation of the forecast may be in the form of a single value or in the form of probability distribution. Forecasting soil moisture would help in deciding irrigation application, enable prediction of the annual evaporation loss from shallow water table and recharge to groundwater storage due to rainfall.

## Modelling of Flow in Saturated Zone

The main objective of saturated groundwater modelling exercise is to forecast the response of the aquifer to a proposed man-made excitation (e.g. pumpage, artificial recharge) acting in conjunction with other forms of natural excitation. A critical evaluation of the computed response can determine whether the proposal is acceptable or unacceptable in respect of prestipulated constraints. The response computed directly through the solution of the governing equations comprises water table or piezometric elevations distributed in space and time. The distributions can provide other forms of the response (e.g. stream-aquifer interaction, maximum and minimum depths to water table and/or saturated thicknesses) that a planner may be interested in.

The flow in the saturated zone is governed by a second order linear or non-linear differential equation, depending upon whether the flow is under confined or unconfined condition respectively. The non-linear equation governing unconfined flow can be linearised either by replacing space-time variant saturated thickness by an average space variant saturated thickness (first method of linearisation) or by change of variable from saturated thickness to the square of saturated thickness (second method of linearisation). The differential equations may account for only the horizontal flow (one or two dimensional) or may account for vertical flows also. The numerical solutions may be based upon either the finite difference or finite elements.

The ultimate objective of saturated flow modelling is to be able to forecast the aquifer response to a variety of excitation patterns. Such a modelling is termed as Direct Problem. However, for solving a direct problem, it is necessary to know spatially distributed estimates of the aquifer parameters. It is generally not feasible to carry out test pumping at large enough number of space points to get an adequate spatial distribution of the parameters. The way out is to arrive at such a distribution of the parameters which yields the best possible reproduction of historical water table/piezometric elevations under historical excitation. The inverse problem can be solved by either repeated solutions of the Direct Problem with varying aquifer parameters (indirect method of solving inverse problem) or by solving the governing differential equation directly for the parameters (direct method of solving inverse problem). The indirect methods can be based upon either simulation or rigorous optimization theory.

Most of the saturated flow modelling studies carried out in India have been related to unconfined alluvial aquifers bounded by rivers. The flow has generally been assumed to be two-dimensional horizontal and the non-linear differential equation governing such a flow linearised by the first method. The resulting equation has been mostly solved by finite differences although there have been a few studies employing finite elements. The inverse problem has been generally solved by the simulation based indirect method (i.e. repeated solution of the direct problem). Keeping in view the large areal extent of hard rock aquifers in

India, the number of model studies on such aquifers has been rather marginal. Similarly, the number of studies on confined aquifers has been very marginal.

The theory of groundwater flow towards partially and fully penetrating wells in unconfined, confined, and leaky aquifers has been developed by numerous authors. Results reported for wells in confined aquifers are derived as solutions to Jacob's differential equation for flow of slightly compressible fluids through deformable porous media. Results in unconfined aquifers are solutions to either the Laplace or Boussinesq equations describing hydraulic head distribution in porous media.

## Spring Flow Modelling

Spring is a ready source of water, a place of natural beauty and a recreational spot. Springs generally provide clean water. They are found in the Himalayas, in the Western Ghats and in other places in India where it is logistically difficult to create storage for water. As such, study of spring flow has relevance to the water supply to rural areas, specially in the hilly region. However, there is no systematic hydrologic study of the springs so far and there is enough scope and need to study springs, particularly in respect of mathematical modelling of spring flow. Exploitation of forested area for food, fibre and minerals and urbanization lead to deforestation and changes in the watershed characteristics. This human interference leads to destruction of the internal hydrological system. As a consequence, the spring flow diminishes which may lead to drying of the spring.

Springs are basically natural outlets of concentrated discharge from an aquifer. A large spring indicates existence of thick transmissive aquifer whereas a small spring indicates an aquifer of low transmissivity. Discharge rate from a spring depends on the extent of the recharge area, the precipitation on the catchment of the spring, aquifer geometry, area of opening of the spring, geology and geomorphology of the area and diffusivity of the aquifer. The discharge of a spring depends on the difference between the elevations of the water table (or piezometric head) in the aquifer in the vicinity of the spring, and the elevation of spring outlet (called as threshold). During dry season, the spring discharge is derived from water stored in the aquifer. Consequently, the water level in the aquifer gradually falls and the spring discharge declines. The recession part of a spring hydrograph, in a semi-log plot (with time on the linear scale), may follow a straight line. Based on such observation, some conceptual linear hydrologic models have been developed to assess the dynamic storage which subsequently appears as spring discharge. These models assume that the spring flow is linearly proportional to the dynamic storage in the spring flow domain. The dynamic storage of groundwater at any time during recession is equal to the product of depletion time and discharge of the spring. It is not yet verified that spring discharge from an aquifer conforming to a linear system would follow strictly an exponential decay curve.

## Groundwater and Surface Water Interaction

Groundwater and surface water contained in the hydrological system are closely interrelated. Therefore, the study of the groundwater/ surface water interrelationship has always been the subject of great attention on the part of hydrogeologists, hydrologists, and water resources specialists. The study of the character and particularities of the ground water/surface water interrelationship is currently one of the main interests of water sciences. The studies resulted in examination of the processes of groundwater flow generation and estimation of groundwater discharge including groundwater discharge to rivers (base flow). There are many methods for quantitative estimation of streamflow components based on the analysis of streamflow depletion curves, separation of streamflow hydrographs, studying groundwater regime and balance under natural conditions, establishing the correlation links between the levels and discharges of groundwater and surface water under different natural conditions. During recent decades, the above studies were appreciably intensified primarily due to requirements of practice - the need or regional quantitative estimation of groundwater discharge and natural resources (recharge) of fresh groundwater.

The separation of streamflow hydrographs, taking into account the character of the hydraulic relationship between ground water and surface water under different hydrogeological conditions and taking into consideration the dynamics of base flow during the seasons of the year, allows us to estimate base flow quantitatively and, therefore, determine the natural groundwater resources of the drainage zone in the case of groundwater discharge to rivers.

In a groundwater basin, it is common to identify several aquifers separated either by less permeable or impermeable layers. A river in general penetrates partially the upper aquifer. When the river stage rises during the passage of a flood, the upper aquifer is recharged through the bed and banks of the river. The lower aquifer is recharged through the intervening aquitard. A single aquifer river interaction problem has been studied analytically by several investigators. A digital model of multi-aquifer system has been developed assuming horizontal flow in the aquifers and vertical flow through the confining layers which separate the aquifers. These assumptions have reduced the mathematical problem to one of solving coupled two-dimensional equation for each aquifer in the system. An interactive, alternating-direction, implicit scheme has been used to solve the system of simultaneous, finite difference equations which describe the response of the aquifer system to applied stresses. The quasi three-dimensional model has been developed to simulate a groundwater system having any number of aquifers.

The studies on the ground water/surface water interrelationship made it possible to solve a number of important scientific and practical problems:

1. to estimate base flow and, therefore, sustained low river discharges of different probabilities;

2. to estimate the groundwater contribution to total water resources and the water balance of regions;

3. to evaluate quantitatively the natural groundwater resources for determining the prospects of their use within large areas and as a component of the safe groundwater yield.

The methods for estimating the groundwater discharge of the upper hydrodynamic zone are fairly well developed and are being widely employed in hydrogeological practice, while studies on the estimation of the groundwater discharge of deep artesian aquifers on their contribution to surface runoff are being carried out insufficiently.

## Conjunctive Use of Surface Water and Groundwater

In a canal command, because of continuous and intensive application of surface water for irrigation, the water table comes up which may lead to water logging and salinity hazards. These problems can be prevented by withdrawing water from aquifer. If the groundwater is not saline, it may be used to irrigate part of the canal command or it may be transported to other areas through the canal conveying surface water to the region. Losses from irrigation schemes to groundwater are often significant. In a typical scheme, as much as 40 % of the water released from the reservoir is lost to the groundwater. These losses occur in the canals and distributories as well as in the fields. Conjunctive use appears to be a realistic approach for the efficient use of the water. If the water lost from the conveyance system and agricultural field is pumped from the aquifer and supplement the canal water for irrigation, high overall efficiencies could be obtained.

Conjunctive utilization of surface and groundwater implies not only their joint use but their coordinated use so that the useful water is more than the sum of the components, and is also used in an efficient economical way. Modern methods of hydrologic analysis can be used for evaluation of surface water resources, safe yield of aquifers and their interactions including the effect of development on availability. Planning models can then be used for efficient economic development and management of these resources in a conjunctive use framework.

While a general conceptual framework is available for the conjunctive utilization problems, case studies are available only at a much simpler level and it is hoped that more realistic studies will be made in the near future because of the importance of conjunctive utilization and the problems of unplanned, uncontrolled and uncoordinated development of different sources of water.

## Artificial Recharge

It has been recognized that aquifers are not only sources of water but also storage reservoirs that require proper management for efficient use. With respect to management, an aquifer may be considered as a reservoir for long term storage artificially produced and as a water quality control tool because of its filtering characteristic that reclaims artificially recharged waste water. Artificial recharge may be viewed as an augmentation of the natural movement of surface water into underground formation by some method of construction, by surface spreading of water or by artificially changing natural conditions. The purpose of artificial recharge of groundwater is to reduce or reverse declining levels of groundwater in a basin, to prevent salt water intrusion from sea to coastal aquifer and to store surplus surface water and reclaimed water for future use. The base flow of a stream can be augmented by recharging ground water at locations far away from the stream so that the recharged water will reach the stream during periods of low flow. The underground fresh water in coastal aquifer can be protected by a hydraulic barrier which can be created by artificial recharge through a line of wells.

The saturated and unsaturated groundwater flow equations provide a means of analysing the impact of artificial recharge on groundwater system. The complex hydrological conditions that develop during certain type of artificial recharge are (i) large change in saturated thickness and (ii) transport of contaminants in an aquifer. In recent years, numerical models have been exclusively used for solving the complex groundwater flow problem.

In a hard rock groundwater basin, it is common to find a weathered zone underlain by massive and fractured zones. The weathered zone and the fractured zone provide opportunity for storing surplus water in them. Both the layers can be recharged economically through a single injection well provided the well intercepts both the layers. Assessment of the quantity of water which is recharged to individual layer and determination of the part of the recharged water that is available in the zone of interest at any time are important tasks.

Construction of percolation tanks is a common practice in several parts of India objective of which is artificial replenishment of groundwater for lift irrigation in small agricultural tracts. A percolation tank is created by constructing a small earthern dam across a natural stream at a suitable location. It is located upstream of an existing cluster of dug wells. The surface runoff during the short monsoon period is collected in the tanks. Under favourable soil and rock conditions, the water percolates and recharge the groundwater. Effectiveness of percolation tanks in recharging groundwater has been studied using the isotope method. In choosing among the several sources of available water for groundwater recharge, increasing importance has been placed in recent years on the use of reclaimed municipal waste water.

**Sea Water Intrusion**

In the coastal margins of groundwater basin, the lowering of water level or potentiometric head can result in the intrusion of sea water, with a resulting degradation of water quality in the basin or a portion of the basin owing to the intruded sea water. For saline intrusion to occur, permeable formation must be in hydraulic connection with sea water, either directly on the ocean floor or along a river estuary or bay which contains sea water. Another necessary condition of saline intrusion is that there be an inland gradient, that is, there must be a tendency for water to move from the sea water sources to the pumping area. Such an inland gradient normally would result from pumping at rate higher than the recharge to the ground water basin. If these conditions exist, there will be a sea water wedge moving inland. The wedge-shaped intrusion results from the fact that sea water is approximately 1.025 times heavier than fresh water. The greater the depth below sea level, the greater is the pressure differential between sea water and fresh water, and hence there is faster movement of the sea water. In most coastal aquifers, the bodies of sea water and fresh water will maintain, for all practical purposes, separate identities because of the difference in density of the water. An exception to this condition occurs where there is a considerable variation in water levels from pumping and tidal fluctuations and where the formation is very permeable. There a transition or mixed zone of fresh water and sea water is known to occur underground.

Several methods to control saline intrusion have been suggested by hydrologists. These are (a) reduction of ground water extraction, (b) artificial recharge by spreading, (c) physical barrier, (d) dumping trough, (e) hydraulic ridge, and f) combination of pumping trough and hydraulic ridge. The relationship between quantity of water flowing to the sea and the length of the wedge is available in literature. Also mathematical modelling of unsteady flow of saline and fresh water in aquifer is documented. The method of controlling saline intrusion must be chosen considering the special condition of the area.

# Groundwater Pollution

In terms of biologic or organic quality, groundwater is generally better than surface water and in terms of chemical quality (dissolved solids) surface water is generally better than ground water. Although it seems that groundwater is more protected than surface water against pollution, it is still subject to pollution, and once pollution of groundwater occurs, the restoration to the original, non-polluted state, is more difficult. Flushing processes in groundwater reservoir are inefficient and expensive. It is easier to prevent the cumulative and complex process of deterioration than to prevent once pollutant concentration have reached problem proportion.

With the increased demand for water and with the intensification of water utilization, the water quality problem becomes the limiting factor in the development of water resources in many parts of the world. Most essential to developing an appropriate contaminant containment/control programme is that a thorough understanding of the subsurface system be developed. A detailed appraisal of the geologic and hydrogeologic setting must be made and the magnitude of the pollution hazard for a specific incident must be evaluated. As the movement of contaminants, and therefore, the methods of containment/control of the contaminants are largely dependent on the hydrogeologic environment, the effectiveness of the various methods will vary with different geologic settings. Designing a programme for containment or control of groundwater contamination is further complicated by the fact that not all the contaminants behave in the same manner in the subsurface. Many contaminants move at different rates in the groundwater system and may occupy different levels in aquifers according to their solubility in water, their density, and other physical properties peculiar to the contaminants. All of the processes of migration and alterations present in ground water are also present in the unsaturated zone. However the flow of water through the unsaturated zone is considerably more complex due to the presence of the air water vapour phases. Nevertheless it is important to note that the attenuation mechanisms in the unsaturated zone can provide a powerful barrier to the passage of contaminants to the saturated zone.

The containment and/or control of contaminated groundwater can generally be accomplished using one or a combination of several available techniques. The alternatives available for remedial action can be classified into three broad categories (i) physical containment measures, including slurry trench cutoff walls, grout curtains, sheet piling, and hydrodynamic control, (ii) aquifer rehabilitation, including withdrawal, treatment, reinjection (or recharge), and in-situ treatment such as chemical neutralization and biological neutralization, (iii) withdrawal, treatment and use.

The objective of groundwater pollution study in a basin is to answer the questions (i) what contaminants are present in the groundwater? (ii) what are harmful levels for specific contaminants? (iii) how will the contamination levels in the groundwater system change with time? (iv) can the source of contamination be controlled effectively? These questions could be answered by a hydrodynamic transport model. A program of study of the quality of groundwater envisages field observations regarding the source and environment of groundwater occurrences, source of pollution and other related aspects having a bearing on the quality of groundwater. In fact, determination of safe yield for a groundwater resource must consider both its quantity and quality.

While exploiting groundwater, it is appropriate to ascertain whether a groundwater abstraction zone is liable to contamination and whether a potential source of pollution is contaminating the aquifer. In either case, the present and/or future conditions applicable to the migration of contaminants are determined. For analysis of pollution potential at a particular site, data describing contaminants, their migration characteristics and the characteristics of groundwater regime are collected. Whilst inspection of the available data can provide a strong insight into a potential pollution hazard, the use of models may provide more appropriate and rigorous method for integrating all the available data together and for evaluation of the response of the aquifer system to a contamination event. The models are generally derived from the expression of the flow and transport processes in terms of mathematical equations which are then solved by incorporating appropriate parameter values and boundary conditions derived from the collected field data. Analytical solutions for solute transport in homogeneous and isotropic saturated porous media, in which the flow velocity is uniform, are available for simple one and two dimensional cases. Contaminant transport problems have been solved by numerical methods which are well documented. The numerical methods exhibit numerical spatial oscillations while solving the advection dominated transport problems.

**Concluding Remarks**

Although in recent years, intensive studies have been carried out on various aspects of groundwater hydrology, there are still many gaps in our knowledge. The following are some of the aspects which are of great relevance for proper utilization of groundwater resources and therefore need immediate attention:

(a) There is a need for establishing an integrated National and State groundwater Data Storage and Retrieval System to collect, store, update, process and disseminate groundwater data to enable planning and management of groundwater resources.

(b) There is a need for constituting a new committee for updating the methodology for groundwater assessment. This will also be in line with the recommendations made in National Water Policy for periodic assessment of groundwater.

(c) There is a need for studying unsaturated and saturated flow through weathered and fractured rocks for finding the recharge components from rainfall and from percolation tanks in hard rock groundwater basin. The irrigation return flow under different irrigation practices for different soils and different crops needs to be quantified. Also user friendly software should be developed for quick assessment of regional groundwater resources.

(d) There is a need for studying soil moisture movement on a continuous basis considering both the storm and interstorm periods. The soil moisture forecast model needs to be made operational on interactive mode.

(e) There is a need for the study of interaction among a stream and several aquifers which are separated by an aquitard considering the changing river width and stage that may occur during passage of flood. A groundwater flow model for a multi-aquifer system to cater the need of groundwater management should be developed. The assessment of induced recharge due to groundwater abstraction in the river basin should be made for planning optimal operation of a groundwater reservoir.

(f) A hydrologic study is required to rejuvenate drying springs. Delineation of recharge area of the spring by remote sensing and nuclear methods needs special efforts for controlling human interference. There is a need for finding the true relationship between dynamic storage of groundwater and the spring discharge.

(g) There is a need for development of methodology to predict quantity of water artificially recharged by different methods and their temporal and spatial availability in the area of interest. Appropriate method of artificial recharge in hard rock basin and method for retaining the recharged water in the subsurface reservoir need to be established. It remains to be investigated how effectiveness of percolation tanks decreases with time because of siltation, what fraction of the water stored joins the groundwater because of evaporation from tank, how much increment in recharge rate could be achieved through construction of recharge shafts. Pretreatment process of the waste water and renovation of waste water with rapid infiltration land treatment system are presently important areas of research.

(h) There is a need to develop numerical scheme to eliminate numerical oscillation while solving the advection dominated transport problems. Assessment of groundwater quality in many groundwater basins remains as a task yet to be performed.

**References**

1.      Engelen, G. B. and G. P. Jones (1986). Developments in the Analysis of Groundwater Flow Systems. IAHS Publication No. 163, A Contribution to the International Hydrological Programme of UNESCO (Project A 2.8), 1986, 356 p.

2.      Kashyap, Deepak (1989). Mathematical Modelling for Groundwater Management - Status in India. Proceedings, Indo-French Seminar on Management of Water Resources, 22-24 September 1989, Festival of France - 1989, Jaipur, pp. IV-59 to IV-75.

3.      Kumar, C. P. (1992). Ground Water Modelling. In. Hydrological Developments in India Since Independence, A Contribution to Hydrological Sciences, National Institute of Hydrology, Roorkee, pp. 235-261.

4.      Mishra, G. C. (1993). Research and Development in Hydrology - Ground Water Hydrology. Proceedings, National Workshop on Thrust Areas of Research in Hydrology, 17-18 June 1993, Roorkee, pp. 72-89.

# 4-Management of Aquifer Recharge

## Introduction

Groundwater recharge is the replenishment of an aquifer with water from the land surface. It is usually expressed as an average rate of mm of water per year, similar to precipitation. In addition to precipitation, other sources of recharge to an aquifer are stream and lake or pond seepage, irrigation return flow (both from canals and fields), inter-aquifer flows, and urban recharge. In contrast to natural recharge (which results from natural causes), artificial recharge is the use of water to artificially replenish the water supply in an aquifer. Of all the factors in the evaluation of groundwater resources, the rate of recharge is one of the most difficult to derive with confidence. Estimates of recharge are normally subject to large uncertainties and spatial and temporal variability.

The increasing demand for water has increased awareness towards the use of artificial recharge to augment ground water supplies. Stated simply, artificial recharge is a process by which excess surface water is directed into the ground – either by spreading on the surface, by using recharge wells, or by altering natural conditions to increase infiltration – to replenish an aquifer. It refers to the movement of water through man-made systems from the surface of the earth to underground water-bearing strata where it may be stored for future use. Artificial recharge (sometimes called planned recharge) is a way to store water underground in times of water surplus to meet demand in times of shortage.

Some factors to consider for Artificial Recharge are (O'Hare et al., 1986)

- Availability of waste water
- Quantity of source water available
- Quality of source water available
- Resulting water quality (reactions with native water and aquifer materials)
- Clogging potential
- Underground storage space available
- Depth to underground storage space
- Transmission characteristics
- Topography/applicable methods (injection or infiltration)
- Legal/institutional constraints
- Costs
- Cultural/social considerations

## Artificial Recharge Projects

The goal of most artificial recharge projects is to convey water to the saturated zone. Evaluation of the viability of proposed projects and of the effectiveness of existing projects requires an understanding and predictive capability of their hydraulic and chemical effects. It focuses on the potential hydraulic consequences of altering the saturated flow system through artificial recharge, which are largely controlled by the geologic and hydrologic characteristics of the aquifer system. A combination of field, laboratory, analytical, and simulation methods generally are used to develop an understanding of the hydrogeologic system as a basis for predicting potential consequences. Optimization techniques may be coupled with predictive models of ground-water flow and other processes to create an effective tool for planning and management of artificial recharge projects. Pre-project and long-term monitoring of key aspects of a flow system is an essential part of a successful management plan.

Artificial recharge projects are undertaken for many purposes in a variety of aquifer systems and associated hydraulic conditions. Regardless of the initial distribution and trend of hydraulic heads in these systems, artificial recharge will alter these heads and associated conditions. Characterization of the geology is important in determining the viability of an artificial recharge project, particularly where significant lateral and (or) vertical ground-water flow is required between recharge and discharge locations.

Hydrologic considerations for the saturated-flow component of an artificial recharge project typically include the distribution of head and stress prior to and during project operations, hydraulic properties, the fate of artificially recharged water, and offsite effects. The prediction of saturated flow during artificial recharge projects requires information on the distribution of stress, or recharge and discharge. These stresses can include a variety of natural and artificial processes that can be measured in a variety of ways. The hydraulic properties of an aquifer system, along with the distribution of stress, determine the direction and rate of saturated flow. Given the distribution of head, stress, and hydraulic properties, simulation models can be developed to help address the fate of artificially recharged water and offsite effects. Monitoring and simulation are both used to address offsite effects; however, simulation can also be used to design an efficient monitoring network prior to full-scale implementation.

Successful planning and management of an artificial recharge project often requires consideration of many water management objectives, water routing capabilities, economics, offsite effects, as well as other factors. Optimization techniques are designed to identify an optimal way to meet an objective given a set of constraints. The linkage of a predictive ground-water flow model with optimization techniques, or a simulation/optimization model, allows for simultaneous consideration of the flow system and physical and (or) economic constraints determined by water-resource managers.

Simulation/optimization models have been applied to ground-water problems for decades and have been used to plan and manage artificial recharge projects. Monitoring of hydraulic conditions prior to and during an artificial recharge project is an essential part of a management plan, and often is an integral part of project operations. Measurement of project performance is clearly one goal of a monitoring program. A second goal is to provide the information needed for future improvement of predictive modeling capabilities and adjustment of optimization constraints. Reduced uncertainty in model results translates directly to increased confidence in management decisions based on these models.

Artificial recharge projects can be a valuable component of a groundwater management and conjunctive use strategy, for long-term reliability of groundwater supply, improvement of basin water quality, and for water banking opportunities.

Artificial Recharge programs are typically conducted in three phases:

## 1. Feasibility

Evaluate dynamics of groundwater flow and basin recharge, and considers options for artificial recharge techniques which can be used. A primary concern is the identification of basin compartmentalization or impermeable layers within the aquifer which inhibit recharge to the basin aquifers. Also important are concerns about chemical mixing of surface waters and native groundwater, hydrologic variability within the aquifers, and the nature of probable migration of recharged water. Different sources of surface water, together with potentially different regulatory concerns are also evaluated as part of the feasibility program. Where applicable, prepare necessary feasibility and hydrologic reports for regulatory oversight and permitting agencies.

## 2. Test Program Design and Operation

Based on results of the feasibility analysis, design a test program, using existing facilities if possible. This work includes chemical and physical modeling of recharge options, detailed chemical analyses of co-mingled waters which have different initial chemical signatures, and measurement of recharge rates in the test program.

## 3. Full-Scale Project Implementation

Test program results are used to recommend final, full-scale program parameters, including sites for additional wells or infiltration ponds (if necessary), potential future options for surface water sources, recharge management planning during regular operations, and necessary monitoring. Focus on keeping the system design flexible, so that changing needs of the client can be integrated with existing recharge operations and facilities.

## Methods of Artificial Recharge

Artificial recharge methods can be classified into two broad groups (i) direct methods, and (ii) indirect methods.

### Direct Methods

(a) Surface Spreading Techniques

The most widely practiced methods of artificial recharge of groundwater employ different techniques of increasing the contact area and resident time of surface water with the soil so that maximum quantity of water can infiltrate and augment the groundwater storage. Areas with gently sloping land without gullies or ridges are most suited for surface water spreading techniques.

1. Flooding

Flooding technique is very useful in selected areas where favourable hydrogeological situation exists for recharging the unconfined aquifer by spreading the surplus surface water from canals/streams over large area for sufficiently long period of time so that it recharges the groundwater body. This technique can be used for gently sloping land of 1 to 3% without gullies and ridges.

2.  Ditch and Furrows

In areas with irregular topography, shallow, flat bottomed and closely spaced ditches or furrows provide maximum water contact area for recharge water from source stream or canal. This technique requires less soil preparation than the recharge basins and is less sensitive to silting.

3.  Recharge Basin

Artificial recharge basins are either excavated or enclosed by dykes or levees. They are commonly built parallel to ephemeral or intermittent stream channels. The water contact area in this method is quite high which typically ranges from 75 to 90% of the total recharge area. In this method, efficient use of space is made and the shape of basins can be made suited to the terrain condition and the available apace.

4.  Runoff Conservation Structures

In areas receiving low to moderate rainfall mostly during a single monsoon season and not having access to water transferred from other areas, the entire effort of water conservation is required to be related to the available 'insitu' precipitation.

*Gully plugs* are the smallest runoff conservation structures built across small gullies and streams rushing down the hill slopes carrying drainage of tiny catchments during rainy season. Usually, the bund is constructed by using local stones, earth and weathered rock, brush wood, and other such local materials.

Sloping lands with surface gradients upto 8% having adequate soil cover can be levelled through *bench terracing* for bringing under cultivation. It helps in soil conservation and holding runoff water on terraced area for longer duration giving rise to increased infiltration recharge.

*Contour bunding* is a watershed management practice to build up soil moisture storages. This technique is adopted generally in low rainfall areas where the monsoon runoff is impounded by putting bunds on the sloping ground all along the contour of equal elevation. Contour bunding is taken up on lands with moderate slopes without involving terracing.

As compared to gully plugs, *nala bunds* are constructed across bigger nalas of second order streams in areas having gentler slopes. A nala bund acts like a mini percolation tank. The total catchment of the nala should be between 40 to 100 hectares. The annual rainfall in the catchment should be less than 1000 mm. The nala bunds should be preferably located in area where contour or graded bunding on lands have been carried out.

In areas where uncultivated land is available in and around the stream channel section, which shows sufficient permeability for sub-surface percolation, small tanks are created by making low elevation stop dam across the stream. The tank can also be located adjacent to the stream by excavation and connecting it to the stream through a delivery canal. These tanks called *"percolation tanks"* are thus artificially created surface water bodies submerging a highly permeable land area so that the surface runoff is made to percolate and recharge the groundwater storage. Normally, a percolation tank should not retain water beyond February. It should be located downstream of runoff zone, preferably towards the edge of piedmont zone or in the upper part of transition zone (land slope between 3 to 5%). There should be adequate area suitable for irrigation near a percolation tank.

5.  Stream Channel Modification

The natural drainage channel can be modified with a view to increase the infiltration by detaining stream flow and increasing the stream bed area in contact with water. This method can be employed in areas having influent streams (stream bed above water table) which are mostly located in piedmont regions and areas with deep water table (semi-arid, arid region and valley fill deposits). Stream channel modification methods are generally applied in alluvial areas.

6.  Surface Irrigation

Surface irrigation aims at increasing agricultural production by providing dependable watering of crops during gaps in monsoon and during non-monsoon period. Wherever adequate drainage is assured, if additional source water becomes available, surface irrigation should be given first priority as it gives a dual benefit of augmenting groundwater resources.

(b) Sub-surface Techniques

When deeper aquifers are overlain by impervious layers, the infiltration from surface can not recharge the sub-surface aquifer under natural conditions. The techniques adopted to recharge the confined aquifers directly from surface water source are grouped under sub-surface recharge techniques.

1.  Injection Wells

Injection wells are structures similar to a tube well but with the purpose of augmenting the groundwater storage of a confined aquifer by "pumping in" treated surface water under pressure. The aquifer to be replenished is generally one which is already over exploited by tube well pumpage and the declining trend of water levels in the aquifer has set in. Artificial recharge of aquifers by injection wells is also done in coastal regions to arrest the ingress of sea water and to combat the problems of land subsidence in areas where confined aquifers are heavenly pumped. Due to higher well losses caused by clogging, the injection wells display lower efficiency (40 to 60%) as compared to a pumping well of similar design in the same situation. The source water and the water in the aquifer should be compatible to avoid any precipitation, causing clogging of well. Injection-cum-pumping wells are more efficient because the well can be cleaned during pumping operation.

## 2. Gravity Head Recharge Wells

In addition to specially designed injection wells, ordinary bore wells and dug wells used for pumping may also be alternatively used as recharge wells, whenever source water becomes available. In certain situations, such wells may also be constructed for effecting recharge by gravity inflow. In areas where water levels are declining due to over development, using available structures for causing recharge may be the immediately available economic option. Studies can be undertaken in selected areas where excess surface water is available during rainy season in rivers and the adjoining dug wells show sufficient unsaturated phreatic aquifer even after rains.

## 3. Connector Wells

Connector wells are special type of recharge wells where, due to difference in potentiometric head in different aquifers, water can be made to flow from one aquifer to other without any pumping. The aquifer horizons having higher heads start recharging aquifer having lower heads.

## 4. Recharge pits

Recharge pits are structures which overcome the difficulty of artificial recharge of phreatic aquifer from surface water sources. Recharge pits are excavated of variable dimensions that are sufficiently deep to penetrate less permeable strata. *Nala trench* is a special case of recharge pit dug across a nala bed. An ideal site for nala trench is influent stretch of a stream which shows up as dry patch. One variation of recharge pit is a *contour trench* extending over long distances across the slope and following topographic contour. This measure is more suitable in piedmont regions and in areas with higher surface gradients. As in case of other water spreading methods, the source water used should be as silt free as possible. In case of hard rock terrain, a nala bed section crossing permeable strata of weathered fractured rock or the nala section coinciding with a prominent lineament or intersection of two lineaments, form ideal sites for nala trench.

## 5. Recharge Shafts

In case, the water table aquifer located deep below land surface, is overlain by poorly permeable strata, a shaft is used for causing artificial recharge. A recharge shaft is similar to a recharge pit but much smaller in cross-section.

Indirect Methods

(a) Induced Recharge

It is an indirect method of artificial recharge involving pumping from aquifer hydraulically connected with surface water, to induce recharge to the groundwater reservoir. It is more a pumpage augmentation rather than artificial recharge measure. In hard rock areas, the abandoned channels often provide good sites for induced recharge. The greatest advantage of this method is that under favourable hydrogeological situations, the quality of surface water generally improves due to its path through the aquifer materials before it is discharged from the pumping well.

## 1. Pumping Wells

Induced recharge system is installed near perennial streams that are hydraulically connected to an aquifer through the permeable rock material of the stream channel. The outer edge of a bend in the stream is favourable for location of well site. The chemical quality of surface water source is one of the most important consideration during induced recharge.

## 2. Collector Wells

For obtaining very large water supplies from river bed, lake bed deposits or water-logged areas, collector wells are constructed. The large discharges and lower lift heads make these wells economical even if initial capital cost is higher as compared to tube well. In areas where the phreatic aquifer adjacent to the river is of limited thickness, horizontal wells may be more appropriate than vertical wells. Collector well with horizontal laterals and infiltration galleries can get more induced recharge from the stream.

## 3. Infiltration Gallery

Infiltration galleries are other structures used for tapping groundwater reservoir below river bed strata. The gallery is a horizontal perforated or porous structure (pipe) with open joints, surrounded by a gravel filter envelope laid in permeable saturated strata having shallow water table and a perennial source of recharge. The galleries are usually laid at depths between 3 to 6 metres to collect water under gravity flow. The galleries can also be constructed across the river bed if the river bed is not too wide. The collector well is more sophisticated and expensive but has higher capacities than the infiltration gallery. Hence, choice should be made by the required yield followed by economic aspects.

## (b) Aquifer Modifications

These techniques modify the aquifer characteristics to increase its capacity to store and transmit water. With such modifications, the aquifer, at least locally, becomes capable of receiving more natural as well as artificial recharge. Hence, in a sense these techniques are artificial yield augmentation measures rather than artificial recharge measures.

## 1. Bore Blasting

These techniques are suited to hard crystalline and consolidated strata. Through hydrogeological investigation, suitable sites are fixed where the aquifer displays limited yield which dwindles or dries in winter or summer months. All the blast holes reach the depth of the aquifer required to be benefitted, whether unconfined or confined. All the charges of row or circle are exploded at a time. G.S.D.A. (Maharashtra) call this technique as 'Jacket well' because a jacket of better aquifer is created around a water supply well.

## 2.  Hydro Fracturing

In many cases, the blasting has given indifferent results. Hydro fracturing is a later technique being used to improve secondary porosity in hard rock strata. The hydro fracturing is a process whereby hydraulic pressure is applied to an isolated zone of bore wells to initiate and propagate fractures and extend existing fractures. The water under high pressure break up the fissures, cleans away clogging and leads to a better contact with adjacent water bearing strata. The yield of the bore well is improved. In hydro fracturing, vertical fractures are initiated which inter-connects aquifers at different levels in addition to extension of existing fractures. This leads to better conditions for artificial recharge. The technique may be applied at bore well sites located in hard crystalline rock or other massive consolidated strata including metamorphic and sedimentary formations. Generally, a bore well giving low or poor yield is treated, but the technique can also benefit other wells.

## (c) Groundwater Conservation Structures

The water artificially recharged into an aquifer is immediately governed by natural groundwater flow regime. It is necessary to adopt groundwater conservation measures so that the recharged water remains available when needed.

## 1.  Groundwater Dams / Underground Bandharas

A groundwater dam is a sub-surface barrier across stream which retards the natural groundwater flow of the system and stores water below ground surface to meet the demands during the period of greatest need. The main purpose of groundwater dam is to arrest the flow of groundwater out of the sub-basin and increase the storage within the aquifer. The sub-surface barriers need not be only across the nala bed. In some micro watersheds, sub-surface dykes can be put to conserve the groundwater flow in larger area in a valley. Sites have to be located in areas where there is a great scarcity of water during the summer months or there is a need for additional water for irrigation. Technical possibilities of constructing the dyke and achieving large storage reservoirs with suitable recharge conditions and low seepage losses are the main criteria for sub-surface dyke. It directly benefits upgradient area and hence care should be taken that a large number of users are not located immediately downstream.

## 2.  Fracture Sealing Cementation Technique

In many hard rock areas, the groundwater circulation to deeper levels is governed by shear, fault or fracture plane indicated by lineaments. The boreholes located on such zones prove productive but due to dissipation of the limited storage along preferred flow planes, in case of adverse topographic situation, these become dry by end of winter or summer. Fracture seal cementation is a suitable water conservation measure in such situations. These measures can also be used to prevent ingress of saline or polluted water from a known source. The groundwater flow system at the site should be adequately known to establish the outflow direction and the preferred fracture planes along which the flow occurs under natural hydraulic gradient.

Under certain hydrogeological situations, a combination of several surface and sub-surface recharge methods and groundwater conservation techniques, can be used in conjunction with one another for an optimal recharge of groundwater resources.

## Artificial Groundwater Recharge in India

Technological developments in well construction and pumping methods have resulted in the large-scale exploitation of groundwater in India and elsewhere. In many parts of India, due to the vagaries of the monsoon, and, in the arid and semi-arid regions, to the lack or scarcity of surface water resources, dependence on groundwater has increased tremendously in recent years. Thus, given the potential for the available groundwater resources to be over-exploited in these areas, it is essential that proper storage and management of available groundwater resources be instituted.

Replenishment of groundwater by artificial recharge of aquifers in the arid and semi-arid regions of India is essential as the intensity of normal rainfalls is grossly inadequate to produce any moisture surplus under normal infiltration conditions. Although artificial groundwater recharge methods have been extensively used in the developed nations for several decades, their use in developing nations, like India, has occurred only during the last ten to twenty years. Techniques such as nala bunding, constructing percolation tanks, trenching along slopes and around hills, etc., have been used for some time, but have typically lacked a scientific basis (e.g., a knowledge of the geological, hydrological and morphological features of the areas) for selecting the sites on which the recharge structures are located.

Various techniques for artificial groundwater recharge have been employed in the states of Maharashtra, Gujarat, Tamil Nadu and Kerala. In Maharashtra, studies were carried out on seven percolation tanks in the Sina and the Main River basins. The average recharge volume of these tanks was 50% of the capacity of the tank, provided the tank bottom was maintained by removing accumulated sediment and debris prior to the annual monsoon. Best results were obtained from systems located in areas of vesicular or fractured basalt. Nala (stream) bunding, where the recharge structure was situated within the course of the nala, was found to be most effective and economical as the surface area exposed to evaporation was, on average, 10% of that of an average-sized percolation tank. Within nala bunds, the rate of infiltration varied from 50% to 70% of the capacity of the reservoir. Infiltration was aided by a connector well linking the phreatic, alluvial aquifer at 6 m depth with the deeper, confined basaltic aquifer at 63 m depth, allowing the free flow of water by gravity from phreatic aquifer to the confined aquifer at the rate of 0.19 million m3/year. The water level in the phreatic aquifer, which was saturated due to infiltration from the surface reservoir, was 3 m below ground level, and the piezometric level in confined aquifer was 30 m below ground level.

In Tamil Nadu and Kerala, studies were carried out on nine percolation tanks in the semi-arid regions of the Noyil Ponani and Vattamalai River basins. Rates of percolation were as high as 163 mm/day at the beginning of the rainy season, but diminished thereafter mainly due to the accumulation of silt in the bottoms of the tanks. Periodic desilting, therefore, was determined to be an essential element in the maintenance of these tanks. In contrast, subsurface dikes of 1 m to 4 m in height were found effective in augmenting groundwater resources, particularly in the hard rock areas underlain by fractured aquifers.

In Punjab, studies of artificial recharge using injection wells were carried out in the Ghaggar River basin, using canal water as the primary surface water source. The injection rate was initially 43.8 l/sec at an injection pressure of one atmosphere (atm). The pressure increased to 2 atm after 5 hours, and remained constant thereafter, although the recharge rate gradually diminished to 3.5 l/sec after few days. The natural, gravity-controlled recharge rate was 5.1 l/sec. Notwithstanding, over time, the reproducable recharge rate obtained using the pressure injection system was found to be about 10 times greater than the rate obtained using gravity flow. The increase in pressure during injection was due to clogging of the interstitial spaces within the aquifer, which can be minimized by careful control of the source water quality. Periodic cleaning of well was also required, whenever the pressure increased beyond 6 atm or showed a sudden rise. Further studies were conducted on induced recharge from the Ghaggar River using a well field, with individual wells spaced at 200 m intervals, within 100 m of the river bank. As with the injection wells, periodic removal of the clay film deposited in the floodplain above the natural recharge areas of the aquifer was required to improve recharge efficiency.

In Gujarat, studies of artificial recharge were carried out in two areas. In the Central Mehsana area of North Gujarat, artificial recharge was carried out using injection wells, connector wells, and infiltration channels and ponds. Surplus groundwater from the floodplain aquifers of the major rivers in Mehsana area and tail-end releases from the Dharoi Canal System were utilized as the water sources. In addition, the injection of water from the phreatic aquifers into the deeper, overexploited aquifers was investigated in the Central Mehsana area. In the coastal areas of Saurashtra, artificial recharge was carried out using injection wells and recharge basins. Stormwater runoff and tail-end releases from the canal system of the Hiran Irrigation Project were used as the water sources, and the studies included an evaluation of the effectiveness of the existing tidal regulators and check dams, designed to limit the extent of seawater intrusion. Of the methods studied in the Central Mehsana area, spreading methods, using techniques such as spreading channels, recharge pits and ponds, were found to be more economical than injection methods, although dual purpose connector wells were found to be more economical for recharging the deep aquifer. The dual purpose connector wells not only supplied water by gravity to the deep aquifer, but also abstracted water by periodic pumping, which reduced the extent of clogging of the wells. In contrast, the coastal Saurashtra area where the aquifers are highly porous and drain to the coastal zone, the rapid outflow of recharged water to the sea did not make artificial recharge a viable proposal. However, the tidal regulators which created barriers of freshwater along the creeks and in coastal depressions effectively prevented seawater intrusion in these areas.

Also in Gujarat, studies of subsurface storage were carried out. In the Jamnagar District, naturally-occurring basaltic dikes were known to retain groundwater. However, it was also known that the surface soils in the District were not waterlogged. The studies indicated that, while the lower portion of the dike acted as a barrier to the passage of groundwater, the top few metres of the dike, composed of fractured basalt, allowed the passage of groundwater through the soil profile, preventing waterlogging in the aquifer area. This design feature was subsequently incorporated into the specifications of subsurface dikes.

Elsewhere in India, watershed management practices adopted in some states to minimize soil loss in erosion gullies also contribute to groundwater recharge. Check dams not only store surface water during portions of the year, but also encourage infiltration into the surfacial aquifers, providing a threefold benefit to communities (i.e., prevention of soil loss, provision of water for livestock watering and human use, and groundwater recharge). Such works have been implemented on an extensive scale in Gujarat, Maharashtra, Madhya Pradesh, and Rajasthan since 1960.

## Advantages and Disadvantages

Artificial recharge has several potential advantages:

- The use of aquifers for storage and distribution of water and removal of contaminants by natural cleaning processes which occur as polluted rain and surface water infiltrate the soil and percolate down through the various geological formations.
- The technology is appropriate and generally well understood by both the technicians and the general population.
- Very few special tools are needed to dig drainage wells.
- In rock formations with high, structural integrity few additional materials may be required (concrete, softstone or coral rock blocks, metal rods) to construct the wells.
- Groundwater recharge stores water during the wet season for use in the dry season, when demand is highest.
- Aquifer water can be improved by recharging with high quality injected water.
- Recharge can significantly increase the sustainable yield of an aquifer.
- Recharge methods are environmentally attractive, particularly in arid regions.
- Most aquifer recharge systems are easy to operate.
- In many river basins, control of surface water runoff to provide aquifer recharge reduces sedimentation problems.
- Recharge with less-saline surface waters or treated effluents improves the quality of saline aquifers, facilitating the use of the water for agriculture and livestock.

Artificial Recharge has some disadvantages too:

- In the absence of financial incentives, laws, or other regulations to encourage landowners to maintain drainage wells adequately, the wells may fall into disrepair and ultimately become sources of groundwater contamination.
- There is a potential for contamination of the groundwater from injected surface water runoff, especially from agricultural fields and roads surfaces. In most cases, the surface water runoff is not pre-treated before injection.
- Recharge can degrade the aquifer unless quality control of the injected water is adequate.
- Unless significant volumes can be injected into an aquifer, groundwater recharge may not be economically feasible.
- The hydrogeology of an aquifer should be investigated and understood before any future full-scale recharge project is implemented. In karstic terrain, dye tracer studies can assist in acquiring this knowledge.
- During the construction of water traps, disturbances of soil and vegetation cover may cause environmental damage to the project area.

## Operation and Maintenance

Periodic maintenance of artificial recharge structures is essential because infiltration capacity is rapidly reduced as a result of silting, chemical precipitation, and accumulation of organic matter. In the case of spreading structures, annual maintenance consists of scraping the infiltration surfaces to remove accumulated silt and organic matter. In the case of injection wells and connector wells, periodic maintenance of the system consists of pumping and/or flushing with a mildly acidic solution to remove encrusting chemical precipitates and bacterial growths on the well tube slots. By converting the injection or connector wells into dual purpose wells, the interval between periodic cleanings can be extended, but, in the case of spreading structures except for subsurface dikes constructed with an overflow or outlet, annual desilting is a must. Unfortunately, because the structures are installed as a drought relief measure, the periodic maintenance is often neglected until a subsequent drought, at which time the structures must be restored (the 5 to 7 year frequency of droughts, however, means that some maintenance does take place). Structural maintenance is normally carried out by several agencies and individuals. Maintenance of minor irrigation tanks is normally carried out by the state irrigation department, maintenance of contour bunds and trenches (along with related afforestation activities) by the state forestry department, and maintenance of farm ponds and related structures by the cultivators.

## Concluding Remarks

- The costs of recharge schemes, in general, depend upon the degree of treatment of the source water, the distance over which source water must be transported, and stability of recharge structure and resistance to siltation and/or clogging.

- Artificial recharge of ground water should be licensed and controlled by competent authorities according to specific requirements laid down in an appropriate permit system which should be flexible so as to adopt to site-specific conditions. The question of ground-water exploitability should be clarified on a case-by-case basis, taking into account all relevant aspects, including ecological ones. The relevant regulations should establish the extent to which exemptions can be allowed.

- Authorization for artificially recharging the aquifer should be granted only if the hydrogeological situation, environmental conditions and the recharge-water quality permit injection, percolation or infiltration of water by artificial means into aquifers for storage and retrieval of good-quality water as well as for restoring over-exploited ground-water resources. For induced recharge from adjacent streams or lakes, appropriate security measures should be applied to forestall accidental pollution.

- Appropriate measures should be taken to combat saltwater encroachment into coastal aquifers. In such areas, special regulations for ground-water abstraction should be enforced to avoid seepage into aquifers owing to overpumping and the resultant lowering of the ground-water table.

# References

1.      Manual on Artificial Recharge of Ground Water (1994). Technical Series – M, No. 3, Central Ground Water Board, Faridabad, March 1994, 215 p.

2.      National Drinking Water Mission and Department of Rural Development (1989). Rain Water Harvesting. Government of India, New Delhi.

3.      Nayantara Nanda Kumar and Niranjan Aiyagari (1997). Artificial Recharge of Groundwater.
        http://www.cee.vt.edu/program_areas/environmental/teach/gwprimer/recharge/recharge.html

4.      O'Hare, M.P., Fairchild, D.M., Hajali, P.A., Canter, L.W. (1986). Artificial Recharge of Groundwater. Proceedings of the Second International Symposium on Artificial Recharge of Groundwater.

5.      Phillips, Steven P. The Role of Saturated Flow in Artifical Recharge Projects (2002). U.S. Geological Survey, Placer Hall, 6000 J Street, Sacramento, California 95819-6129.

6.      Sophocleous, M.A. and J. A. Schloss. Estimated Annual Groundwater Recharge.
        http://www.kgs.ukans.edu/HighPlains/atlas/atrch.htm

# 5-Assessment of Groundwater Potential

## Introduction

Rapid industrial development, urbanisation and increase in agricultural production have led to freshwater shortages in many parts of the world. In view of increasing demand of water for various purposes like agricultural, domestic and industrial etc., a greater emphasis is being laid for a planned and optimal utilisation of water resources. The water resources of the basins remain almost constant while the demand for water continues to increase. The utilisable water resources of India are estimated to be 1121 BCM out of which 690 BCM is surface water resources and 431 BCM is groundwater resources.

Due to uneven distribution of rainfall both in time and space, the surface water resources are unevenly distributed. Also, increasing intensities of irrigation from surface water alone may result in alarming rise of water table creating problems of water-logging and salinisation, affecting crop growth adversely and rendering large areas unproductive. This has resulted in increased emphasis on development of groundwater resources. The simultaneous development of groundwater, specially through dug wells and shallow tubewells, will lower water table, provide vertical drainage and thus can prevent water-logging and salinisation. Areas, which are already waterlogged, can also be reclaimed.

On the other hand, continuous increased withdrawals from a groundwater reservoir in excess of replenishable recharge may result in regular lowering of water table. In such a situation, a serious problem is created resulting in drying of shallow wells and increase in pumping head for deeper wells and tubewells. This has led to emphasis on planned and optimal development of water resources. An appropriate strategy will be to develop water resources with planning based on conjunctive use of surface water and groundwater.

For sustainable development of water resources, it is imperative to make a quantitative estimation of the available water resources. For this, the first task would be to make a realistic assessment of the surface water and groundwater resources and then plan their use in such a way that full crop water requirements are met and there is neither water-logging nor excessive lowering of groundwater table. It is necessary to maintain the groundwater reservoir in a state of dynamic equilibrium over a period of time and the water level fluctuations have to be kept within a particular range over the monsoon and non-monsoon seasons.

Groundwater is a dynamic system. The total annual replenishable groundwater resource of India is around 431 BCM. Inspite of the national scenario on the availability of groundwater being favourable, there are many areas in the country facing scarcity of water. This is because of the unplanned groundwater development resulting in fall of water levels, failure of wells, and salinity ingress in coastal areas. The development and over-exploitation of groundwater resources in certain parts of the country have raised the concern and need for judicious and scientific resource management and conservation.

The '*National Water Policy*' adopted by the Government of India in 1987 and revised in 2002 and 2012, regards water as one of the most crucial elements in developmental planning. Regarding groundwater, it recommends that

♦ A portion of river flows should be kept aside to meet ecological needs ensuring that the low and high flow releases are proportional to the natural flow regime, including base flow contribution in the low flow season through regulated ground water use.

♦ The anticipated increase in variability in availability of water because of climate change should be dealt with by increasing water storage in its various forms, namely, soil moisture, ponds, ground water, small and large reservoirs and their combination. States should be incentivized to increase water storage capacity, which inter-alia should include revival of traditional water harvesting structures and water bodies.

♦ There is a need to map the aquifers to know the quantum and quality of ground water resources (replenishable as well as non-replenishable) in the country. This process should be fully participatory involving local communities. This may be periodically updated.

♦ Declining ground water levels in over-exploited areas need to be arrested by introducing improved technologies of water use, incentivizing efficient water use and encouraging community based management of aquifers. In addition, where necessary, artificial recharging projects should be undertaken so that extraction is less than the recharge. This would allow the aquifers to provide base flows to the surface system, and maintain ecology.

♦ Water saving in irrigation use is of paramount importance. Recycling of canal seepage water through conjunctive ground water use may also be considered.

♦ There should be concurrent mechanism involving users for monitoring if the water use pattern is causing problems like unacceptable depletion or building up of ground waters, salinity, alkalinity or similar quality problems, etc., with a view to planning appropriate interventions.

♦ The over-drawal of groundwater should be minimized by regulating the use of electricity for its extraction. Separate electric feeders for pumping ground water for agricultural use should be considered.

♦ Quality conservation and improvements are even more important for ground waters, since cleaning up is very difficult. It needs to be ensured that industrial effluents, local cess pools, residues of fertilizers and chemicals, etc., do not reach the ground water.

♦ Industries in water short regions may be allowed to either withdraw only the make up water or should have an obligation to return treated effluent to a specified standard back to the hydrologic system. Tendencies to unnecessarily use more water within the plant to avoid treatment or to pollute ground water need to be prevented.

♦ Appropriate institutional arrangements for each river basin should be developed to collect and collate all data on regular basis with regard to rainfall, river flows, area irrigated by crops and by source, utilizations for various uses by both surface and ground water and to publish water accounts on ten daily basis every year for each river basin with appropriate water budgets and water accounts based on the hydrologic balances. In addition, water budgeting and water accounting should be carried out for each aquifers.

♦ Appropriate institutional arrangements for each river basin should also be developed for monitoring water quality in both surface and ground waters.

A complexity of factors - hydrogeological, hydrological and climatological, control the groundwater occurrence and movement. The precise assessment of recharge and discharge is rather difficult, as no techniques are currently available for their direct measurements. Hence, the methods employed for groundwater resource estimation are all indirect. Groundwater being a dynamic and replenishable resource is generally estimated based on the component of annual recharge, which could be subjected to development by means of suitable groundwater structures.

For quantification of groundwater resources, proper understanding of the behaviour and characteristics of the water bearing rock formation, known as aquifer, is essential. An aquifer has two main functions - (i) to transit water (conduit function) and (ii) to store it (storage function). The groundwater resources in unconfined aquifers can be classified as static and dynamic. The static resources can be defined as the amount of groundwater available in the permeable portion of the aquifer below the zone of water level fluctuation. The dynamic resources can be defined as the amount of groundwater available in the zone of water level fluctuation. The replenishable groundwater resource is essentially a dynamic resource which is replenished annually or periodically by precipitation, irrigation return flow, canal seepage, tank seepage, influent seepage, etc.

The methodologies adopted for computing groundwater resources, are generally based on the hydrologic budget techniques. The hydrologic equation for groundwater regime is a specialized form of water balance equation that requires quantification of the components of inflow to and outflow from a groundwater reservoir, as well as changes in storage therein. Some of these are directly measurable, few may be determined by differences between measured volumes or rates of flow of surface water, and some require indirect methods of estimation.

Water balance techniques have been extensively used to make quantitative estimates of water resources and the impact of man's activities on the hydrological cycle. The study of water balance requires the systematic presentation of data on the water supply and its use within a given study area for a specific period. The water balance of an area is defined by the hydrologic equation, which is basically a statement of the law of conservation of mass as applied to the hydrological cycle. With water balance approach, it is possible to evaluate quantitatively individual contribution of sources of water in the system, over different time periods, and to establish the degree of variation in water regime due to changes in components of the system.

A basinwise approach yields the best results where the groundwater basin can be characterized by prominent drainages. A thorough study of the topography, geology and aquifer conditions should be taken up. The limit of the groundwater basin is controlled not only by topography but also by the disposition, structure and permeability of rocks and the configuration of the water table.

Generally, in igneous and metamorphic rocks, the surface water and groundwater basins are coincident for all practical purposes, but marked differences may be encountered in stratified sedimentary formations. Therefore, the study area for groundwater balance study is preferably taken as a doab which is bounded on two sides by two streams and on the other two sides by other aquifers or extension of the same aquifer. Once the study area is identified, comprehensive studies can be undertaken to estimate for selected period of time, the input and output of water, and change in storage to draw up water balance of the basin.

The estimation of groundwater balance of a region requires quantification of all individual inflows to or outflows from a groundwater system and change in groundwater storage over a given time period. The basic concept of water balance is:

Input to the system - outflow from the system =  change in storage of the system
(over a period of time)

The general methodology of computing groundwater balance consists of the following:

- Identification of significant components,
- Evaluating and quantifying individual components, and
- Presentation in the form of water balance equation.

The groundwater balance study of an area may serve the following purposes:

- As a check on whether all flow components involved in the system have been quantitatively accounted for, and what components have the greatest bearing on the problem under study.
- To calculate one unknown component of the groundwater balance equation, provided all other components are quantitatively known with sufficient accuracy.
- As a model of the hydrological processes under study, which can  be used to predict the effect that changes imposed on certain components will have on the other components of groundwater system.

**Groundwater Balance Equation**

Considering the various inflow and outflow components in a given study area, the groundwater balance equation can be written as:

$$R_r + R_c + R_i + R_t + S_i + I_g = E_t + T_p + S_e + O_g + \Delta S \qquad \qquad ...(1)$$

where,

$R_r$ = recharge from rainfall;
$R_c$ = recharge from canal seepage;
$R_i$ = recharge from field irrigation;
$R_t$ = recharge from tanks;

$S_i$ = influent seepage from rivers;
$I_g$ = inflow from other basins;
$E_t$ = evapotranspiration from groundwater;
$T_p$ = draft from groundwater;
$S_e$ = effluent seepage to rivers;
$O_g$ = outflow to other basins; and
$\Delta S$ = change in groundwater storage.

Preferably, all elements of the groundwater balance equation should be computed using independent methods. However, it is not always possible to compute all individual components of the groundwater balance equation separately. Sometimes, depending on the problem, some components can be lumped, and account only for their net value in the equation.

Computations of various components usually involve errors, due to shortcomings in the estimation techniques. The groundwater balance equation therefore generally does not balance, even if all its components are computed by independent methods. The resultant discrepancy in groundwater balance is defined as a residual term in the balance equation, which includes errors in the quantitative determination of various components as well as values of the components which have not been accounted in the equation.

The water balance may be computed for any time interval. The complexity of the computation of the water balance tends to increase with increase in area. This is due to a related increase in the technical difficulty of accurately computing the numerous important water balance components.

## Data Requirements for a Groundwater Balance Study

For carrying out a groundwater balance study, following data may be required over a given time period:

*Rainfall data*: Monthly rainfall data of sufficient number of rainguage stations lying within or around the study area, along with their locations, should be available.

*Land use data and cropping patterns*: Land use data are required for estimating the evapotranspiration losses from the water table through forested area. Cropping pattern data are necessary for estimating the spatial and temporal distributions of groundwater withdrawals, if required. Monthly pan evaporation rates should also be available at few locations for estimation of consumptive use requirements of different crops.

*River data*: Monthly river stage and discharge data along with river cross-sections are required at few locations for estimating the river-aquifer interflows.

*Canal data*: Monthwise water releases into the canal and its distributaries along with running days during each month are required. To account for the seepage losses through the canal system, the seepage loss test data are required in different canal reaches and distributaries.

*Tank data*: Monthly tank gauges and water releases should be available. In addition, depth vs. area and depth vs. capacity curves should also be available for computing the evaporation and seepage losses from tanks. Field test data are required for computing infiltration capacity to be used to evaluate the recharge from depression storage.

*Water table data*: Monthly water table data (or at least pre-monsoon and post-monsoon data) from sufficient number of well-distributed observation wells along with their locations are required. The available data should comprise reduced level (R.L.) of water table and depth to water table.

*Groundwater draft*: For estimating groundwater withdrawals, the number of each type of wells operating in the area, their corresponding running hours each month and discharge are required. If a complete inventory of wells is not available, then this can be obtained by carrying out sample surveys.

*Aquifer parameters*: Data regarding the storage coefficient and transmissivity are required at sufficient number of locations in the study area.

## Groundwater Resource Estimation Methodology

The Groundwater Estimation Committee (GEC) was constituted by the Government of India in 1982 to recommend methodologies for estimation of the groundwater resource potential in India. It was recommended by the committee that the groundwater recharge should be estimated based on groundwater level fluctuation method. However, in areas, where groundwater level monitoring is not being done regularly, or where adequate data about groundwater level fluctuation is not available, adhoc norms of rainfall infiltration may be adopted. In order to review the recommended methodology, the committee was reconstituted in 1995, which released its report in 1997. This committee proposed several improvements in the existing methodology based on groundwater level fluctuation approach. Salient features of their recommendations are given below.

(a)     Watershed may be used as the unit for groundwater resource assessment in hard rock areas, which occupies around $2/3^{rd}$ part of the country. The size of the watershed as a hydrological unit could be of about 100 to 300 sq. km. area. The assessment made for watershed as unit may be transferred to administrative unit such as block, for planning development programmes.

(b)     For alluvial areas, the present practice of assessment based on block/taluka/mandal-wise basis is retained. The possibility of adopting doab as the unit of assessment in alluvial areas needs further detailed studies.

(c)     The total geographical area of the unit for resource assessment is to be divided into subareas such as hilly regions (slope > 20%), saline groundwater areas, canal command areas and non-command areas, and separate resource assessment may be made for these subareas. Variations in geomorphological and hydrogeological characteristics may be considered within the unit.

(d)     For hard rock areas, the specific yield value may be estimated by applying the water level fluctuation method for the dry season data, and then using this specific yield value in the water level fluctuation method for the monsoon season to get recharge. For alluvial areas, specific yield values may be estimated from analysis of pumping tests. However, norms for specific yield values in different hydrogeological regions may still be necessary for use in situations where the above methods are not feasible due to inadequacy of data.

(e)     There should be at least 3 spatially well-distributed observation wells in the unit, or one observation well per 100 sq. km. whichever is more.

(f)     The problem of accounting for groundwater inflow/outflow and base flow from a region is difficult to solve. If watershed is used as a unit for resource assessment in hard rock areas, the groundwater inflow/outflow may become negligible. The base flow can be estimated if one stream gauging station is located at the exit of the watershed.

(g)     Norms for return flow from groundwater and surface water irrigation are revised taking into account the source of water (groundwater/surface water), type of crop (paddy/non-paddy) and depth of groundwater level.

In subsequent section, the recommended GEC norms for estimation of various inflow/outflow components of the groundwater balance equation have been mentioned at appropriate places, along with other methodologies/formulae in use.

**Estimation of Groundwater Balance Components**

The various inflow/outflow components of the groundwater balance equation may be estimated through appropriate empirical relationships suitable for a region, Groundwater Estimation Committee norms (1997), field experiments or other methods, as discussed below.

**Recharge from Rainfall ($R_r$)**

Rainfall is the major source of recharge to groundwater. Part of the rain water, which falls on the ground, is infiltrated into the soil. A part of this infiltrated water is utilized in filling the soil moisture deficiency while the remaining portion percolates down to reach the water table, which is termed as *rainfall recharge* to the aquifer. The amount of rainfall recharge depends on various hydrometeorological and topographic factors, soil characteristics and depth to water table. The methods for estimation of rainfall recharge involve the empirical relationships established between recharge and rainfall developed for different regions, Groundwater Resource Estimation Committee norms, groundwater balance approach, and soil moisture data based methods.

**Empirical Methods**

Several empirical formulae have been worked out for various regions in India on the basis of detailed studies. Some of the commonly used formulae are:

*(a) Chaturvedi formula*: Based on the water level fluctuations and rainfall amounts in Ganga-Yamuna doab, Chaturvedi in 1936, derived an empirical relationship to arrive at the recharge as a function of annual precipitation.

$$R_r = 2.0 (P - 15)^{0.4} \qquad \qquad ...(2)$$

where,

$R_r$ = net recharge due to precipitation during the year, in inches; and
$P$ = annual precipitation, in inches.

This formula was later modified by further work at the U.P. Irrigation Research Institute, Roorkee and the modified form of the formula is

$$R_r = 1.35 (P - 14)^{0.5} \qquad \qquad ...(3)$$

The Chaturvedi formula has been widely used for preliminary estimations of groundwater recharge due to rainfall. It may be noted that there is a lower limit of the rainfall below which the recharge due to rainfall is zero. The percentage of rainfall recharged commences from zero at $P = 14$ inches, increases upto 18% at $P = 28$ inches, and again decreases. The lower limit of rainfall in the formula may account for the soil moisture deficit, the interception losses and potential evaporation. These factors being site specific, one generalized formula may not be applicable to all the alluvial areas. Tritium tracer studies on groundwater recharge in the alluvial deposits of Indo-Gangetic plains of western U.P., Punjab, Haryana and alluvium in Gujarat state have indicated variations with respect to Chaturvedi formula.

*(b) Kumar and Seethapathi (2002)*: They conducted a detailed seasonal groundwater balance study in Upper Ganga Canal command area for the period 1972-73 to 1983-84 to determine groundwater recharge from rainfall. It was observed that as the rainfall increases, the quantity of recharge also increases but the increase is not linearly proportional. The recharge coefficient (based upon the rainfall in monsoon season) was found to vary between 0.05 to 0.19 for the study area. The following empirical relationship (similar to Chaturvedi formula) was derived by fitting the estimated values of rainfall recharge and the corresponding values of rainfall in the monsoon season through the non-linear regression technique.

$$R_r = 0.63 (P - 15.28)^{0.76} \qquad \qquad ...(4)$$

where,

$R_r$ = Groundwater recharge from rainfall in monsoon season (inch);
$P$ = Mean rainfall in monsoon season (inch).

The relative errors (%) in the estimation of rainfall recharge computed from the proposed empirical relationship was compared with groundwater balance study. In almost all the years, the relative error was found to be less than 8%. On the other hand, relative errors (%) computed from Chaturvedi formula (equations 2 and 3) were found to be quite high. Therefore, equation (4) can conveniently be used for better and quick assessment of natural groundwater recharge in Upper Ganga Canal command area.

*(c) Amritsar formula*: Using regression analysis for certain doabs in Punjab, the Irrigation and Power Research Institute, Amritsar, developed the following formula in 1973.

$$R_r = 2.5 \, (P - 16)^{0.5} \qquad\qquad ...(5)$$

where, $R_r$ and P are measured in inches.

*(d) Krishna Rao*: Krishna Rao gave the following empirical relationship in 1970 to determine the groundwater recharge in limited climatological homogeneous areas:

$$R_r = K \, (P - X) \qquad\qquad ...(6)$$

The following relation is stated to hold good for different parts of Karnataka:

$R_r = 0.20 \, (P - 400)$ for areas with annual normal rainfall (P) between 400 and 600 mm
$R_r = 0.25 \, (P - 400)$ for areas with P between 600 and 1000 mm
$R_r = 0.35 \, (P - 600)$ for areas with P above 2000 mm

where, $R_r$ and P are expressed in millimetres.

The relationships indicated above, which were tentatively proposed for specific hydrogeological conditions, have to be examined and established or suitably altered for application to other areas.

**Groundwater Resource Estimation Committee Norms**

If adequate data of groundwater levels are not available, rainfall recharge may be estimated using the rainfall infiltration method. The same recharge factor may be used for both monsoon and non-monsoon rainfall, with the condition that the recharge due to non-monsoon rainfall may be taken as zero, if the rainfall during non-monsoon season is less than 10% of annual rainfall. Groundwater Resource Estimation Committee (1997) recommended the following rainfall infiltration factors:

(a) *Alluvial areas*

| | | |
|---|---|---|
| Indo-Gangetic and inland areas | - | 22 % |
| East coast | - | 16 % |
| West coast | - | 10 % |

(b) *Hard rock areas*

| | | |
|---|---|---|
| Weathered granite, gneiss and schist with low clay content | - | 11 % |
| Weathered granite, gneiss and schist with significant clay content | - | 8 % |
| Granulite facies like charnockite etc. | - | 5 % |
| Vesicular and jointed basalt | - | 13 % |
| Weathered basalt | - | 7 % |

| Laterite | - | 7 % |
|---|---|---|
| Semi-consolidated sandstone | - | 12 % |
| | | |
| Consolidated sandstone, Quartzites, Limestone (except cavernous limestone) | - | 6 % |
| | | |
| Phyllites, Shales | - | 4 % |
| Massive poorly fractured rock | - | 1 % |

An additional 2% of rainfall recharge factor may be used in areas where watershed development with associated soil conservation measures is implemented. This additional factor is separate from contribution due to water conservation structures such as check dams, nalla bunds, percolation tanks etc., for which the norms are defined separately.

**Groundwater Balance Approach**

In this method, all components of the groundwater balance equation (1), except the rainfall recharge, are estimated individually. The algebraic sum of all input and output components is equated to the change in groundwater storage, as reflected by the water table fluctuation, which in turn yields the single unknown in the equation, namely, the rainfall recharge. A pre-requisite for successful application of this technique is the availability of very extensive and accurate hydrological and meteorological data. The groundwater balance approach is valid for the areas where the year can be divided into monsoon and non-monsoon seasons with the bulk of rainfall occurring in former.

Groundwater balance study for monsoon and non-monsoon periods is carried out separately. The former yields an estimate of recharge coefficient and the later determines the degree of accuracy with which the components of water balance equation have been estimated. Alternatively, the average specific yield in the zone of fluctuation can be determined from a groundwater balance study for the non-monsoon period and using this specific yield, the recharge due to rainfall can be determined using the groundwater balance components for the monsoon period.

**Soil Moisture Data Based Methods**

Soil moisture data based methods are the lumped and distributed model and the nuclear methods. In the lumped model, the variation of soil moisture content in the vertical direction is ignored and any effective input into the soil is assumed to increase the soil moisture content uniformly. Recharge is calculated as the remainder when losses, identified in the form of runoff and evapotranspiration, have been deducted from the precipitation with proper accounting of soil moisture deficit. In the distributed model, variation of soil moisture content in the vertical direction is accounted and the method involves the numerical solution of partial differential equation (Richards equation) governing one-dimensional flow through unsaturated medium, with appropriate initial and boundary conditions.

## (a)    Soil Water Balance Method

Water balance models were developed in the 1940s by Thornthwaite (1948) and revised by Thornthwaite and Mather (1955). The method is essentially a book-keeping procedure which estimates the balance between the inflow and outflow of water. When applying this method to estimate the recharge for a catchment area, the calculation should be repeated for areas with different precipitation, evapotranspiration, crop type and soil type. The soil water balance method is of limited practical value, because evapotranspiration is not directly measurable. Moreover, storage of moisture in the unsaturated zone and the rates of infiltration along the various possible routes to the aquifer form important and uncertain factors. Another aspect that deserves attention is the depth of the root zone which may vary in semi-arid regions between 1 and 30 meters. Results from this model are of very limited value without calibration and validation, because of the substantial uncertainty in input data.

## (b)    Nuclear Methods

Nuclear techniques can be used for the determination of recharge by measuring the travel of moisture through a soil column. The technique is based upon the existence of a linear relation between neutron count rate and moisture content (% by volume) for the range of moisture contents generally occurring in the unsaturated soil zone.

### Neutron soil moisture probe

Neutron soil moisture probe is used to determine soil moisture contents at different depths in the unsaturated zone. The neutron probe consists of a fast neutron source of radioactive isotopes (Radium and Beryllium) and a slow neutron detector adjacent to the neutron source. The neutron source of the probe emits fast neutrons into the surrounding soil. In soil, hydrogen is present mainly in the form of moisture. Hence, fast neutrons when projected into the soil, are scattered by the hydrogen nuclei after collision and are slowed down to the thermal energy level by losing their kinetic energy. The number of neutrons slowed down thus is proportional to the hydrogen atoms available in the soil and hence, to the moisture content of the soil. These slow down neutrons are detected by a slow neutron detector (thermal neutron detector). The measurement of number of counts per second gives the measure of the soil moisture content, which is converted into actual values of soil moisture content after applying suitable calibration equation. By repeated measurements of moisture contents through neutron soil moisture probe, at different times during the rainy season, groundwater recharge can be estimated.

### Tritium tagging technique

The basic principle of estimation of recharge to groundwater by using this method assumes that soil water in the unsaturated zone moves downward 'layer by layer' similar to a piston flow, i.e. if any amount of water is added to the ground surface due to precipitation or irrigation, it will percolate by pushing equal amount of water beneath it further downwards such that an equal amount of the moisture of the last layer in the unsaturated zone is added to the groundwater. If the tritium radioisotope is tagged below the active root zone and sun-heating zone, it will be mixed with the soil moisture available at that depth and act as an impermeable layer. Therefore, if any water is added to the top soil, it will be infiltrated into the soil strata by pushing down the older water, this shift in the tritium can be observed after a specified period depending upon the input supply of water. However, the injected tritium will be found in the form of a broadened peak due to molecular diffusion, stream line dispersion, asymmetrical flow and other heterogeneities of the soil media.

## Recharge from Canal Seepage ($R_c$)

Seepage refers to the process of water movement from a canal into and through the bed and wall material. Seepage losses from irrigation canals often constitute a significant part of the total recharge to groundwater system. Hence, it is important to properly estimate these losses for recharge assessment to groundwater system. Recharge by seepage from canals depends upon the size and cross-section of the canal, depth of flow, characteristics of soils in the bed and sides, and location as well as level of drains on either side of the canal. A number of empirical formulae and formulae based on theoretical considerations have been proposed to estimate the seepage losses from canals.

Recharge from canals that are in direct hydraulic connection with a phreatic aquifer underlain by a horizontal impermeable layer at shallow depth, can be determined by Darcy's equation, provided the flow satisfies Dupuit assumptions.

$$R_c = K \frac{h_s - h_1}{L} A \qquad \qquad ...(7)$$

where, $h_s$ and $h_1$ are water-level elevations above the impermeable base, respectively, at the canal, and at distance L from it. For calculating the area of flow cross-section, the average of the saturated thickness $(h_s + h_1)/2$ is taken. The crux of computation of seepage depends on correct assessment of the hydraulic conductivity, K. Knowing the percentage of sand, silt and clay, the hydraulic conductivity of undisturbed soil can be approximately determined using the soil classification triangle showing relation of hydraulic conductivity to texture for undisturbed sample (Johnson, 1963).

A number of investigations have been carried out to study the seepage losses from canals. The following formulae/values are in vogue for the estimation of seepage losses:

(a)     As reported by the Indian Standard (IS: 9452 Part 1, 1980), the loss of water by seepage from unlined canals in India varies from 0.3 to 7.0 cumec per million square meter of wetted area depending on the permeability of soil through which the canal passes, location of water table, distance of drainage, bed width, side slope, and depth of water in the canal. Transmission loss of 0.60 cumec per million square meter of wetted area of lined canal is generally assumed (IS: 10430, 1982).

(b)     For unlined channels in Uttar Pradesh, it has been proposed that the losses per million square meter of  wetted area are 2.5 cumec for ordinary clay loam to about 5 cumec for sandy loam with an average of 3 cumec. Empirically, the seepage losses can be computed using the following formula:

$$Losses\ in\ cumecs\ /\ km = \frac{C}{200} (B + D)^{2/3} \qquad ...(8)$$

67

where, B and D are the bed width and depth, respectively, of the channel in meters, C is a constant with a value of 1.0 for intermittent running channels and 0.75 for continuous running channels.

(c) For lined channels in Punjab, the following formula is used for estimation of seepage losses:

$$R_c = 1.25 \, Q^{0.56} \qquad \qquad \ldots(9)$$

where, $R_c$ is the seepage loss in cusec per million square foot of wetted area and Q, in cusec, is the discharge carried by the channel. In unlined channels, the loss rate on an average is four times the value computed using the above formula.

(d) U. S. B. R. recommended the channel losses based on the channel bed material as given below:

| Material | | Seepage Losses (cumec per million square meter of wetted area) |
|---|---|---|
| Clay and clay loam | : | 1.50 |
| Sandy loam | : | 2.40 |
| Sandy and gravely soil | : | 8.03 |
| Concrete lining | : | 1.20 |

(e) Groundwater Resource Estimation Committee (1997) has recommended the following norms:

(i) Unlined canals in normal soil with some clay content along with sand
- 1.8 to 2.5 cumec per million square meter of wetted area.

(ii) Unlined canals in sandy soil with some silt content
- 3.0 to 3.5 cumec per million square meter of wetted area.

(iii) Lined canals and canals in hard rock areas
- 20% of the above values for unlined canals.

These values are valid if the water table is relatively deep. In shallow water table and water logged areas, the recharge from canal seepage may be suitably reduced. Specific results from case studies may be used, if available. The above norms take into consideration the type of soil in which the canal runs while computing seepage. However, the actual seepage will also be controlled by the width of canal (B), depth of flow (D), hydraulic conductivity of the bed material (K) and depth to water table.

Knowing the values of B and D, the range of seepage losses ($R_{c\_max}$ and $R_{c\_min}$) from the canal may be obtained as

$$R_{c\_max} = K (B + 2D) \text{ (in case of deeper water table)} \qquad \dots(10a)$$

$$R_{c\_min} = K (B - 2D) \text{ (in case of water table at the level of channel bed)} \qquad \dots(10b)$$

However, the various guidelines for estimating losses in the canal system, as given above, are at best approximate. Thus, the seepage losses may best be estimated by conducting actual tests in the field. The methods most commonly adopted are:

*Inflow - outflow method*: In this method, the water that flows into and out of the section of canal, under study, is measured using current meter or Parshall flume method. The difference between the quantities of water flowing into and out of the canal reach is attributed to seepage. This method is advantageous when seepage losses are to be measured in long canal reaches with few diversions.

*Ponding method*: In this method, bunds are constructed in the canal at two locations, one upstream and the other downstream of the reach of canal with water filled in it. The total change in storage in the reach is measured over a period of time by measuring the rate of drop of water surface elevation in the canal reach. Alternatively, water may be added to maintain a constant water surface elevation. In this case, the volume of water added is measured along with the elapsed time to compute the rate of seepage loss. The ponding method provides an accurate means of measuring seepage losses and is especially suitable when they are small (e.g. in lined canals).

*Seepage meter method*: The seepage meter is a modified version of permeameter developed for use under water. Various types of seepage meters have been developed. The two most important are seepage meter with submerged flexible water bag and falling head seepage meter. Seepage meters are suitable for measuring local seepage rates in canals or ponds and used only in unlined or earth-lined canals. They are quickly and easily installed and give reasonably satisfactory results for the conditions at the test site but it is difficult to obtain accurate results when seepage losses are low.

The total losses from the canal system generally consist of the evaporation losses ($E_c$) and the seepage losses ($R_c$). The evaporation losses are generally 10 to 15 percent of the total losses. Thus the $R_c$ value is 85 to 90 percent of the losses from the canal system.

## Recharge from Field Irrigation ($R_i$)

Water requirements of crops are met, in parts, by rainfall, contribution of moisture from the soil profile, and applied irrigation water. A part of the water applied to irrigated field crops is lost in consumptive use and the balance infiltrates to recharge the groundwater. The process of re-entry of a part of the water used for irrigation is called return flow. Percolation from applied irrigation water, derived both from surface water and groundwater sources, constitutes one of the major components of groundwater recharge. The irrigation return flow depends on the soil type, irrigation practice and type of crop. Therefore, irrigation return flows are site specific and will vary from one region to another.

For a correct assessment of the quantum of recharge by applied irrigation, studies are required to be carried out on experimental plots under different crops in different seasonal conditions. The method of estimation comprises application of the water balance equation involving input and output of water in experimental fields.

The recharge due to irrigation return flow may also be estimated, based on the source of irrigation (groundwater or surface water), the type of crop (paddy, non-paddy) and the depth of water table below ground surface, using the norms provided by Groundwater Resource Estimation Committee (1997), as given below (as percentage of water application):

| Source of Irrigation | Type of Crop | Water table below ground surface | | |
|---|---|---|---|---|
| | | <10m | 10-25m | >25m |
| Groundwater | Non-paddy | 25 | 15 | 5 |
| Surface water | Non-paddy | 30 | 20 | 10 |
| Groundwater | Paddy | 45 | 35 | 20 |
| Surface water | Paddy | 50 | 40 | 25 |

For surface water, the recharge is to be estimated based on water released at the outlet from the canal/distribution system. For groundwater, the recharge is to be estimated based on gross draft. Where continuous supply is used instead of rotational supply, an additional recharge of 5% of application may be used. Specific results from case studies may be used, if available.

## Recharge from Tanks ($R_t$)

Studies have indicated that seepage from tanks varies from 9 to 20 percent of their live storage capacity. However, as data on live storage capacity of large number of tanks may not be available, seepage from tanks may be taken as 44 to 60 cm per year over the total water spread, taking into account the agro-climatic conditions in the area. The seepage from percolation tanks is higher and may be taken as 50 percent of its gross storage. In case of seepage from ponds and lakes, the norms as applied to tanks may be taken. Groundwater Resource Estimation Committee (1997) has recommended that based on the average area of water spread, the recharge from storage tanks and ponds may be taken as 1.4 mm/day for the period in which tank has water. If data on the average area of water spread is not available, 60% of the maximum water spread area may be used instead of average area of water spread.

In case of percolation tanks, recharge may be taken as 50% of gross storage, considering the number of fillings, with half of this recharge occurring in monsoon season and the balance in non-monsoon season. Recharge due to check dams and nala bunds may be taken as 50% of gross storage (assuming annual desilting maintenance exists) with half of this recharge occurring in the monsoon season and the balance in the non-monsoon season.

## Influent and Effluent Seepage ($S_i$ & $S_e$)

The river-aquifer interaction depends on the transmissivity of the aquifer system and the gradient of water table in respect to the river stage. Depending on the water level in the river and in the aquifer (in the vicinity of river), the river may recharge the aquifer (influent) or the aquifer may contribute to the river flow (effluent). The effluent or influent character of the river may vary from season to season and from reach to reach. The seepage from/to the river can be determined by dividing the river reach into small sub-reaches and observing the discharges at the two ends of the sub-reach along with the discharges of its tributaries and diversions, if any. The discharge at the downstream end is expressed as:

$$Q_d. \Delta t = Q_u. \Delta t + Q_g. \Delta t + Q_t. \Delta t - Q_o. \Delta t - E. \Delta t \pm S_{rb} \qquad ...(11)$$

where,

$Q_d$ =     discharge at the downstream section;
$Q_u$ =     discharge at the upstream section;
$Q_g$ =     groundwater contribution (unknown quantity; -ve computed value indicates influent conditions);
$Q_t$ =     discharge of tributaries;
$Q_o$ =     discharge diverted from the river;
E =     rate of evaporation from river water surface and flood plain (for extensive bodies of surface water and for long time periods, evaporation from open water surfaces can not be neglected);
$S_{rb}$ =     change in bank storage ( + for decrease and - for increase); and
$\Delta t$ =     time period.

The change in bank storage can be determined by monitoring the water table along the cross-section normal to the river. Thus, using the above equation, seepage from/to the river over a certain period of time $\Delta t$ can be computed. However, this would be the contribution from aquifers on both sides of the stream. The contribution from each side can be separated by the following method:

$$Contribution\ from\ left\ bank\ =\ \frac{I_L T_L}{I_L T_L + I_R T_R} . Q_g \qquad ...(12)$$

$$Contribution\ from\ right\ bank\ =\ \frac{I_R T_R}{I_L T_L + I_R T_R} . Q_g \qquad ...(13)$$

where, $I_L$ and $T_L$ are gradient and transmissivity respectively on the left side and $I_R$ and $T_R$ are those on the right.

## Inflow from and Outflow to Other Basins ($I_g$ & $O_g$)

For the estimation of groundwater inflow/outflow from/to other basins, regional water table contour maps are drawn based on the observed water level data from wells located within and outside the study area. The flows into and out of a region are governed mainly by the hydraulic gradient and transmissivity of the aquifer. The gradient can be determined by taking the slope of the water table normal to water table contour. The length of the section, across which groundwater inflow/outflow occurs, is determined from contour maps, the length being measured parallel to the contour. The inflow/outflow is determined as follows:

$$I_g\ or\ O_g\ =\ \sum^{L} T\ I\ \Delta L \qquad ...(14)$$

where, T is the transmissivity and I is the hydraulic gradient averaged over a length $\Delta L$ of contour line.

## Evapotranspiration from Groundwater ($E_t$)

Evapotranspiration is the combined process of transpiration from vegetation and evaporation from both soil and free water surfaces. Potential evapotranspiration is the maximum loss of water through evapotranspiration. Evapotranspiration from groundwater occurs in waterlogged areas or in forested areas with roots extending to the water table. From the land use data, area under forests is available while the waterlogged areas may be demarcated from depth to water table maps. The potential evapotranspiration from such areas can be computed using standard methods.

Depth to water table maps may be prepared based on well inventory data to bring into focus the extensiveness of shallow water table areas. During well inventory, investigation should be specifically oriented towards accurately delineating water table depth for depths less than 2 meters. The evapotranspiration can be estimated based on the following equations:

$$E_t = PE_t * A \qquad \text{if } h > h_s \qquad \qquad ...(15a)$$
$$E_t = 0 \qquad \text{if } h < (h_s - d) \qquad \qquad ...(15b)$$
$$E_t = PE_t * A (h - (h_s - d))/d \qquad \text{if } (h_s - d) \le h \le h_s \qquad \qquad ...(15c)$$

where,

$E_t$ = evapotranspiration in volume of water per unit time $[L^3 \, T^{-1}]$;

$PE_t$ = maximum rate of evapotranspiration in volume of water per unit area per unit time $[L^3 \, L^{-2} \, T^{-1}]$;

$A$ = surface area $[L^2]$;

$h$ = water table elevation $[L]$;

$h_s$ = water table elevation at which the evapotranspiration loss reaches the maximum value; and

$d$ = extinction depth. When the distance between $h_s$ and $h$ exceeds $d$, evapotranspiration from groundwater ceases $[L]$.

## Draft from Groundwater ($T_p$)

Draft is the amount of water lifted from the aquifer by means of various lifting devices. To estimate groundwater draft, an inventory of wells and a sample survey of groundwater draft from various types of wells (state tubewells, private tubewells and open wells) are required. For state tubewells, information about their number, running hours per day, discharge, and number of days of operation in a season is generally available in the concerned departments. To compute the draft from private tubewells, pumping sets and rahats etc., sample surveys have to be conducted regarding their number, discharge and withdrawals over the season.

In areas where wells are energised, the draft may be computed using power consumption data. By conducting tests on wells, the average draft per unit of electricity consumed can be determined for different ranges in depth to water levels. By noting the depth to water level at each distribution point and multiplying the average draft value with the number of units of electricity consumed, the draft at each point can be computed for every month.

In the absence of sample surveys, the draft can be indirectly estimated from the net crop water requirement which is based upon the cropping pattern and irrigated areas under various crops. The consumptive use requirements of crops are calculated using the consumptive use coefficient and effective rainfall. The consumptive use coefficient for crops is related to percentage of crop growing season (Table 1). The consumptive use for each month can be evaluated by multiplying consumptive use coefficient with monthly pan evaporation rates. For the computation of net irrigation requirement, the effective rainfall has to be evaluated. Effective rainfall is the portion of rainfall that builds up the soil moisture in the root zone after accounting for direct runoff and deep percolation. The normal monthly effective rainfall, as related to average monthly consumptive use, is given in Table 2. Net crop water requirement is obtained after subtracting effective rainfall from consumptive use requirement. The groundwater draft can thus be estimated by subtracting canal water released for the crops from the net crop water requirement.

Table 1: Crop Consumptive Use Coefficients

| Percent of crop growing season | Consumptive use (Evapotranspiration) coefficient (to be multiplied by Class 'A' Pan Evaporation) | | | | | | | |
|---|---|---|---|---|---|---|---|---|
| | Group A | Group B | Group C | Group D | Group E | Group F | Group G | Rice |
| 0 | 0.20 | 0.15 | 0.12 | 0.08 | 0.90 | 0.60 | 0.50 | 0.80 |
| 5 | 0.20 | 0.15 | 0.12 | 0.08 | 0.90 | 0.60 | 0.55 | 0.90 |
| 10 | 0.36 | 0.27 | 0.22 | 0.25 | 0.90 | 0.60 | 0.60 | 0.95 |
| 15 | 0.50 | 0.38 | 0.30 | 0.19 | 0.90 | 0.60 | 0.65 | 1.00 |
| 20 | 0.64 | 0.48 | 0.38 | 0.27 | 0.90 | 0.60 | 0.70 | 1.05 |
| 25 | 0.75 | 0.56 | 0.45 | 0.33 | 0.90 | 0.60 | 0.75 | 1.10 |
| 30 | 0.84 | 0.63 | 0.50 | 0.40 | 0.90 | 0.60 | 0.80 | 1.14 |
| 35 | 0.92 | 0.69 | 0.55 | 0.46 | 0.90 | 0.60 | 0.86 | 1.17 |
| 40 | 0.97 | 0.73 | 0.58 | 0.52 | 0.90 | 0.60 | 0.90 | 1.21 |
| 45 | 0.99 | 0.74 | 0.60 | 0.58 | 0.90 | 0.60 | 0.95 | 1.25 |
| 50 | 1.00 | 0.75 | 0.60 | 0.65 | 0.90 | 0.60 | 1.00 | 1.30 |
| 55 | 1.00 | 0.75 | 0.60 | 0.71 | 0.90 | 0.60 | 1.00 | 1.30 |
| 60 | 0.99 | 0.74 | 0.60 | 0.77 | 0.90 | 0.60 | 1.00 | 1.30 |
| 65 | 0.96 | 0.72 | 0.58 | 0.82 | 0.90 | 0.60 | 0.95 | 1.25 |
| 70 | 0.91 | 0.68 | 0.55 | 0.88 | 0.90 | 0.60 | 0.90 | 1.20 |
| 75 | 0.85 | 0.64 | 0.51 | 0.90 | 0.90 | 0.60 | 0.85 | 1.15 |
| 80 | 0.75 | 0.56 | 0.45 | 0.90 | 0.90 | 0.60 | 0.80 | 1.00 |
| 85 | 0.60 | 0.45 | 0.36 | 0.80 | 0.90 | 0.60 | 0.75 | 1.00 |
| 90 | 0.46 | 0.35 | 0.28 | 0.70 | 0.90 | 0.60 | 0.70 | 0.90 |
| 100 | 0.20 | 0.20 | 0.17 | 0.20 | 0.90 | 0.60 | 0.50 | 0.20 |

Group A - beans, maize, cotton, potatoes, sugar beets, jowar and peas
Group B - dates, olives, walnuts, tomatoes, and hybrid jowar
Group C - melons, onions, carrots, and grapes
Group D - barley, flex, wheat, and other small grains
Group E - pastures, orchards, and cover crops

Group F -   citrus crops, oranges, limes and grape fruit
Group G -   sugarcane and alfalfa

Table 2:   Normal Monthly Effective Rainfall as related to Normal Monthly Rainfall and Average Monthly Consumptive Use

| Normal Monthly Rainfall (mm) | Average Monthly Consumptive Use (mm) | | | | | | | | | | | | | |
| | 25 | 50 | 75 | 100 | 125 | 150 | 175 | 200 | 225 | 250 | 275 | 300 | 325 | 350 |
| | Normal Monthly Effective Rainfall (mm) | | | | | | | | | | | | | |
| 25 | 15 | 17 | 18 | 18 | 19 | 20 | 21 | 22 | 25 | 25 | 25 | 25 | 25 | 25 |
| 50 | 25 (42) | 33 | 35 | 36 | 37 | 40 | 41 | 44 | 48 | 50 | 50 | 50 | 50 | 50 |
| 75 | | 47 | 51 | 54 | 56 | 58 | 61 | 65 | 69 | 74 | 75 | 75 | 75 | 75 |
| 100 | | 50 (81) | 65 | 69 | 75 | 75 | 79 | 83 | 89 | 96 | 100 | 100 | 100 | 100 |
| 125 | | | 75 (124) | 85 | 89 | 91 | 93 | 102 | 108 | 110 | 119 | 119 | 125 | 125 |
| 150 | | | | 97 | 104 | 108 | 113 | 120 | 127 | 138 | 144 | 150 | 150 | 150 |
| 175 | | | | 100 (162) | 117 | 120 | 128 | 136 | 143 | 154 | 163 | 172 | 175 | 175 |
| 200 | | | | | 125 (200) | 131 | 140 | 149 | 158 | 169 | 184 | 191 | 197 | 200 |
| 225 | | | | | | 142 | 152 | 162 | 175 | 189 | 200 | 210 | 220 | 225 |
| 250 | | | | | | 145 | 164 | 175 | 192 | 206 | 215 | 228 | 236 | 245 |
| 275 | | | | | | 150 (260) | 175 | 188 | 205 | 229 | 233 | 242 | 256 | 265 |
| 300 | | | | | | | 175 (290) | 195 | 215 | 235 | 246 | 258 | 275 | 288 |
| 325 | | | | | | | | 199 | 220 | 242 | 256 | 275 | 290 | 304 |
| 350 | | | | | | | | 200 (330) | 224 | 245 | 265 | 285 | 305 | 320 |
| 375 | | | | | | | | | 225 (360) | 248 | 270 | 292 | 310 | 328 |
| 400 | | | | | | | | | | 250 (390) | 275 | 296 | 317 | 335 |
| 425 | | | | | | | | | | | 275 (420) | 300 | 320 | 340 |
| 450 | | | | | | | | | | | | 300 (450) | 322 | 343 |
| 475 | | | | | | | | | | | | | 324 | 346 |
| 500 | | | | | | | | | | | | | 325 (485) | 348 |
| 525 | | | | | | | | | | | | | | 350 |

The above table is based on 75 mm net depth of application. For other net depth of application, multiply by the factors shown below.

| Net depth of application (mm) | 25 | 38 | 50 | 63 | 75 | 100 | 125 | 150 | 175 |
| --- | --- | --- | --- | --- | --- | --- | --- | --- | --- |
| Factor | 0.77 | 0.85 | 0.93 | 0.98 | 1.00 | 1.02 | 1.04 | 1.05 | 1.07 |

**Change in Groundwater Storage (ΔS)**

To estimate the change in groundwater storage, the water levels are observed through a network of observation wells spread over the area. The water levels are highest immediately after monsoon in the month of October or November and lowest just before rainfall in the month of May or June. During the monsoon season, the recharge is more than the extraction; therefore, the change in groundwater storage between the beginning and end of monsoon season indicates the total volume of water added to the groundwater reservoir. While the change in groundwater storage between the beginning and end of the non-monsoon season indicates the total quantity of water withdrawn from groundwater storage. The change in storage (ΔS) is computed as follows:

$$\Delta S = \sum \Delta h\, A\, S_y \qquad\qquad \ldots(16)$$

where, $\Delta h$ = change in water table elevation during the given time period;
$A$ = area influenced by the well; and
$S_y$ = specific yield.

Groundwater Resource Estimation Committee (1997) recommended that the size of the watershed as a hydrological unit could be of about 100 to 300 sq. km. area and there should be at least three spatially well-distributed observation wells in the unit, or one observation well per 100 sq. km., whichever is more. However, as per IILRI (1974), the following specification may serve as a rough guide:

| Size of the Area (ha) | Number of Observation Points | Number of Observation Points per 100 hectares |
|---|---|---|
| 100 | 20 | 20 |
| 1,000 | 40 | 4 |
| 10,000 | 100 | 1 |
| 1,00,000 | 300 | 0.3 |

The specific yield may be computed from pumping tests. Groundwater Resource Estimation Committee (1997) recommended the following values of specific yield for different geological formations:

(a) **Alluvial areas**

| | | |
|---|---|---|
| Sandy alluvium | - | 16.0 % |
| Silty alluvium | - | 10.0 % |
| Clayey alluvium | - | 6.0 % |

(b) **Hard rock areas**

| | | |
|---|---|---|
| Weathered granite, gneiss and schist with low clay content | - | 3.0 % |

| Weathered granite, gneiss and schist with significant clay content | - | 1.5 % |
| Weathered or vesicular, jointed basalt | - | 2.0 % |
| Laterite | - | 2.5 % |
| Sandstone | - | 3.0 % |
| Quartzites | - | 1.5 % |
| Limestone | - | 2.0 % |
| Karstified limestone | - | 8.0 % |
| Phyllites, Shales | - | 1.5 % |
| Massive poorly fractured rock | - | 0.3 % |

The values of specific yield in the zone of fluctuation of water table in different parts of the basin can also be approximately determined from the soil classification triangle showing relation between particle size and specific yield (Johnson, 1967).

**Establishment of Recharge Coefficient**

Groundwater balance study is a convenient way of establishing the rainfall recharge coefficient, as well as to cross check the accuracy of the various prevalent methods for the estimation of groundwater losses and recharge from other sources. The steps to be followed are:

1.  Divide the year into monsoon and non-monsoon periods.

2.  Estimate all the components of the water balance equation other than rainfall recharge for monsoon period using the available hydrological and meteorological information and employing the prevalent methods for estimation.

3.  Substitute these estimates in the water balance equation and thus calculate the rainfall recharge and hence recharge coefficient (recharge/rainfall ratio). Compare this estimate with those given by various empirical relations valid for the area of study.

4.  For non-monsoon season, estimate all the components of water balance equation including the rainfall recharge which is calculated using recharge coefficient value obtained through the water balance of monsoon period. The rainfall recharge ($R_r$) will be of very small order in this case. A close balance between the left and right sides of the equation will indicate that the net recharge from all the sources of recharge and discharge has been quantified with a good degree of accuracy.

By quantifying all the inflow and outflow components of a groundwater system, one can determine which particular component has the most significant effect on the groundwater flow regime. Alternatively, a groundwater balance study may be used to compute one unknown component (e.g. the rainfall recharge) of the groundwater balance equation, when all other components are known. The balance study may also serve as a model of the area under study, whereby the effect of change in one component can be used to predict the effect of changes in other components of the groundwater system. In this manner, the study of groundwater balance has a significant role in planning a rational groundwater development of a region.

**Concluding Remarks**

- Water balance approach, essentially a lumped model study, is a viable method of establishing the rainfall recharge coefficient and for evaluating the methods adopted for the quantification of discharge and recharge from other sources. For proper assessment of potential, present use and additional exploitability of water resources at optimal level, a water balance study is necessary. It has been reported that the groundwater resource estimation methodology recommended by Groundwater Resource Estimation Committee (1997) is being used by most of the organisations in India.

- Groundwater exploitation should be such that protection from depletion is provided, protection from pollution is provided, negative ecological effects are reduced to a minimum and economic efficiency of exploitation is attained. Determination of exploitable resources should be based upon hydrological investigations. These investigations logically necessitate use of a mathematical model of groundwater system for analysing and solving the problems. The study of water balance is a pre-requisite for groundwater modelling.

- There is a need for studying unsaturated and saturated flow through weathered and fractured rocks for finding the recharge components from rainfall and from percolation tanks in hard rock groundwater basins. The irrigation return flow under different soils, crops and irrigation practices need to be quantified. Assessment of groundwater quality in many groundwater basins is a task yet to be performed. A hydrological database for groundwater assessment should be established. Also user friendly software should be developed for quick assessment of regional groundwater resources. Tata Infotech Limited (India) has developed a proprietary software 'GEMS' (Groundwater Estimation & Management System) based upon the recommendations of Groundwater Resource Estimation Committee (1997).

- Non-conventional methods for utilisation of water such as through inter-basin transfers, artificial recharge of groundwater and desalination of brackish or sea water as well as traditional water conservation practices like rainwater harvesting, including roof-top rainwater harvesting, need to be practiced to further increase the utilisable water resources.

**References**

1. Chandra, Satish and R. S. Saksena (1975). Water Balance Study for Estimation of Groundwater Resources. Journal of Irrigation and Power, India, October 1975, pp. 443-449.

2.    Central Ground Water Board (2011). Dynamic Ground Water Resources of India (as on 31 March 2009). Ministry of Water Resources, Government of India, November 2011, 225 p.

3.    Groundwater Resource Estimation Methodology - 1997. Report of the Groundwater Resource Estimation Committee, Ministry of Water Resources, Government of India, New Delhi, June 1997.

4.    IILRI (1974). Drainage Principles and Applications, Survey and Investigation. Publication 16, Vol. III.

5.    Johnson, A. I. (1963). Application of Laboratory Permeability Data. Open File Report, U.S.G.S., Water Resources Division, Denver, Colorado, 34 p.

6.    Johnson, A. I. (1967). Specific Yield - Compilation of Specific Yields for Various Materials. Water Supply Paper, U.S.G.S., 1962-D, 74 p.

7.    Karanth, K. R. (1987). Groundwater Assessment, Development and Management. Tata McGraw-Hill Publishing Company Limited, New Delhi, pp. 576-657.

8.    Kumar, C. P. and P. V. Seethapathi (1988). Effect of Additional Surface Irrigation Supply on Groundwater Regime in Upper Ganga Canal Command Area, Part I - Groundwater Balance. National Institute of Hydrology, Case Study Report No. CS-10 (Secret/Restricted), 1987-88.

9.    Kumar, C. P. and P. V. Seethapathi (2002). Assessment of Natural Groundwater Recharge in Upper Ganga Canal Command Area. Journal of Applied Hydrology, Association of Hydrologists of India, Vol. XV, No. 4, October 2002, pp. 13-20.

10.   Mishra, G. C. (1993). Current Status of Methodology for Groundwater Assessment in the Country in Different Region. National Institute of Hydrology, Technical Report No. TR-140, 1992-93, 25 p.

11.   National Water Policy (2012). Ministry of Water Resources, Government of India.

12.   Sokolov, A. A. and T. G. Chapman (1974). Methods for Water Balance Computations. The UNESCO Press, Paris.

13.   Sophocleous, Marios A. (19910. Combining the Soilwater Balance and Water-Level Fluctuation Methods to Estimate Natural Groundwater Recharge: Practical Aspects. Journal of Hydrology, Vol. 124, pp. 229-241.

14.   Thornthwaite, C. W. (1948). An Approach towards a Rational Classification of Climate. Geogr. Rev., Vol. 38, No. 1, pp. 55-94.

15.   Thornthwaite, C. W. and J. W. Mather (19550. The Water Balance. Publ. Climatol. Lab. Climatol. Drexel Inst. Technol., Vol. 8, No. 1, pp. 1-104.

# 6-Groundwater Data Requirement and Analysis

## Introduction

Groundwater is used for a variety of purposes, including irrigation, drinking and manufacturing. Groundwater is also the source of a large percentage of surface water. To verify that groundwater is suited for its purpose, its quality can be evaluated (i.e., monitored) by collecting samples and analyzing them. In simplest terms, the purpose of groundwater monitoring is to define the physical, chemical, and biological characteristics of groundwater.

Accurate and reliable groundwater resource information (including quality) is critical to planners and decision-makers. Huge investment in the areas of groundwater exploration, development and management at state and national levels aims to meet the groundwater requirement for drinking and irrigation and generates enormous amount of data. We need to focus on improved data management, precise analysis and effective dissemination of data.

Numerical models are capable of solving large and complex groundwater problems varying widely in size, nature and real life situations. With the advent of high speed computers, spatial heterogeneities, anisotropy and uncertainties can be tackled easily. However, the success of any modelling study, to a large measure depends upon the availability and accuracy of measured/recorded data required for that study. Therefore, identifying the data needs of a particular modelling study and collection/monitoring of required data form an integral part of any groundwater modelling exercise.

## Data Requirement for Groundwater Studies

The first phase of any groundwater study consists of collecting all existing geological and hydrological data on the groundwater basin in question. This will include information on surface and subsurface geology, water tables, precipitation, evapotranspiration, pumped abstractions, stream flows, soils, land use, vegetation, irrigation, aquifer characteristics and boundaries, and groundwater quality. If such data do not exist or are very scanty, a program of field work must first be undertaken, for no model whatsoever makes any hydrological sense if it is not based on a rational hydrogeological conception of the basin. All the old and newly-found information is then used to develop a conceptual model of the basin, with its various inflow and outflow components.

A conceptual model is based on a number of assumptions that must be verified in a later phase of the study. In an early phase, however, it should provide an answer to the important question: does the groundwater basin consist of one single aquifer (or any lateral combination of aquifers) bounded below by an impermeable base? If the answer is yes, one can then proceed to the next phase: developing the numerical model. This model is first used to synthesize the various data and then to test the assumptions made in the conceptual model. Developing and testing the numerical model requires a set of quantitative hydrogeological data that fall into two categories:

- Data that define the physical framework of the groundwater basin

- Data that describe its hydrological stress

These two sets of data are then used to assess a groundwater balance of the basin. The separate items of each set are listed below.

*Physical framework*

1. Topography
2. Geology
3. Types of aquifers
4. Aquifer thickness and lateral extent
5. Aquifer boundaries
6. Lithological variations within the aquifer
7. Aquifer characteristics

*Hydrological stress*

1. Water table elevation
2. Type and extent of recharge areas
3. Rate of recharge
4. Type and extent of discharge areas
5. Rate of discharge

It is common practice to present the results of hydrogeological investigations in the form of maps, geological sections and tables - a procedure that is also followed when developing the numerical model. The only difference is that for the model, a specific set of maps must be prepared. These are:

- Contour maps of the aquifer's upper and lower boundaries
- Maps of the aquifer characteristics
- Maps of the aquifer's s net recharge
- Water table contour maps

Some of these maps cannot be prepared without first making a number of auxiliary maps. A map of the net recharge, for instance, can only be made after topographical, geological, soil, land use, cropping pattern, rainfall, and evaporation maps have been made.

The data needed in general for a groundwater flow modelling study can be grouped into two categories: (a) Physical framework and (b) Hydrogeologic framework (Moore, 1979). The data required under physical framework are:

1. Geologic map and cross section or fence diagram showing the areal and vertical extent and boundaries of the system.
2. Topographic map at a suitable scale showing all surface water bodies and divides. Details of surface drainage system, springs, wetlands and swamps should also be available on map.
3. Land use maps showing agricultural areas, recreational areas etc.
4. Contour maps showing the elevation of the base of the aquifers and confining beds.
5. Isopach maps showing the thickness of aquifers and confining beds.
6. Maps showing the extent and thickness of stream and lake sediments.

These data are used for defining the geometry of the groundwater domain under investigation, including the thickness and areal extent of each hydrostratigraphic unit.

Under the hydrogeologic framework, the data requirements are:

1. Water table and potentiometric maps for all aquifers.
2. Hydrographs of groundwater head and surface water levels and discharge rates.
3. Maps and cross sections showing the hydraulic conductivity and/or transmissivity distribution.
4. Maps and cross sections showing the storage properties of the aquifers and confining beds.
5. Hydraulic conductivity values and their distribution for stream and lake sediments.
6. Spatial and temporal distribution of rates of evaporation, groundwater recharge, surface water - groundwater interaction, groundwater pumping, and natural groundwater discharge.

## Groundwater Data Acquisition

Obtaining all the information necessary for modelling is not an easy task. In fact, a modeller may have to devote considerable effort and time in data acquisition, especially when the database for the study area is non-existent. Some data may be obtained from existing reports of various agencies/departments, but in most cases, additional field work is required. Moreover, the data is not readily available in the format required by the model, and requires additional work to process it. The observed raw data obtained from the field may also contain inconsistencies and errors. Before proceeding with data processing, it is essential to carry out data validation in order to correct errors in recorded data and assess the reliability of a record. In addition, as the modelling exercise progresses, certain gaps in the database get identified. In such cases, the field monitoring program may undergo some revision including installation of new piezometers/monitoring wells.

Amongst the hydrologic stresses including groundwater pumping, evapotranspiration and recharge, groundwater pumpage is the easiest to estimate. Field information for estimating evapotranspiration is likely to be sparse and can be estimated from information about the land use and potential evapotranspiration values. Recharge is one of the most difficult parameters to estimate. Recharge refers to the volume of infiltrated water that crosses the water table and becomes part of the groundwater flow system. This infiltrated water may be a certain percentage of rainfall, irrigation return flow, seepage from surface water bodies etc., depending upon topography, soil characteristics, depth to groundwater level and other factors.

Values of transmissivity and storage coefficient are usually obtained from data generated during pumping tests and subsequent data processing. For modelling at a local scale, values of hydraulic conductivity may be determined by pumping tests if volume-averaged values are required or by slug-tests if point values are desired. For unconsolidated sand-size sediment, hydraulic conductivity may be obtained from laboratory permeability tests using permeameters. However, due to rearrangement of grains during repacking the sample into permeameter, the obtained hydraulic conductivity values are typically several orders of magnitude smaller than values measured *in situ*. Furthermore, in a laboratory column sample, large-scale features such as fractures and gravel lenses that may impart transmission characteristics to the hydrogeologic unit as a whole are not captured. Due to this reason, laboratory analyses of core samples tend to give lower values of hydraulic conductivity than are measured in the field. In the absence of site-specific field or laboratory measurements, initial estimates for aquifer properties may be taken from standard tables.

When simulating anisotropic media, information is required on principal components of hydraulic conductivity tensor $K_x$, $K_y$ and $K_z$. Vertical anisotropy is defined by the anisotropy ratio between $K_x$ and $K_z$. For most groundwater problems, it is impossible to model geologic units at the isotropic scale. When the thickness of the model layer ($B_{ij}$) is much larger than the thickness of isotropic layer ($b_{ijk}$) (assuming this thickness can be identified from bedding information), the hydrologically equivalent horizontal and vertical hydraulic conductivities for the model layer may be calculated as:

$$(K_x)_{ij} = \sum_{k=1}^{m} \frac{K_{ijk} b_{ijk}}{B_{ij}} \qquad (K_z)_{ij} = \frac{B_{ij}}{\sum_{k=1}^{m} \frac{b_{ijk}}{K_{ijk}}} \qquad B_{ij} = \sum_{k=1}^{m} b_{ijk}$$

where $i$ represents column, $j$ represents row and $k$ represents layer number. Vertical anisotropy for each hydrogeologic unit or model layer may be computed using above equations, if sufficient stratigraphic information is available. The anisotropy ratio may also be estimated during model calibration. The thickness and vertical hydraulic conductivity of stream and lake sediments are required for estimating seepage. These values may be obtained from field measurements or during model calibration.

*Monitoring of Groundwater Levels*

To obtain data on the depth and configuration of the water table, the direction of groundwater movement, and the location of recharge and discharge areas, a network of observation wells and/or piezometers has to be established. The objectives of the groundwater level monitoring are to:

- detect impact of groundwater recharge and abstractions,
- monitor the groundwater level changes,
- assess depth to water level,
- detect long term trends,
- compute the groundwater resource availability,
- assess the stage of development,
- design management strategies at regional level.

The water table reacts to various recharge and discharge components that characterize a groundwater system and is therefore constantly changing. Important in any drainage investigation are the (mean) highest and the (mean) lowest water table positions, as well as the mean water table of a hydrological year. For this reason, water level measurements should be made at frequent intervals for at least a year. The interval between readings should not exceed one month, but a fortnight may be better. All measurements should, as far as possible, be made on the same day because this gives a complete picture of the water table.

*Monitoring of Groundwater Quality*

For various reasons, a knowledge of the groundwater quality is required. These are:

- Any lowering of the water table may provoke the intrusion of salty groundwater from adjacent areas, or from the deep underground, or from the sea. The drained area and its surface water system will then be charged daily with considerable amounts of dissolved salts;

- The disposal of the salty drainage water into fresh-water streams may create environmental and other problems, especially if the water is used for irrigation and/or drinking;

- In arid and semi-arid regions, soil salinization is directly related to the depth of the groundwater and to its salinity;

- Groundwater quality dictates the type of cement to be used for hydraulic structures, especially when the groundwater is rich in sulphates.

Groundwater is sampled to assess its quality for a variety of purposes. Whatever the purpose, it can only be achieved if results are representative of actual site conditions and are interpreted in the context of those conditions. Substantial costs are incurred to obtain and analyze samples. Field costs for drilling, installing, and sampling monitoring wells and laboratory costs for analyzing samples are not trivial. The utility of such expenditures can be jeopardized by the manner in which reported results are interpreted as well as by problems in how samples were obtained and analyzed. Considerable attention has been given to standardizing procedures for sampling and analyzing groundwater. Although following such standard procedures is important and provides a necessary foundation for understanding results, it neither guarantees that reported results will be representative nor necessarily have any real relationship to actual site conditions.

Comprehensive data analysis and evaluation by a knowledgeable professional should be the final quality assurance step, it may indeed help to find errors in field or laboratory work that went otherwise unnoticed, and provides the best chance for real understanding of the meaning of reported results. To facilitate interpretation, the following steps should be included:

1. Collection, analysis and evaluation of background data on regional and site-specific geology, hydrology and potential anthropogenic factors that could influence groundwater quality and collection of background information on the environmental chemistry of the analytes of concern.

2. Planning and carrying out of field activities using accepted standard procedures capable of producing data of known quality.

3. Selection of a laboratory to analyze groundwater samples based on careful evaluation of laboratory qualifications.

4. The use of appropriate quality control/quality assurance (QC/QA) checks of field and laboratory work (including field blank, duplicate, and performance evaluation samples).

5. Comprehensive interpretation of reported analytical data by a knowledgeable professional. The analytical data must be accompanied by appropriate QC/QA data, be cross-checked

using standard water quality checks and relationships where possible, and be correlated with information on regional and site-specific geology and hydrology, environmental chemistry, and potential anthropogenic influences.

The objectives of the water quality monitoring network are to:

- establish the bench mark for different water quality parameters, and compare the different parameters against the national standards,
- detect water quality changes with time,
- identify potential areas that show rising trend,
- detect potential pollution sources,
- study the impact of land use and industrialization on groundwater quality,
- Data collection for the reporting year.

The frequency of sampling required in a groundwater-quality monitoring program is dictated by the expected rate of change in the concentrations of chemical constituents and the physio-chemical properties of the water being measured. Groundwater moves slowly, perhaps only a few centimeters to a few decimeters per day, so that day-to-day fluctuations in concentrations of constituents and properties at a point (or well) commonly are too small to be detected. For monitoring concentrations of major ions and nutrients, and values of physical properties of groundwater, twice yearly sampling should be sufficient, and by varying the season selected for sampling, conditions during all the seasons could be documented over a 2-year cycle. A second group of constituents, trace inorganic and organic compounds, could be adequately monitored by collecting samples once every 2 years from wells in background areas (those areas unaffected by human activities), but more frequent sampling should be considered if the types and conditions of any upgradient sources of these compounds are changing.

Consideration of several factors suggests that monitoring of groundwater quality should be a long-term activity. Not only does the structure of the program described herein mandate long-term monitoring, but also the scales over which groundwater quality is likely to fluctuate are long. Because of the slow rate of groundwater movement, any changes in factors that affect the quality of the water in recharge areas can take a long time to be reflected in surface water bodies that are discharge areas for the groundwater. The duration of a groundwater-quality monitoring program also is affected by the time scale of changes in the source area for the chemical constituents of interest.

**Processing of Groundwater Data**

Before any conclusions can be drawn about the cause, extent and severity of an area's groundwater related problems, the raw groundwater data on water levels and water quality have to be processed. They then have to be related to the geology and hydrogeology of the area. The results, presented in graphs, maps, and cross-sections, will enable a diagnosis of the problems. We shall assume that such basic maps as topographic, geological, and pedological maps are available.

The following graphs and maps have to be prepared that are discussed hereunder:

- ❖ Groundwater hydrographs;

- ❖ Water table-contour map;
- ❖ Depth-to-water table map;
- ❖ Water table-fluctuation map;
- ❖ Head-differences map;
- ❖ Groundwater-quality map.

It must be emphasized that a proper interpretation of groundwater data, hydrographs, and maps requires a coordinate study of a region's geology, soils, topography, climate, hydrology, land use, and vegetation. If the groundwater conditions in irrigated areas are to be properly understood and interpreted, cropping patterns, water distribution and supply, and irrigation efficiencies should be known too.

Hydrological Information System (HIS) developed under Hydrology Project provides easy access to different variety of data. The monitoring data are systematically organized in the HIS data base, including:

- well inventory
- exploratory drilling
- pumping test data
- logging
- water level
- water quality
- rainfall data
- meteorological data

*Groundwater Hydrographs*

When the amount of groundwater in storage increases, the water table rises; when it decreases, the water table falls. This response of the water table to changes in storage can be plotted in a hydrograph (Figure 1). Groundwater hydrographs show the water-level readings, converted to water levels below ground surface, against their corresponding time. A hydrograph should be plotted for each observation well or piezometer. It is important to know the rate of rise of the water table, and even more important, that of its fall. If the groundwater is not being recharged, the fall of the water table will depend on:

- transmissivity of the water-transmitting layer, KH;
- storativity of this layer, S;
- hydraulic gradient, dh/dx.

After a period of rain (or irrigation) and an initial rise in groundwater levels, they then decline, rapidly at first, and then more slowly as time passes because both the hydraulic gradient and the transmissivity decrease. The graphical representation of the water table decline is known as the natural recession curve. It can be shown that the logarithm of the water table height decreases linearly with time. Hence, a plot of the water table height against time on semi-logarithmic paper gives a straight line. Groundwater recession curves are useful in studying changes in groundwater storage and in predicting future groundwater levels.

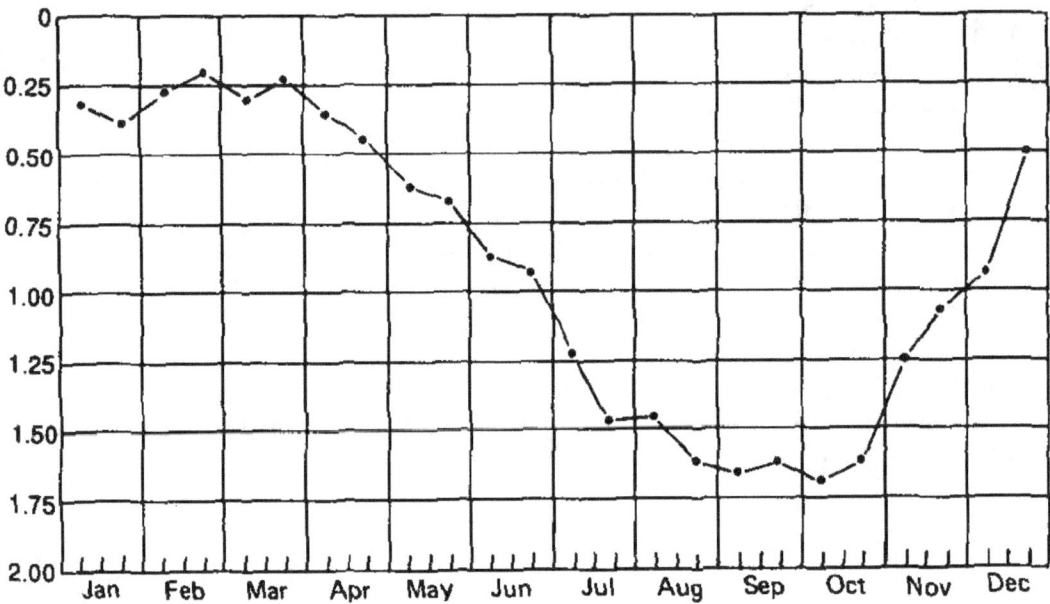

Figure 1:     Hydrograph of a Water Table Observation Well

The groundwater hydrographs of all the observation points should be systematically analyzed. A comparison of these hydrographs enables us to distinguish different groups of observation wells. Each well, belonging to a certain group, shows a similar response to the recharge and discharge pattern of the area. By a similar response, we mean that the water level in these wells starts rising at the same time, attains its maximum value at the same time, and after recession starts, reaches its minimum value at the same time. The amplitude of the water level fluctuation in the various wells need not necessarily be exactly the same, but should show a great similarity. Areas where such wells are sited can then be regarded as hydrological units (i.e. sub-areas in which the watertable reacts to recharge and discharge everywhere in the same way).

Groundwater hydrographs also offer a means of estimating the annual groundwater recharge from rainfall. This, however, requires several years of records on rainfall and water tables. An average relationship between the two can be established by plotting the annual rise in water table against the annual rainfall (Figure 2). Extending the straight line until it intersects the abscissa gives the amount of rainfall below which there is no recharge of the groundwater. Any quantity less then this amount is lost by surface runoff and evapotranspiration.

*Water Table-Contour Map*

A water table-contour map shows the elevation and configuration of the water table on a certain date. To construct it, we first have to convert the water-level data from the form of depth below surface to the form of water table elevation (= water level height above a datum plane, e.g. mean sea level). These data are then plotted on a topographic base map and lines of equal water table elevation are drawn. A proper contour interval should be chosen, depending on the slope of the water table. For a flat water table, 0.25 to 0.50 m may suit; in steep water table areas, intervals of 1 to 5 m or even more may be needed to avoid overcrowding the map with contour lines. The topographic base map should contain contour lines of the land surface and should show all natural drainage channels and open water bodies. For the given date, the water levels of these surface waters should also be plotted on the map. Only with these data and data on the land surface elevation can water table contour lines be drawn correctly (Figure 3).

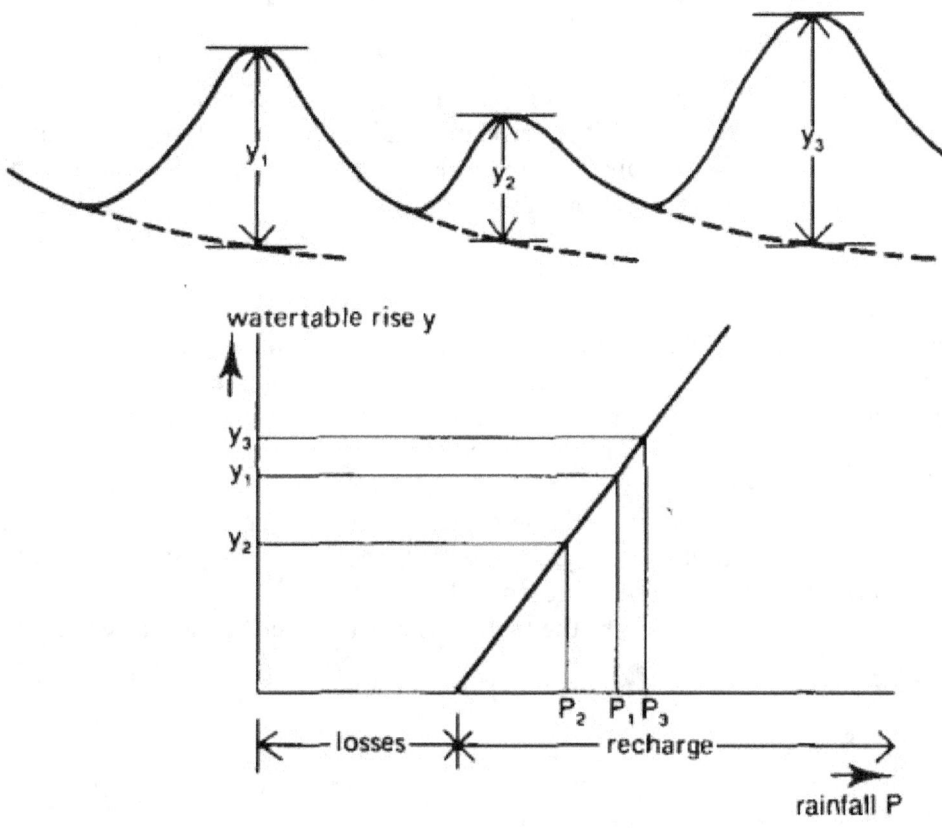

Figure 2:      Relationship between Annual Groundwater Recharge and Rainfall

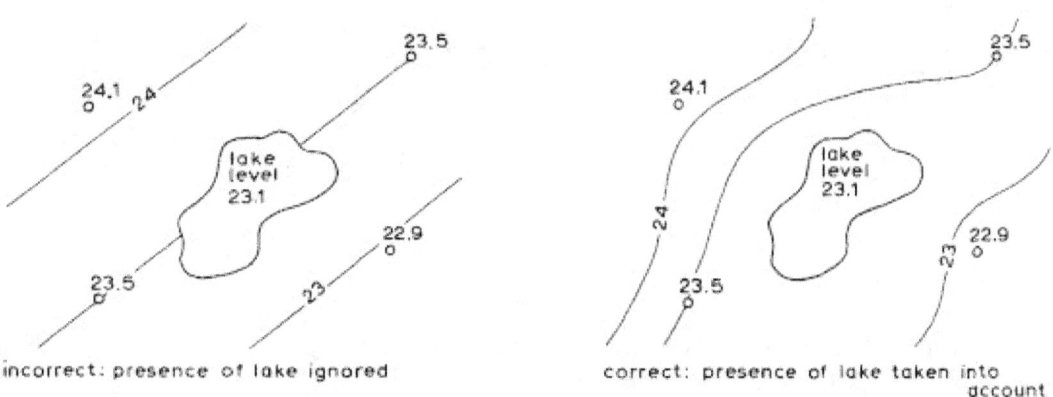

incorrect: presence of lake ignored

correct: presence of lake taken into account

Figure 3:      Example of Water Table-Contour Lines

To draw the water table-contour lines, we have to interpolate the water levels between the observation points, using the linear interpolation method. Instead of preparing the map for a certain date, we could also select a period (e.g. a season or a whole year) and calculate the mean water table elevation of each well for that period. This has the advantage of smoothing out local or occasional anomalies in water levels. A water table-contour map is an important tool in groundwater investigations because, from it, one can derive the gradient of the water table (dh/dx) and the direction of groundwater flow, which is perpendicular to the water table-contour lines.

For a proper interpretation of a water table-contour map, one has to consider not only the topography, natural drainage pattern, and local recharge and discharge patterns, but also the subsurface geology. More specifically, one should know the spatial distribution of permeable and less permeable layers below the water table. For instance, a clay lens impedes the downward flow of excess irrigation water or, if the area is not irrigated, the downward flow of excess rainfall. A groundwater mound will form above such a horizontal barrier (Figure 4).

Water table-contour maps are graphic representations of the hydraulic gradient of the water table. The velocity of groundwater flow (v) varies directly with the hydraulic gradient (dh/dx) and, at constant flow velocity, the gradient is inversely related to the hydraulic conductivity (K), or v = -K(dh/dx) (Darcy's law). This is a fundamental law governing the interpretation of hydraulic gradients of water tables. Suppose the flow velocity in two cross-sections of equal depth and width is the same, but one cross-section shows a greater hydraulic gradient than the other, then its hydraulic conductivity must be lower. A steepening of the hydraulic gradient may thus be found at the boundary of fine-textured and
coarse-textured material, or at a fault where the thickness of the water bearing layers changes abruptly.

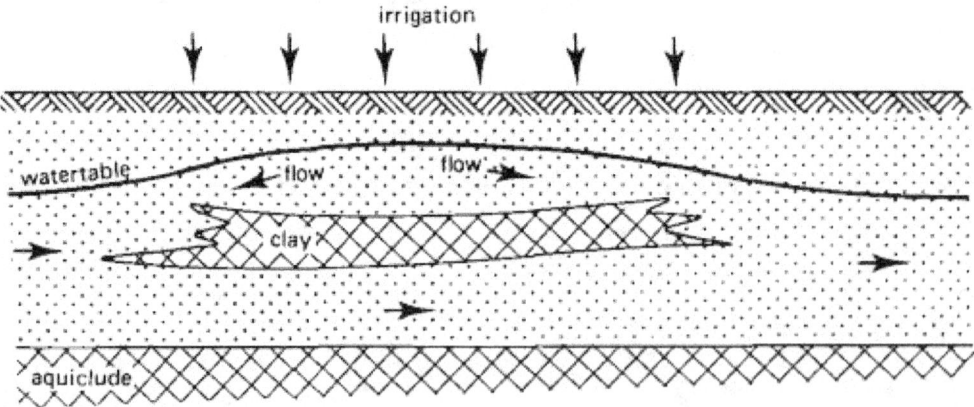

Figure 4:     A Clay Lens under an Irrigated Area impedes the Downward Flow of Excess Irrigation Water

*Depth-to-Water Table Map*

A depth-to-water table map or isobath map, as these names imply, shows the spatial distribution of the depth of the water table below the land surface. It can be prepared in two ways. The water level data from all the observation wells for a certain date should first be converted to water levels below land surface (the reference point from which the readings are taken needs not necessarily to be the land surface). One then plots the transformed data on the topographical base map near each observation point and draws isobaths or lines of equal depth to groundwater. A suitable contour interval may be 50 cm (Boonstra, 1988). Another way of preparing an isobath map is made by superimposing a water table contour map for a special date on the topographical map showing contour lines of the land surface. From the two families of contour lines, read the differences in elevation at contour intersections, plot these data on a clean topographical map and draw the isobaths.

According to the observation results for a year, the map drawn using the highest water table levels indicates the highest water table levels for a year in irrigated lands. The regions of map where the groundwater level is between 0-2 m depicts the area having drainage problems. On the other hand, based on measurement results for a year, the map drawn using the lowest water table levels indicates to which extent the groundwater falls in a year in irrigated area. The section where the water table level is between 0-1 m determines the areas in which groundwater exists in the root-zone throughout a year. In these areas, farm drainage systems also need to be established. The map drawn based on water table measurements done in the month of most intensive irrigation indicates to which extent irrigation activities influence the water table. Due to most intensive month in terms of irrigation application differing from one irrigation scheme to another, this month is defined as month in which most of water is released to the system.

A variety of factors must be considered if one is to interpret a depth-to-groundwater or isobath map properly. Shallow water tables may occur temporarily, which means that the natural groundwater runoff cannot cope with an incidental precipitation surplus or irrigation percolation. They may occur (almost) permanently because the inflow of groundwater exceeds the outflow, or groundwater outflow is lacking as in topographic depressions. The depth and shape of the first impermeable layer below the water table strongly affect the height of the water table. To explain differences and variations in the depth to water table, one has to consider topography, surface and subsurface geology, climate, direction and rate of groundwater flow, land use, vegetation, irrigation, and the abstraction of groundwater by wells.

*Water Table-Fluctuation Map*

A water table-fluctuation map is a map that shows the magnitude and spatial distribution of the change in water table over a period (e.g. a season or a whole hydrological year). Using such graphs, we calculate the difference between the highest and the lowest water table height (or preferably the difference between the mean highest and the mean lowest water table height for the two seasons). We then plot these data on a topographic base map and draw lines of equal change in water table, using a convenient contour interval. A water table-fluctuation map is a useful tool in the interpretation of drainage problems in areas with large water table fluctuations, or in areas with poor natural drainage (or upward seepage) and permanently high water tables (i.e. areas with minor water table fluctuations).

The water table in topographic highs is usually deep, whereas in topographic lows it is shallow. This means that on topographic highs there is sufficient space for the water table to change. This space is lacking in topographic lows where the water table is often close to the surface. Water table fluctuations are therefore closely related to depth to groundwater.

Another factor to consider in interpreting water table-fluctuation maps is the drainable pore space of the soil. The change in water table in fine-textured soils will differ from that in coarse-textured soils, for the same recharge or discharge.

## Head-Differences Map

A head-differences map is a map that shows the magnitude and spatial distribution of the differences in hydraulic head between two different soil layers. We calculate the difference in water level between the two piezometers, and plot the result on a map. After choosing a proper contour interval (e.g. 0.10 or 0.20 m), we draw lines of equal head difference. Another way of drawing such a map is to superimpose a water table-contour map on a contour map of the piezometric surface of the underlying layer. We then read the head differences at contour line intersections, plot these on a base map, and draw lines of equal head difference. The map is a useful tool in estimating upward or downward seepage.

The difference in hydraulic head between the shallow and the deep groundwater is directly related to the hydraulic resistance of the low-permeable layer(s). Because such layers are seldom homogeneous and equally thick throughout an area, the hydraulic resistance of these layers varies from one place to another. Consequently, the head difference between shallow and deep groundwater varies. Local 'leaks' in low-permeable layers may result in anomalous differences in hydraulic heads. The hydraulic resistance is especially of interest when one is defining upward seepage or natural drainage or the possibilities for tubewell drainage.

## Groundwater-Quality Map

A groundwater-quality map e.g. an electrical-conductivity (EC) map is a map that shows the magnitude and spatial variation in the salinity of the groundwater. The EC values of all representative wells (or piezometers) are used for this purpose. Groundwater salinity varies not only horizontally but also vertically; a zonation of groundwater salinity is common in many areas (e.g. in delta and coastal plains, and in arid plains). It is therefore advisable to prepare an electrical-conductivity map not only for the shallow groundwater but also for the deep groundwater.

In electrical conductivity maps, critical groundwater salinity is taken as 5000 micro-mhos/cm, although it changes according to species of the crop to be grown. However, it must be examined whether salt accumulation risks occur in the root-zone through plotting of all the areas having that salinity level on the critical highest depth-to-water table map. By plotting all the EC values on a map, lines of equal electrical conductivity (equal salinity) can be drawn. Preferably the following limits should be taken: less than 100 micro-mhos/cm, 100 to 250; 250-750; 750 to 2500; 2500 to 5000; and more than 5000. Other limits may, of course, be chosen depending on the salinity found in the waters. Other types of groundwater-quality maps can be prepared by plotting different quality parameters e.g. Sodium Adsorption Ratio (SAR) values.

Spatial variations in groundwater quality are closely related to topography, geological environment, direction and rate of groundwater flow, residence time of the groundwater, depth to water table, and climate. Topographic highs, especially in the humid zones, are areas of recharge if their permeability is fair to good. The quality of groundwater in such areas almost resembles that of rainwater. On its way to topographic lows (areas of discharge), groundwater becomes more mineralized because of the dissolution of minerals. Although the water may be still fresh in discharge areas, its electrical conductivity can be several times higher than in recharge areas.

The groundwater in the lower portions of coastal and delta plains may be brackish to extremely salty, because of sea-water encroachment and the marine environment in which all or most of the mass of sediments was deposited. Their upper parts, which are usually topographic highs, are now-a-days recharge areas and consequently contain fresh groundwater.

In the arid and semi-arid zones, shallow water table areas, as can be found in the lower parts of alluvial fans, coastal plains, and delta plains, may contain very salty groundwater because of high rates of evaporation. Irrigation in such areas may contribute to the salinity of the shallow groundwater through the dissolution of salts accumulated in the soil layers. Sometimes, however, irrigated land can have groundwater of much better quality than adjacent non-irrigated land. Because of the irrigation percolation losses, the water table under the irrigated land is usually higher than in the adjacent non-irrigated land. Consequently, there is a continuous transport of salt-bearing groundwater from the irrigated to the non-irrigated land. This causes the water table in the non-irrigated land to rise to close to the surface, where evapotranspiration further contributes to the salinization of groundwater and soil.

## Interpretation of Hydraulic Head and Groundwater Conditions

*Groundwater flow direction:* Measurements of hydraulic head, normally achieved by the installation of a piezometer or well point, are useful for determining the directions of groundwater flow in an aquifer system. In Figure 5(a), three piezometers installed to the same depth enable the determination of the direction of groundwater flow and, with the application of Darcy's law, the calculation of the horizontal component of flow. In Figure 5(b), two examples of piezometer nests are shown that allow the measurement of hydraulic head and the direction of groundwater flow in the vertical direction to be determined either at different levels in the same aquifer formation or in different formations.

## Interpolation of Field Data by Kriging

Assigning field data as input to a groundwater model is difficult because the model requires values for each grid node or cell and field data are typically sparse. To define spatial variation of a parameter over the study area, interpolation of measured data points may need to be carried out using suitable interpolation techniques like inverse distance to a power method, least square fitting of a polynomial, kriging etc.

Kriging: The most often used method for interpolation is kriging. It is a statistical interpolation method that chooses the best linear unbiased estimate for the variable in question (DeMarsily, 1986). The variable is assumed to be a random function with some kind of correlation (structure) in its spatial distribution, that is defined by a variogram (Figure 6). A variogram is a measure of the change in the variable with changes in distance.

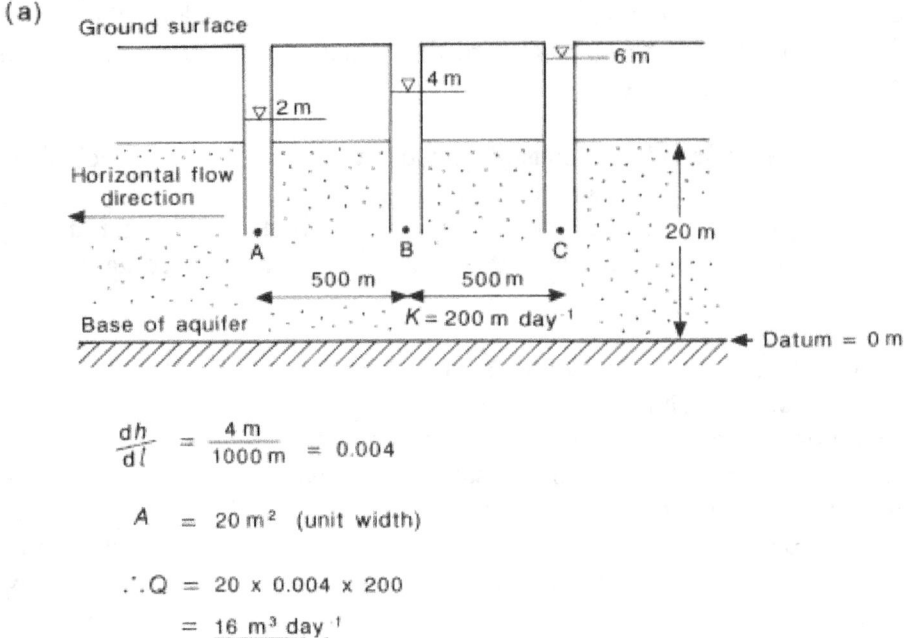

(a)

$$\frac{dh}{dl} = \frac{4\,m}{1000\,m} = 0.004$$

$$A = 20\,m^2 \ \ (unit\ width)$$

$$\therefore Q = 20 \times 0.004 \times 200$$
$$= \underline{16\ m^3\ day^{-1}}$$

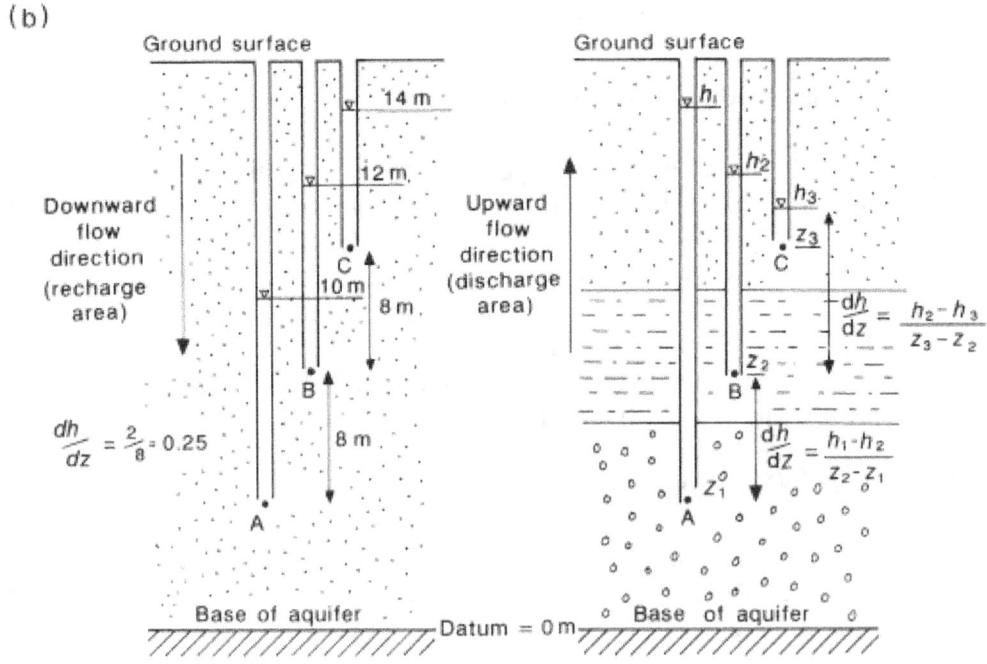

Figure 5: Determination of groundwater flow direction and hydraulic head gradient from piezometer measurements for (a) horizontal flow and (b) vertical flow. The elevation of the water level indicating the hydraulic head at each of the points A, B and C is noted adjacent to each piezometer.

Higher correlation is expected between points measured at small separation distances. For example, hydraulic head, transmissivity, permeability, thickness of a layer, storage coefficient, rainfall, effective recharge, etc., are all functions of space and are very often highly variable. This spatial variability is, in general, not purely random; if measurements are made at two different locations, the closer the measurements points are to each other, the closer are the measured values.

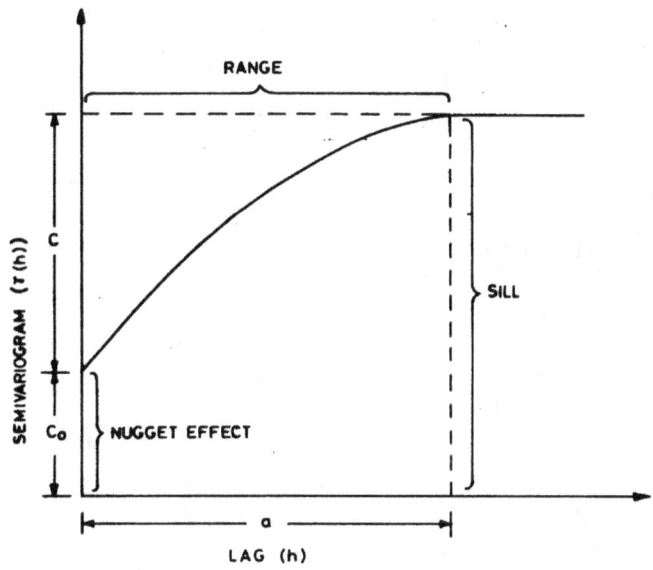

Figure 6: Plot of Variogram

For the purpose of analysis, the spatial continuity is expressed as a variogram, which is a plot of the semi-variance ($\gamma$) against the separation distance ($h$). The semi-variance is defined as half the mean of squares of the differences of the recorded values at a pair of stations spaced at a given distance. According to the definition of $\gamma$, its value at $h$ equal to zero is half the variance of a data from itself, which is zero. Thus, the variogram has to pass through the origin. Further, for a distance less than the minimum distance between the observation points, $\gamma$ is not defined. The minimum distance may be quite small in comparison with the total range of $h$, and $\gamma$ corresponding to this distance may not be close to zero. Since, the variogram has to pass through the origin, this leads to a nugget effect, which is basically a vertical rise of the plotted curve at the origin (Figure 6). The value of $\gamma$ rises gradually upto a limiting distance beyond which it becomes a constant. This distance is known as the range of the variable. The maximum value of $\gamma$ reached at the range is known as the sill. The range can be viewed as the effective neighbourhood within which the continuity of the variable holds.

The structural information as represented in the variogram is then fitted with a model. The commonly used models are linear, spherical, exponential and Gaussian. These are illustrated in Figure 7.

The model fitting involves choosing an appropriate model and estimating its parameters, such as the range and nugget. Thus, kriging is the process of estimating the value of a spatially distributed variable from adjacent values while considering the interdependence expressed in the fitted variogram model.

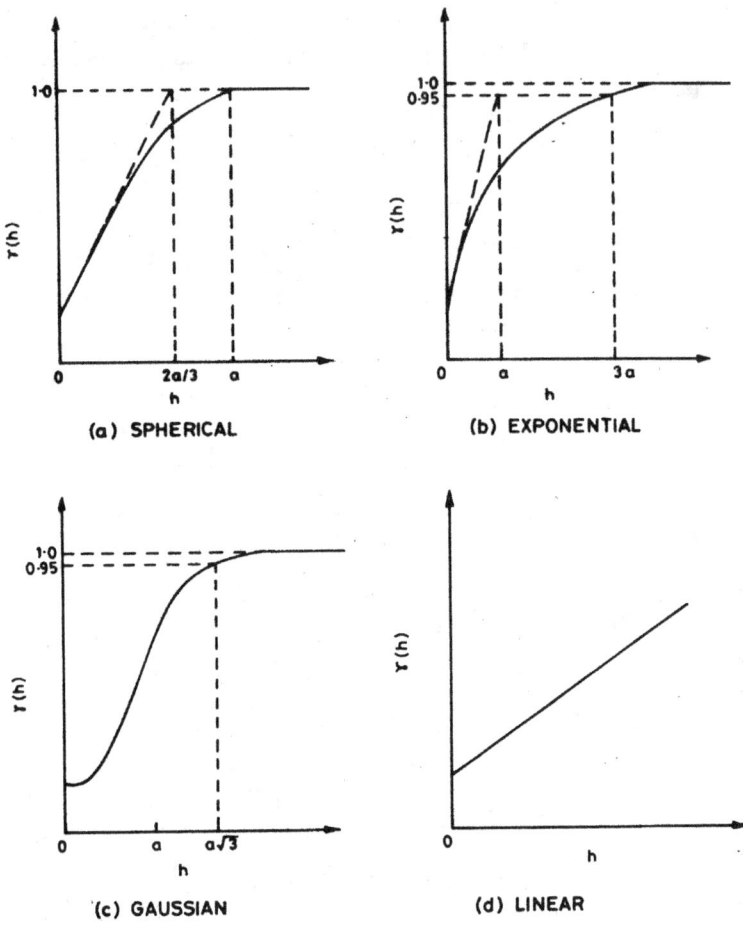

Figure 7:      Theoretical Models of Variogram

Kriging differs from other interpolation methods because it considers the spatial structure of a variable, and also preserves the field value at measurement points (unlike some other interpolation schemes such as least square fitting of a polynomial). Irrespective of the method used to assign parameter values to nodal points of grid, care should be taken to check that the resulting parameter distribution is reasonable in hydrogeologic terms and that the values fall within appropriate range for the geologic setting.

**GIS for Groundwater Studies**

For handling groundwater data, the GIS-technology is aptly suited, for the following main reasons:

1. *Concurrent handling of locational and attribute data:* In groundwater studies, one has to deal with information comprising locational data (where it is?) and attribute data (what it is?). GIS packages have the unique capability to handle locational and attribute data; such a capability is not available in other groups of packages (Figure 8).

94

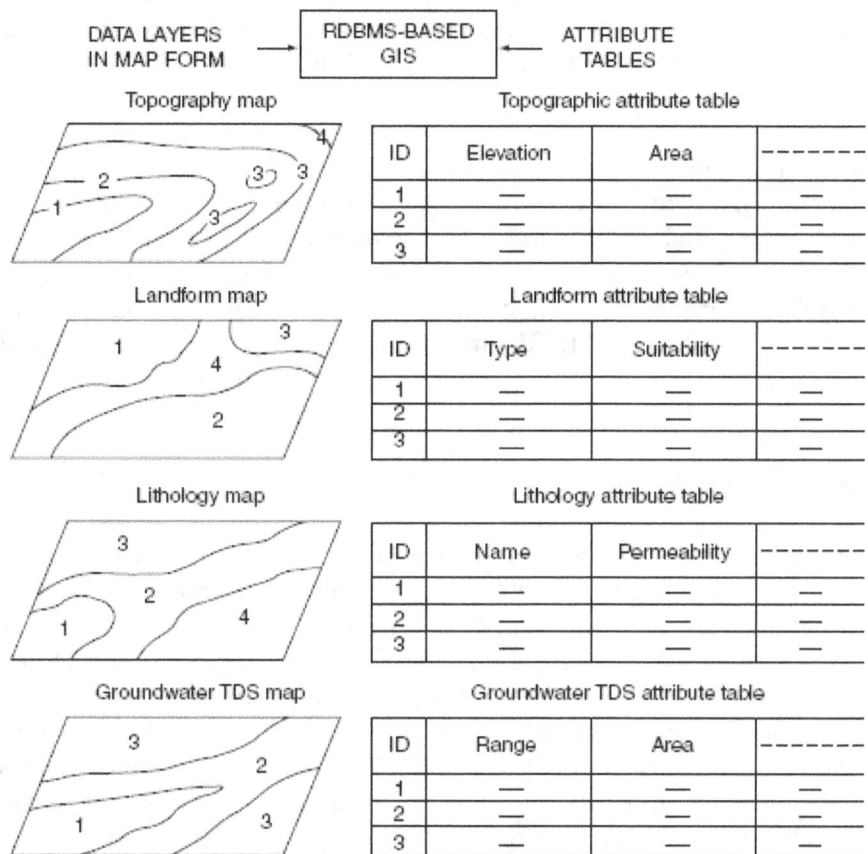

Figure 8:     Schematic Representation of GIS-Working

2. *Variety of data:* Groundwater investigations often comprise diverse forms and types of data, such as: (a) topographic contour maps, (b) landform maps, (c) lithological maps, (d) structural geological maps, (e) isobath map (contour map of equal depth of water-table), (f) isogram (isocone) maps depicting groundwater characteristics by contours of equal concentration of dissolved solids (TDS) or ions, (g) drainage density and other geomorphic maps, (h) tables of various observations and data sets, and (i) point data, say locations and water-levels in observations wells, or spring discharge etc. In these, some of the variables are of continuous type, e.g. TDS content, water-level data etc., and some others are of categorical type, such as low/medium/high drainage density, or gravel/marble/granite lithology. It is essential to integrate the spatial information for coherent and meaningful interpretation, and to avoid compartmentalization of data. GIS offers technological avenues for integrating the variety of data sets in both qualitative and quantitative terms, hitherto not available through any other route.

3. *Flexibility of operations and concurrent display:* Modern GIS packages are endowed with numerous functions for computing, searching for and classifying data, which allow processing and analysis of spatial information in a highly flexible manner and concurrent display, interactively.

95

4. *Speed, time and costs of processing:* Advances in micro-electronics and computer technology have made it possible that modern GIS can store, process and analyze large volumes of data which otherwise would be too expensive, tedious and time-consuming to carry out by other methods.

5. *Higher accuracy and repeatability of results:* The technique being digital computer-based, yields higher accuracy, in comparison to manual cartographic products. The results are amenable to re-checking and confirmation. In a typical hydrogeological investigation, the sources of data could be satellite or aerial sensing, field surveys, geochemical laboratory analyses, geophysical exploration data, etc. and may be available in the form of maps, profiles, point data, tables, lists etc. If GIS methodology is not used, then integrating such a variety of data sets would involve elaborate manual exercises in order to deduce the relevant information.

*Arc Hydro - Geographic Data Model*

The Arc Hydro groundwater data model is a geographic data model for representing spatial and temporal groundwater information within a geographic information system (GIS). The data model is a standardized representation of groundwater systems within a spatial database that provides a public domain template for GIS users to store, document and analyze commonly used spatial and temporal groundwater data sets. It includes two-dimensional (2D) and three-dimensional (3D) object classes for representing aquifers, wells, and borehole data and 3D geospatial context in which these data exist. The framework data model also includes tabular objects for representing temporal information such as water levels and water quality samples that are related with spatial features.

*Geostatistical Analysis using ArcGIS*

GIS is designed to support a range of different kinds of analysis of geographic information: techniques to examine and explore data from a geographic perspective, to develop and test models, and to present data in ways that lead to greater insight and understanding. A linkage between GIS and spatial data analysis is considered to be an important aspect in the development of GIS into a research tool to explore and analyze spatial relationships. The GIS methodology for the spatial analysis of the groundwater levels (as illustrated in Figure 9) involves the following steps:

(a) Exploratory spatial data analysis (ESDA) using ArcGIS software for the water level to study the following:

> ➢ Data distribution.
> ➢ Global and local outliers.
> ➢ Trend analysis.

(b) Spatial interpolation for water level data using ArcGIS software, while ordinary kriging is applied by involving the following procedures:

> ➢ Semi-variogram and covariance modelling.

➤ Model validation using cross-validation.
➤ Surfaces generation of the groundwater level data.

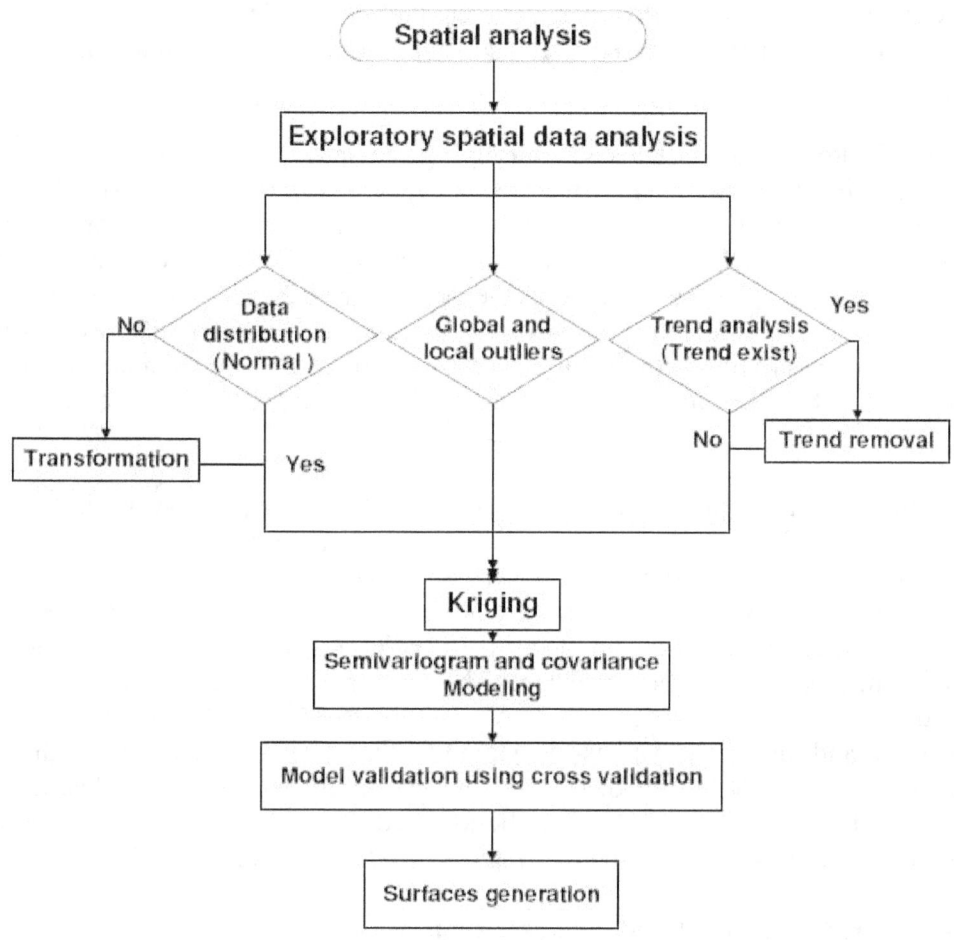

Figure 9:    Flow Chart of the Geostatistical Analysis Steps

There are some methods of spatial modelling that work better if the experimental distribution of available data is close to normal one, therefore it is necessary to check for normality before performing any spatial modelling. Transformations are necessary to drive the data to normal distribution in case of non-normality, where several transformations including Box–Cox also known as power transformations, arcsine, and logarithmic one, can be used to make the data more normally distributed. The groundwater levels data are plotted in the histogram, where groundwater level data is not normally distributed.

Trend is a surface that may be made up of two main components: a fixed global trend and random short-range variation. The global trend is sometimes referred to as the fixed mean structure. Somewhat different approach to ESDA for continuous data represented as a point set with z-values is to examine whether any simple trends are present. The spatial trend can be further analyzed using a three-dimensional perspective using ArcGIS geostatistical analyst trend analysis tool.

A global outlier is a measured sample point that has a very high or a very low value relative to all of the values in a dataset. For example, if 99 out of 100 points have value between 300 and 400, but the 100th point has a value of 750, the 100th point may be a global outlier. Local outlier is a measured sample point that has a value that is within the normal range for the entire dataset, but when compared to the surrounding points, it is unusually high or low.

Kriging is divided into two distinct tasks: quantifying the spatial structure of the data and producing a prediction. Quantifying the structure, known as variography, is where spatial-dependence model fit the data. For making a prediction for an unknown value for a specific location, kriging will use the fitted model from variography, spatial data configuration, and values of the measured sample points around the prediction location. Variography is the process of estimating the theoretical semi-variogram. The semi-variogram and covariance function quantify the assumption that things nearby tend to be more similar than things that are farther apart. They both measure the strength of statistical correlation as a function of distance.

Validation should be carried out before producing the final surface, where it helps in making an informed decision as to which model provides the best predictions. The most popular methods for verifying predictions are cross-validation and validation provided in ArcGIS Geostatistical Analyst. If the mean prediction error is near zero, predictions are centred on the measurement values. The closer the predictions are to their true values, the smaller the root-mean-square prediction errors. The average root-mean-square prediction errors are computed as the square root of the average of the squared difference between observed and predicted values. For a model that provides accurate predictions, the root-mean-squared prediction error should be as small as possible. The average standard error and the mean standardized prediction error should be as small as possible. After model validation, the surface can be generated to produce the water level map.

**Groundwater Data Management and Analysis Tools**

Following are some of the software packages used for groundwater data management and analysis.

*AQTESOLV*: Design and analysis of aquifer tests including pumping tests, step-drawdown tests, variable-rate tests, recovery tests, single-well tests, slug tests and constant-head tests.

*AquaChem*: AquaChem is an integrated software package developed specifically for graphical and numerical analysis of geochemical data sets. It features a database of common geochemical parameters that can be customized and configured to include an unlimited number of attributes per sample. The built-in analysis tools include many common calculations used for analyzing, interpreting and comparing aqueous geochemical data. The built-in graphics include many common geochemical plotting methods such as Piper, Stiff, Durov, Radial, Schoeller, Langlier-Ludwig and more. AquaChem also includes a direct link to the popular PHREEQC program for geochemical modeling.

*AquiferTest Pro*: Graphical analysis and reporting of pumping test and slug test data.

*AquiferWin32*: A software system for analysis and display of aquifer test results.

*Arc Hydro Groundwater*: Arc Hydro Groundwater (AHGW) tools include Groundwater Analyst, MODFLOW Analyst, and Subsurface Analyst. It can access and visualize your groundwater, time series, and geologic data within ArcGIS.

*ChemStat*: RCRA compliant statistical analysis of groundwater data. It includes most methods described in 1989 and 1992 USEPA statistical guidance documents.

*EnviroInsite*: EnviroInsite is a desktop, groundwater visualization package for analysis and communication of spatial and temporal trends in multi-analyte, environmental groundwater data. It facilitates the generation of data queries to generate time history graphs, pie charts, radial diagrams, data tables and dot-plots in plan, on vertical profiles, and in 3D views.

*Environmental Visualization System (EVS)*: Unites advanced gridding, geostatistical analysis, and fully three-dimensional visualization tools into a software system developed to address the needs of all earth science disciplines. The graphical user interface is integrated with modular analysis and graphics routines. The more advanced versions allow these modules to be customized and combined. The software can be used to analyze all types of analytes and geophysical data in any environment (e.g. soil, groundwater, surface water, air, etc.). It includes integrated geostatistics.

*EQWin Data Manager*: EQWin Data Manager is a database application used to validate, store, analyze, and report on environmental data. Using either MS Access or SQL Server databases, EQWin manages many frequently monitored entities, including ground/surface water, soil, air, and meteorological data. EQWin is integrated with Microsoft Excel as its default import/reporting tool.

*GW Contour*: Data interpolation and contouring program for groundwater professionals that also incorporates mapping velocity vectors and particle tracks.

*GW-Base*: GW-Base was developed for expedient evaluation of groundwater information for use in water management, monitoring, investigation and remediation of damage cases. It combines a powerful database with easy-to-use evaluation tools like contour maps, bar-charts, pie-charts, statistical tools, query options and GIS-functionalities.

*HydroGeo Analyst*: Groundwater and borehole data management and visualization technology.

*LogPlot*: Log plotting software for the environmental, petroleum, mining, and academic geoscientist.

*PUMPTEST (IGWMC)*: The PUMPTEST program package is a menu-driven set of independently run programs. It includes three different methods to analyze pumping test data: the time-drawdown method (JACOBFIT); distance-drawdown method (DISTANCE); and recovery method (RECOVERY).

*RockWorks*: Geological data management, analysis and visualization.

*Strater*: Strater is a well log and borehole plotting software program that imports data from a multitude of sources (database files, data files, LAS files, ODBC, and OLE DB data sources). Strater provides innumerable ways to graphically display the data. All the logs are fully customizable.

*WinLoG*: WinLoG can be used to quickly create, edit and print geotechnical, environmental, mining, and oil & gas borehole and well logs. The graphical windows interface displays the log as it is changed and shows exactly how the log will look when it is printed.

## Review of Groundwater Scenario

*Review of Groundwater Level Changes*

➤ Describe the typical long-term water level hydrographs. Show typical examples of villages showing rising water level trends and declining water level trends.
➤ Describe the typical long-term water level hydrographs for typical areas (coastal areas, irrigation commands, over-exploited areas, delta areas, areas close to riverbeds).
➤ Study multiple hydrographs from a number of wells within a watershed to understand the groundwater dynamics. Assess the water level changes in multi-aquifer system.
➤ Delineate areas showing typical long-term water level trends (with list of villages).
➤ Describe typical high frequency water level monitoring hydrographs and their significance.
➤ Explain the recharge – rainfall response for different rainfall intensities.

*Review of Groundwater Flow System Characteristics*

➤ Generate water level fluctuation contour maps using data for the reporting period and the last year (pre/post monsoon) from all the monitoring wells tapping a single aquifer in the network.
➤ Generate water level elevation contour maps using data for the reporting period (pre/post monsoon) from all the monitoring wells tapping a single aquifer in the network.
➤ From the generated map, assess the gradient of groundwater flow, determine the flow path, and delineate the recharge and discharge areas. Detect any change in flow gradient or path, as compared to the previous years.
➤ Generate maps for all the different aquifers and assess the gradient of groundwater flow for the different aquifers. Assess nature of contact/mixing between the aquifers.
➤ Assess the groundwater flow through the aquifer system using supporting data (rainfall, runoff, recharge and draft).
➤ Generate Groundwater worthy map.

*Review of the Groundwater Quality Changes*

➤ Describe the groundwater quality monitoring network, frequency of monitoring, list of laboratories and parameters analysed.
➤ Show sample hydrograph depicting the changing trends in water quality for different parameters. List the parameters that show higher levels of concentration or show increasing trend.
➤ Describe the chemical quality of groundwater in different aquifers; assess the parameters that show higher levels of concentration or show increasing trend.

- Generate water quality contour maps for specific parameters using analysed results for the reporting period (pre/post monsoon) from all the monitoring wells tapping a single aquifer in the network.
- From the generated maps, assess the pattern of contaminant transport, delineate areas showing high concentration, identify polluting sources, if any (natural/industrial).
- Generate water quality maps and diagrams for different aquifers. Assess nature of contact/mixing of contaminants, if any, between aquifers.
- Assess the rate of dilution or increasing concentration. Study the impact of rainfall, surface water bodies, recharge and excess draft on groundwater quality.

*Estimation of Groundwater Resource Availability*

- Carry out groundwater resource estimation for the reporting period based on the GEC (Groundwater Resource Estimation Committee, 1997) norms.
- Identify watersheds/administrative units subjected to overexploitation, as compared to previous years.
- Identify areas that are showing heavy increase in draft.
- Estimate stage of groundwater development.
- Generate notified area map.

*Recommendation for Sustainable Development of Groundwater*

- Based on different analyses, list the administrative blocks showing declining and rising water levels.
- Delineate recharge and discharge areas.
- Identify the technically appropriate programs that need to be considered for containing the declining water level trend and increasing contamination, containing depletion of resources, reducing erosion and increasing recharge.
- Recommend the appropriate designs for efficient wells, artificial recharge structures/water harvesting ponds that can be taken up in different areas.
- Recommend sustainable groundwater development programs for ecolgically fragile areas like coastal/urban/ industrial areas.
- Identify specific research projects that need to be considered for tackling serious groundwater related issues.
- Recommend to administrators/planners the appropriate groundwater policies and legislation that can ensure equity and ensure groundwater sustainability.

## References

1.  Boonstra, J. (1988). Groundwater Survey. Groundwater Balances. Lecture Note. Twenty -Ninth International Course on Land Drainage (1990), ILRI, Wageningen, The Netherlands.

2.  Cheng, Alexander H.-D. And Ouazar, Driss (2004). Coastal Aquifer Management Monitoring, Modeling, and Case Studies. Lewis Publishers.

3.  De Marsily, G. (1986). Quantitative Hydrogeology. Academic Press.

4.      de Ridder, N. A. (1974). Groundwater Investigations, In: Drainage Principles and Applications. ILRI Publication 16.

5.      DHV Consultants. Model Yearbook – Groundwater. Hydrology Project (India).

6.      Groundwater Resource Estimation Methodology - 1997. Report of the Groundwater Resource Estimation Committee, Ministry of Water Resources, Government of India, New Delhi, June 1997.

7.      Gundogdu, K. S., Demir, A. O., Degirmenci, H., Buyukcangaz, H., Akkaya, T. Preparation and Interpretation of Groundwater Maps using Geographical Information System (Arc/Info).

8.      Moore, J. E. (1979). Contributions of Groundwater Modelling to Planning. Journal of Hydrology, Vol. 43, pp. 121-128.

9.      Salah, Hamad (2009). Geostatistical Analysis of Groundwater Levels in the South Al Jabal Al Akhdar Area using GIS. GIS Ostrava 2009.

10.     Singhal, B. B. S. and Gupta, R. P. (2010). Applied Hydrogeology of Fractured Rocks. Springer.

11.     Zemansky, G.M. (1998). Interpretation of Groundwater Chemical Quality Data. Proceedings, 14th Annual Waste Testing & Quality Assurance Symposium, July 13-15, 1998, Arlington, Virginia, pp. 192-201.

# 7-Introduction to Groundwater Modelling

## Introduction

Groundwater lies almost everywhere below the earth's surface. It is an important source of drinking water supply and irrigation and caters to more than 45% of the total irrigation in the country. More than 90% of the world's total supply of drinkable water is groundwater. The contribution of ground water irrigation to achieve self-sufficiency in food grains production in the past three decades is phenomenal. In the coming years, the ground water utilization is likely to increase manifold for expansion of irrigated agriculture and to achieve targets of food production. Although the groundwater is annually replenishable resource, its availability is non-uniform in space and time. Hence, precise estimation of ground water resource and irrigation potential is a prerequisite for planning its development.

*What is Groundwater?*

When rain falls to the ground, the water does not stop moving. Some of it flows along the surface in streams or lakes, some of it is used by plants, some evaporates and returns to the atmosphere, and some sinks into the ground. Imagine pouring a glass of water onto a pile of sand. Where does the water go? The water moves into the spaces between the particles of sand. Groundwater is water that is found underground in cracks and spaces in soil, sand and rocks. The area where water fills these spaces is called the saturated zone. The top of this zone is called the water table...just remember the top of the water is the table. The water table may be only a meter below the ground's surface or it may be hundreds of meters down.

Groundwater can be found almost everywhere. The water table may be deep or shallow; and may rise or fall depending on many factors. Heavy rains or melting snow may cause the water table to rise, or an extended period of dry weather may cause the water table to fall. Groundwater is stored in and moves slowly through layers of soil, sand and rocks called aquifers. The speed, at which groundwater flows, depends on the size of the spaces in the soil or rock and how well the spaces are connected. Aquifers typically consist of gravel, sand, sandstone or fractured rock, like limestone. These materials are permeable because they have large connected spaces that allow water to flow through.

Water in aquifers is brought to the surface naturally through a spring or can be discharged into lakes and streams. This water can also be extracted through a well drilled into the aquifer. A well is a pipe in the ground that fills with groundwater. This water then can be brought to the surface by a pump. Shallow wells may go dry if the water table falls below the bottom of the well. Some wells called artesian wells do not need a pump because of natural pressures that force the water up and out of the well.

Groundwater supplies are replenished or recharged by rain and snow melt. In some areas of the world, people face serious water shortages because groundwater is used faster than it is naturally replenished. In other areas groundwater is polluted by human activities. In areas where material above the aquifer is permeable, pollutants can sink into the groundwater. Groundwater can be polluted by landfills, septic tanks, leaky underground gas tanks and from overuse of fertilizers and pesticides. If groundwater becomes polluted, it will no longer be safe to drink. It is important for all of us to learn to protect our groundwater.

## Contamination and Concerns

A large population of India depends daily on groundwater for their drinking water. Groundwater is also one of our most important sources of irrigation water. Unfortunately, groundwater is susceptible to pollutants. Groundwater is generally a safe source of drinking water; however, there are concerns that contamination may increase as toxins dumped on the ground in the past make their way into groundwater supplies.

## Excessive Ground Water Exploitation

More water is being pumped out of a number of aquifers than is being replaced by the natural recharge. Groundwater levels in some aquifers have declined by tens of meters because of over-pumping, making it more difficult and expensive to abstract more water. This has resulted in land subsidence. Declining groundwater levels have reduced the dry weather flow and have caused some to disappear completely. This also causes deterioration in the quality of the groundwater.

## Pollutants Leach

Pollutants that contaminate groundwater may be some of the same pollutants that contaminate surface water. Compounds from the surface can move through the soil and end up in the groundwater. For example, pesticides and fertilizers can find their way into groundwater supplies over time. Road salt, toxic substances from mining sites and used motor oil also may seep into groundwater. In addition, it is possible for untreated waste from septic tanks and toxic chemicals from underground storage tanks to contaminate groundwater.

## Dangers of Contaminated Groundwater

Drinking contaminated groundwater can have serious health effects. Diseases such as hepatitis and dysentery may be caused by contamination from septic tank waste. Toxins that have leached into well water supplies may cause poisoning. And it is important not to forget that wildlife too can be harmed by contaminated groundwater.

## Sources of Contamination

Groundwater contamination occurs when man-made products and chemicals get into the groundwater and cause it to become unsafe and unfit for human use. Some of the major sources of these products called contaminants are storage tanks, septic systems, hazardous waste sites, landfills and the widespread use of road salts and chemicals.

## Groundwater Protection

Unfortunately, contaminated groundwater is very difficult and expensive to clean up. Solutions can be found after groundwater has been contaminated but this isn't always easy.

The best thing to do is adopt pollution prevention and conservation practices in order to protect important groundwater supplies from being contaminated in the first place.

**Applications of Groundwater Models**

The use of groundwater models is prevalent in the field of environmental science. Models have been applied to investigate a wide variety of hydrogeologic conditions. More recently, groundwater models are being applied to predict the transport of contaminants for risk evaluation.

In general, models are conceptual descriptions or approximations that describe physical systems using mathematical equations; they are not exact descriptions of physical systems or processes. By mathematically representing a simplified version of a hydrogeological system, reasonable alternative scenarios can be predicted, tested, and compared. The applicability or usefulness of a model depends on how closely the mathematical equations approximate the physical system being modeled. In order to evaluate the applicability or usefulness of a model, it is necessary to have a thorough understanding of the physical system and the assumptions embedded in the derivation of the mathematical equations.

Groundwater models describe the groundwater flow and transport processes using mathematical equations based on certain simplifying assumptions. These assumptions typically involve the direction of flow, geometry of the aquifer, the heterogeneity or anisotropy of sediments or bedrock within the aquifer, the contaminant transport mechanisms and chemical reactions. Because of the simplifying assumptions embedded in the mathematical equations and the many uncertainties in the values of data required by the model, a model must be viewed as an approximation and not an exact duplication of field conditions. Groundwater models, however, even as approximations, are a useful investigation tool that groundwater hydrologists may use for a number of applications.

Application of existing groundwater models include water balance (in terms of water quantity), gaining knowledge about the quantitative aspects of the unsaturated zone, simulating of water flow and chemical migration in the saturated zone including river-groundwater relations, assessing the impact of changes of the groundwater regime on the environment, setting up/optimising monitoring networks, and setting up groundwater protection zones.

The modeling studies in India have so far been confined to academic and research organisations. The practising professionals mostly still prefer to employ lumped models for planning of groundwater development and recharge. Such models completely ignore the distributed character of the groundwater regime. Thus, they are based upon rather conservative concepts like safe yields and are incapable of accounting for the stream-aquifer interaction and the dependence of lateral recharge on the water table pattern. Consequently, permissible mining (i.e. withdrawals in excess of vertical recharge) and perennial yield can not be arrived at. The objectives of modeling studies in India have been mainly (i) groundwater recharge, (ii) dynamic behaviour of the water table, (iii) stream-aquifer interaction, and (iv) sea-water intrusion etc.

It is important to understand general aspects of both groundwater flow and transport models so that application or evaluation of these models may be performed correctly.

## Model Development

A groundwater model application can be considered to be two distinct processes (Figure 1). The first process is model development resulting in a software product, and the second process is application of that product for a specific purpose. Groundwater models are most efficiently developed in a logical sequence.

### Model Objectives

Model objectives should be defined which explain the purpose of using a groundwater model. The modeling objectives will profoundly impact the modeling effort required.

### Hydrogeological Characterization

Proper characterization of the hydrogeological conditions at a site is necessary in order to understand the importance of relevant flow or solute transport processes. Without proper site characterization, it is not possible to select an appropriate model or develop a reliably calibrated model.

### Model Conceptualization

Model conceptualization is the process in which data describing field conditions are assembled in a systematic way to describe groundwater flow and contaminant transport processes at a site. The model conceptualization aids in determining the modeling approach and which model software to use.

### Modeling Software Selection

After hydrogeological characterization of the site has been completed and the conceptual model developed, a computer model software is selected. The selected model should be capable of simulating conditions encountered at a site. For example, analytical models can be used where field data show that groundwater flow or transport processes are relatively simple. Similarly, one-dimensional/ two-dimensional/ three-dimensional groundwater flow and transport models should be selected based upon the hydrogeological characterization and model conceptualization.

### Model Design (Input Parameters)

Model design includes all parameters that are used to develop a calibrated model. The input parameters include model grid size and spacing, layer elevations, boundary conditions, hydraulic conductivity/transmissivity, recharge, any additional model input, transient or steady state modeling, dispersion coefficients, degradation rate coefficients etc.

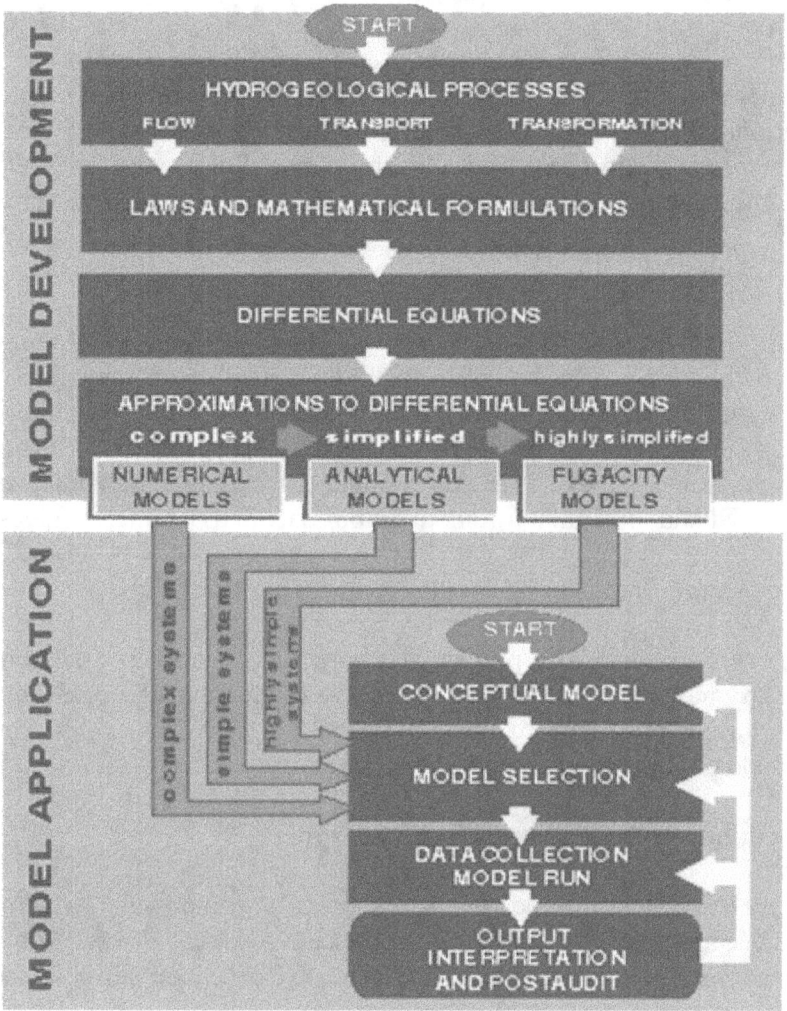

Figure 1:    Development Process of a Model

*Model Calibration*

Model calibration consists of changing values of model input parameters in an attempt to match field conditions within some acceptable criteria. Model calibration requires that field conditions at a site be properly characterized. Lack of proper site characterization may result in a model calibrated to a set of conditions that are not representative of actual field conditions.

*Sensitivity Analysis*

A sensitivity analysis is the process of varying model input parameters over a reasonable range (range of uncertainty in value of model parameter) and observing the relative change in model response. Typically, the observed change in hydraulic head, flow rate or contaminant transport are noted. Data for which the model is relatively sensitive would require future characterization, as opposed to data for which the model is relatively insensitive.

## Model Verification

A calibrated model uses selected values of hydrogeologic parameters, sources and sinks and boundary conditions to match historical field conditions. The process of model verification may result in further calibration or refinement of the model. After the model has successfully reproduced measured changes in field conditions, it is ready for predictive simulations.

## Predictive Simulations

A model may be used to predict some future groundwater flow or contaminant transport condition. The model may also be used to evaluate different remediation alternatives. However, errors and uncertainties in a groundwater flow analysis and solute transport analysis make any model prediction no better than an approximation. For this reason, all model predictions should be expressed as a range of possible outcomes that reflect the assumptions involved and uncertainty in model input data and parameter values.

## Performance Monitoring Plan

Groundwater models are used to predict the migration pathway and concentrations of contaminants in groundwater. Errors in the predictive model, even though small, can result in gross errors in solutions projected forwarded in time. Performance monitoring is required to compare future field conditions with model predictions.

## Modelling of Groundwater Flow and Mass Transport

A groundwater model is a simplified representation of a groundwater system. Groundwater models can be classified as physical or mathematical. A *physical model* (e.g. a sand tank) replicates physical processes, usually on a smaller scale than encountered in the field.

A *mathematical model* describes the physical processes and boundaries of a groundwater system using one or more governing equations. An *analytical model* makes simplifying assumptions (e.g. properties of the aquifer are considered to be constant in space and time) to enable solution of a given problem. Analytical models are usually solved rapidly, sometimes using a computer, but sometimes by hand.

A *numerical model* divides space and/or time into discrete pieces. Features of the governing equations and boundary conditions (e.g. aquifer geometry, hydrogeologogical properties, pumping rates or sources of solute) can be specified as varying over space and time. This enables more complex, and potentially more realistic, representation of a groundwater system than could be achieved with an analytical model. Numerical models are usually solved by a computer and are usually more computationally demanding than analytical models.

Groundwater modelling begins with a conceptual understanding of the physical problem. The next step in modelling is translating the physical system into mathematical terms. In general, the results are the familiar groundwater flow equation and transport equations. The governing flow equation for three-dimensional saturated flow in saturated porous media is:

$$\frac{\partial}{\partial x}(K_{xx}\frac{\partial h}{\partial x})+\frac{\partial}{\partial y}(K_{yy}\frac{\partial h}{\partial y})+\frac{\partial}{\partial z}(K_{zz}\frac{\partial h}{\partial z})-Q=S_s\frac{\partial h}{\partial t} \qquad \text{... (1)}$$

where,

| | | |
|---|---|---|
| $K_{xx}, K_{yy}, K_{zz}$ | = | hydraulic conductivity along the x,y,z axes which are assumed to be parallel to the major axes of hydraulic conductivity; |
| $h$ | = | piezometric head; |
| $Q$ | = | volumetric flux per unit volume representing source/sink terms; |
| $S_s$ | = | specific storage coefficient defined as the volume of water released from storage per unit change in head per unit volume of porous material. |

The transport of solutes in the saturated zone is governed by the advection-dispersion equation which for a porous medium with uniform porosity distribution is formulated as follows:

$$\frac{\partial c}{\partial t}=-\frac{\partial}{\partial x_i}(cv_i)+\frac{\partial}{\partial x_i}\left(D_{ij}\frac{\partial c}{\partial x_j}\right)+R_c \qquad i,j=1,2,3 \qquad \text{... (2)}$$

where,

| | | |
|---|---|---|
| $c$ | = | concentration of the solute; |
| $R_c$ | = | sources or sinks; |
| $D_{ij}$ | = | dispersion coefficient tensor; |
| $v_i$ | = | velocity tensor. |

An understanding of these equations and their associated boundary and initial conditions is necessary before a modelling problem can be formulated. Basic processes, that are considered, include groundwater flow, solute transport and heat transport. Most groundwater modelling studies are conducted using either deterministic models, based on precise description of cause-and-effect or input-response relationships or stochastic models reflecting the probabilistic nature of a groundwater system.

The governing equations for groundwater systems are usually solved either analytically or numerically. Analytical models contain analytical solution of the field equations, continuously in space and time. In numerical models, a discrete solution is obtained in both the space and time domains by using numerical approximations of the governing partial differential equation. Various numerical solution techniques are used in groundwater models. Among the most used approaches in groundwater modelling, three techniques can be distinguished: Finite Difference Method, Finite Element Method, and Analytical Element Method. All techniques have their own advantages and disadvantages with respect to availability, costs, user friendliness, applicability, and required knowledge of the user.

**Groundwater Flow Models**

The most widely used numerical groundwater flow model is MODFLOW which is a three-dimensional model, originally developed by the U.S. Geological Survey (McDonald and Harbaugh, 1988). It uses block-centred finite difference scheme for saturated zone. The advantages of MODFLOW include numerous facilities for data preparation, easy exchange of data in standard form, extended worldwide experience, continuous development, availability of source code, and relatively low price. However, surface runoff and unsaturated flow are not included, hence in case of transient problems, MODFLOW can not be applied if the flux at the groundwater table depends on the calculated head and the function is not known in advance. Salient features of the frequently used unsaturated flow and groundwater models have been presented below.

1. **FEFLOW**

   (Finite Element Subsurface Flow System)

   FEFLOW is a finite-element package for simulating 3D and 2D fluid density-coupled flow, contaminant mass (salinity) and heat transport in the subsurface. It is capable of computing:

   - Groundwater systems with and without free surfaces (phreatic aquifers, perched water tables, moving meshes);
   - Problems in saturated-unsaturated zones;
   - Both salinity-dependent and temperature-dependent transport phenomena (thermohaline flows);
   - Complex geometric and parametric situations.

   The package is fully graphics-based and interactive. Pre-, main- and post-processing are integrated. There is a data interface to GIS (Geographic Information System) and a programming interface. The implemented numerical features allow the solution of large problems. Adaptive techniques are incorporated.

2. **HST3D**

   (3-D Heat and Solute Transport Model)

   HST3D is a powerful user-friendly interface for HST3D integrated within the Argus Open Numerical Environments (Argus ONE) modeling environment. HST3D allows the user to enter all spatial data, graphically run HST3D, and visualize the results. Argus ONE integrates CAD, GIS, Database, Conceptual Modeling, Geostatistics, Automatic Grid and Mesh Generation, and Scientific Visualization within one comprehensive graphical user interface (GUI). The Heat and Solute Transport Model HST3D simulates ground-water flow and associated heat and solute transport in three dimensions. The HST3D model may be used for analysis of problems such as those related to subsurface-waste injection, landfill leaching, saltwater intrusion, freshwater recharge and recovery, radioactive waste disposal, water geothermal systems, and subsurface energy storage. The Argus ONE GIS and Grid Modules are required to run HST3D.

## 3. MODFLOW

(Three-Dimensional Finite-Difference Ground-Water Flow Model)

MODFLOW is the name that has been given the USGS Modular Three-Dimensional Ground-Water Flow Model. Because of its ability to simulate a wide variety of systems, its extensive publicly available documentation, and its rigorous USGS peer review, MODFLOW has become the worldwide standard ground-water flow model. MODFLOW is used to simulate systems for water supply, containment remediation and mine dewatering. When properly applied, MODFLOW is the recognized standard model.

The main objectives in designing MODFLOW were to produce a program that can be readily modified, is simple to use and maintain, can be executed on a variety of computers with minimal changes, and has the ability to manage the large data sets required when running large problems. The MODFLOW report includes detailed explanations of physical and mathematical concepts on which the model is based and an explanation of how those concepts were incorporated in the modular structure of the computer program. The modular structure of MODFLOW consists of a Main Program and a series of highly-independent subroutines called modules. The modules are grouped in packages. Each package deals with a specific feature of the hydrologic system which is to be simulated such as flow from rivers or flow into drains or with a specific method of solving linear equations which describe the flow system such as the Strongly Implicit Procedure or Preconditioned Conjugate Gradient. The division of MODFLOW into modules permits the user to examine specific hydrologic features of the model independently. This also facilitates development of additional capabilities because new modules or packages can be added to the program without modifying the existing ones. The input/output system of MODFLOW was designed for optimal flexibility.

Ground-water flow within the aquifer is simulated in MODFLOW using a block-centered finite-difference approach. Layers can be simulated as confined, unconfined, or a combination of both. Flows from external stresses such as flow to wells, areal recharge, evapotranspiration, flow to drains, and flow through riverbeds can also be simulated.

## 4. MT3D

(A Modular 3D Solute Transport Model)

MT3D is a comprehensive three-dimensional numerical model for simulating solute transport in complex hydrogeologic settings. MT3D has a modular design that permits simulation of transport processes independently or jointly. MT3D is capable of modeling advection in complex steady-state and transient flow fields, anisotropic dispersion, first-order decay and production reactions, and linear and nonlinear sorption. It can also handle bioplume-type reactions, monad reactions, and daughter products. This enables MT3D to do multi-species reactions and simulate or assess natural attenuation within a contaminant plume. MT3D is linked with the USGS groundwater flow simulator, MODFLOW, and is designed specifically to handle advectively-dominated transport problems without the need to construct refined models specifically for solute transport.

5.    **SEAWAT**

(Three-Dimensional Variable-Density Ground-Water Flow)

The SEAWAT program was developed to simulate three-dimensional, variable-density, transient ground-water flow in porous media. The source code for SEAWAT was developed by combining MODFLOW and MT3DMS into a single program that solves the coupled flow and solute-transport equations. The SEAWAT code follows a modular structure, and thus, new capabilities can be added with only minor modifications to the main program. SEAWAT reads and writes standard MODFLOW and MT3DMS data sets, although some extra input may be required for some SEAWAT simulations. This means that many of the existing pre- and post-processors can be used to create input data sets and analyze simulation results. Users familiar with MODFLOW and MT3DMS should have little difficulty applying SEAWAT to problems of variable-density ground-water flow.

6.    **SUTRA**

(2-D/3-D Saturated/Unsaturated Transport Model)

SUTRA is a groundwater saturated-unsaturated transport model, a complete saltwater intrusion and energy transport model. SUTRA simulates fluid movement and transport of either energy or dissolved substances in a subsurface environment. SUTRA employs a two-dimensional hybrid finite-element and integrated finite-difference method to approximate the governing equations that describe the two interdependent processes that are simulated: (1) fluid density-dependent saturated or unsaturated groundwater flow and either (2a) transport of a solute in the groundwater, in which the solute may be subject to equilibrium adsorption on the porous matrix and both first-order and zero-order production or decay, or (2b) transport of thermal energy in the groundwater and solid matrix of the aquifer.

7.    **SWIMv1/SWIMv2**

(Soil water infiltration and movement model - simulate soil water balances)

SWIMv1 (Soil Water Infiltration and Movement model version 1) is a software package for simulating water infiltration and movement in soils. SWIMv1 consists of a menu-driven suite of three programs that allow the user to simulate soil water balances using numerical solutions of the basic soil water flow equations. As in the real world, SWIMv1 allows addition of water to the system as precipitation and removal by runoff, drainage, evaporation from the soil surface and transpiration by vegetation. SWIMv1 helps researchers and consultants understand the soil water balance so they can assess possible effects of such practices as tree clearing, strip mining and irrigation management. SWIMv1 is valuable for scientists and consultants involved in land planning and land management. For example, if a development is being considered which involves tree clearing, SWIMv1 can be used to indicate salinity or surface runoff problems that could result from a change in the soil water balance associated with the removal of the trees.

SWIMv2 (Soil Water Infiltration and Movement model version 2) is a mechanistically-based model designed to address soil water and solute balance issues associated with both production and the environmental consequences of production. SWIMv2 employs fast, numerically-efficient techniques for solving Richards' equation for water flow and the convection-dispersion equation for solute transport and is suitable for personal computer applications. The model deals with a one-dimensional vertical soil profile which may be vertically inhomogeneous but is assumed to be horizontally uniform. It can be used to simulate runoff, infiltration, redistribution, solute transport and redistribution of solutes, plant uptake and transpiration, evaporation, deep drainage and leaching.

## 8.    VISUAL HELP

(Modeling Environment for the U.S. EPA HELP Model for Evaluating and Optimizing Landfill Designs)

Visual HELP is an advanced hydrological modeling environment available for designing landfills, predicting leachate mounding and evaluating potential leachate contamination. Visual HELP combines the latest version of the HELP model with an easy-to user interface and powerful graphical features for designing the model and evaluating the modeling results. Visual HELP's user-friendly interface and flexible data handling procedures provides convenient access to both the basic and advanced features of the HELP model. This completely-integrated modeling HELP environment allows the user to graphically create several profiles representing different parts of a landfill; automatically generate statistically-reliable weather data (or create your own); run complex model simulations; visualize full-color, high-resolution results; and prepare graphical and document materials for your report.

## 9.    Visual MODFLOW

(Integrated Modeling Environment for MODFLOW, MODPATH, MT3D)

Visual MODFLOW provides professional 3D groundwater flow and contaminant transport modeling using MODFLOW-2000, MODPATH, MT3DMS and RT3D. Visual MODFLOW Pro seamlessly combines the standard Visual MODFLOW package with WinPEST and the Visual MODFLOW 3D-Explorer to give the most complete and powerful graphical modeling environment available. This fully-integrated groundwater modeling environment allows to:

- Graphically design the model grid, properties and boundary conditions,
- Visualize the model input parameters in two or three dimensions,
- Run the groundwater flow, pathline and contaminant transport simulations,
- Automatically calibrate the model using WinPEST or manual methods, and
- Display and interpret the modeling results in three-dimensional space using the Visual MODFLOW 3D-Explorer

Considering the large variability and the quick development of groundwater models, a new, more sophisticated model can often replace a previously applied model. Additionally, the reconsideration of the conceptual model and the regeneration of the mesh may need a new allocation of the parameters. Therefore, it is important that model data (information) are stored independently from a given model, with a preference for GIS-based databases. Considerable development in the field of user-friendly GIS and data base servers makes the set-up and the modification of models easier and more time-effective. One such model is FEFLOW which incorporates mathematical modeling with GIS-based data exchange interfaces.

The input data for a groundwater model include natural and artificial stress, and parameters, dimensions, and physico-chemical properties of all aquifers considered in the model. A finer level of detail of the numerical approximation (solution) greatly increases the data requirements. Input data for aquifers are common values such as transmissivities, aquitard resistances, abstraction rates, groundwater recharges, surface water levels etc. The most common output data are groundwater levels, fluxes, velocities and changes in these parameters due to stress put into the model.

**Concluding Remarks**

Mathematical models are tools, which are frequently used in studying groundwater systems. In general, mathematical models are used to simulate (or to predict) the groundwater flow and in some cases the solute and/or heat transport. Predictive simulations must be viewed as estimates, dependent upon the quality and uncertainty of the input data. Models may be used as predictive tools, however field monitoring must be incorporated to verify model predictions. The best method of eliminating or reducing modeling errors is to apply good hydrogeological judgement and to question the model simulation results. If the results do not make physical sense, find out why.

**References**

1. Anderson, M.P. and W.W. Woessner (1992). Applied Groundwater Modeling. Academic Press, Inc., San Diego, CA., 381 p.

2. American Society for Testing and Materials (1993). Standard Guide for Application of a Ground-Water Flow Model to a Site-Specific Problem. ASTM Standard D 5447-93, West Conshohocken, PA, 6 p.

3. American Society for Testing and Materials (1995). Standard Guide for Subsurface Flow and Transport Modeling. ASTM Standard D 5880-95, West Conshohocken, PA, 6 p.

4.    Bear, J., and A. Verruijt (1987). Modeling Groundwater Flow and Pollution. D. Reidel Publishing Company, 414 p.

5.    Franke, O.L., Bennett, G.D., Reilly, T.E., Laney, R.L., Buxton, H.T., and Sun, R.J. (1991). Concepts and Modeling in Ground-Water Hydrology -- A Self-Paced Training Course. U.S. Geological Survey Open-File Report 90-707.

6.    Kashyap, Deepak (1989). Mathematical Modelling for Groundwater Management – Status in India. Indo-French Seminar on Management of Water Resources, 22-24 September, 1989, Festival of France-1989, Jaipur, pp. IV-59 to IV-75.

7.    Kinzelbach, W. (1986). Groundwater Modeling: An Introduction with Sample Programs in BASIC. Elsevier, New York, 333 p.

8.    Kumar, C. P. (1992). Groundwater Modelling – In. Hydrological Developments in India Since Independence. A Contribution to Hydrological Sciences, National Institute of Hydrology, Roorkee, pp. 235-261.

9.    Kumar, C. P. (2001). Common Ground Water Modelling Errors and Remediation. Journal of Indian Water Resources Society, Volume 21, Number 4, October 2001, pp. 149-156.

10.   McDonald, M.G. and A.W. Harbaugh (1988). A Modular Three-Dimensional Finite-Difference Ground-Water Flow Model, USGS TWRI Chapter 6-A1, 586 p.

11.   Pinder, G.F., and J.D. Bredehoeft (1968). Application of the Digital Computer for Aquifer Evaluation, Water Resources Research, Vol. 4, pp. 1069-1093.

12.   Wang, H.F. and M.P. Anderson (1982). Introduction to Groundwater Modeling. W.H. Freeman and Company, San Francisco, CA, 237 p.

13.   Website - Kumar Links to Hydrology Resources:
      http://www.angelfire.com/nh/cpkumar/hydrology.html

14.   Website - Scientific Software Group:
      http://www.scisoftware.com/

15.   Website - USGS Ground-Water Software:
      http://water.usgs.gov/nrp/gwsoftware/

# 8-Basic Concepts for Groundwater Modelling

## Introduction

Groundwater modelling is only one management tool available to catchment managers for developing solutions to complex catchment issues. They are rarely the only component of a large catchment or resource availability study, and are often linked to other socio-economic models and extensive community consultation initiatives.

Groundwater models describe the groundwater flow and transport processes using mathematical equations based on certain simplifying assumptions. These assumptions typically involve the direction of flow, geometry of the aquifer, the heterogeneity or anisotropy of sediments or bedrock within the aquifer, the contaminant transport mechanisms and chemical reactions. Because of the simplifying assumptions embedded in the mathematical equations and the many uncertainties in the values of data required by the model, a model must be viewed as an approximation and not an exact duplication of field conditions. Groundwater models, however, even as approximations, are a useful investigation tool that groundwater hydrologists may use for a number of applications.

Application of existing groundwater models include water balance (in terms of water quantity), gaining knowledge about the quantitative aspects of the unsaturated zone, simulating of water flow and chemical migration in the saturated zone including river-groundwater relations, assessing the impact of changes of the groundwater regime on the environment, setting up/optimising monitoring networks, and setting up groundwater protection zones.

For any modelling study, accurate and reliable data must be available, together with a modelling team with proven skills in the local hydrogeology/hydrology and the designated modelling package. The required resources of time, budget, data and technical expertise will be greater for models with more complexity and where there are higher expectations of outcomes for resource management.

A model review framework is another key element, with reviews recommended at all stages throughout the study, consistent with the objectives, scope, scale and budget of the project. A model review provides a process by which the end-user can check consistently that a model meets the project objectives. It also provides the model developer with a specification against which the modelling study will be evaluated.

Model refinements are based on the data quality and volume, hydrogeological understanding, modelling study scope, and on clientele/community expectations. A hierarchical approach sets out three main model classifications --Basic (Simple), Impact Assessment (Moderate) and Aquifer Simulator (Complex), in order of increasing complexity, and with the associated capability to provide for more complex simulations of hydrogeological process and/or to address resource management issues more comprehensively. The study purpose and objectives must be carefully considered and clearly stated at the outset of any modelling study to develop an adequate tool with the appropriate complexity.

## Conceptualisation

It is necessary to collate and analyse the available data in order to develop an understanding of the important aspects of the physical system, and of the hydrological processes that control or impact the groundwater flow system. The selected modelling approach should be consistent with the available data sets and the current conceptual model of the groundwater system.

Development of a valid conceptual model is the most important step in a computer modelling study. A conceptual model is a simplified representation of the essential features of the physical hydrogeological system, and its hydrological behaviour, to an adequate degree of detail. It forms the foundation upon which the interactive, site-specific model is built, and is itself based on an initial literature review, data collation and hydrogeological interpretation. While the conceptual model is an idealised summary of the current understanding of catchment conditions, and the key aspects of how the flow system works, it is subject to some simplifying assumptions.

Boundary conditions are constraints imposed on the model grid to represent the interface between the model calculation domain and the surrounding environment. The type of boundary selected should be consistent with the conceptual model and the water budget, and should be located and oriented consistent with the physical features it represents. In particular, model domain boundaries should be set far from the area of interest so that imposed stresses on the grid interior do not reach the boundaries. As more layers can be accommodated in a profile model (than an areal model) for the same level of computational demand, profile models are suitable for situations in which detailed simulation of vertical flow components is essential.

Surface-groundwater interaction can form a critical component of the water budget, as well as an essential feature of the conceptual model, and often forms the most complex, sensitive and uncertain parts of a model. This is particularly so because the relevant flow processes commonly involve consideration of unsaturated flow, recharge and evapotranspiration processes. Dynamic simulation of surface-groundwater interaction requires mathematical description of transient effects within complex surface and groundwater flow systems, such as: rainfall-runoff processes, surface water flow routing, and evapotranspiration; fluctuations in the stage (elevation) and volume of a surface water body, and leakage to groundwater; infiltration through the ground surface and flow in the unsaturated (vadose) zone; flow in the saturated (aquifer zone) and discharge to the ground surface. As most groundwater models utilise a one month stress period, this approximation is usually valid.

The simplest approaches involve the use of specified head and specified flow boundaries, while more complex approaches involve head-dependent flow boundaries, which also sometimes account for surface flow volumes. When a head-dependent flow boundary is used, flow is computed at the surface-groundwater interface as a function of the relative water levels at any time and a conductance term at the boundary interface, with the conductance term becoming a calibration parameter. For leakage from a stream, the head difference can and should be limited to the sum of the depth of water in the stream, and the streambed thickness.

Two types of time interval are used in flow models: stress periods during which boundary conditions and stresses (e.g. hydrologic conditions and pumping) are constant, and between which boundary conditions and stresses can vary; and, time steps during which model calculations are made to simulate the effect of stresses on the system.

## Model Calibration

Calibration is defined as the process of refining the model representation of the hydrogeological framework, hydraulic properties and boundary conditions to achieve a desired degree of correspondence between simulations and the groundwater flow system. Medium to high complexity models should be calibrated to measured data before they are used for prediction simulations, and the calibration performance should be presented in qualitative and quantitative terms in comparison to agreed target criteria. In automated model calibration process, an objective function is used to compare how the model simulation matches the historical groundwater system behaviour. Therefore, the formulation of the objective function is a critical step in automated model calibration.

The non-uniqueness problem arises because many different possible sets of model inputs can produce nearly identical model outputs. In other words, multiple calibrations of the same system are possible using different combinations of boundary conditions and aquifer properties, because exact ("unique") solutions cannot be computed when many variables are involved in the calibration approach.

The main methods that should be employed in conjunction to reduce the non-uniqueness problem comprise: (a) calibrating the model using hydraulic conductivity (and other) parameters that are consistent with measured values; and (b) calibrating to multiple distinct hydrological conditions with that parameter set. The first method is designed to restrict the possible range of parameters to values that are consistent with the actual ("unique") values of the aquifer. The second method provides an indication of the predictive performance of a model by demonstrating that a given set of input model parameters (consistent with field measurements) are capable of reproducing system behaviour through a range of distinct hydrological conditions.

Initial conditions refer to the head distribution everywhere in the system at the beginning of the simulation, and thus are boundary conditions in time (Anderson and Woessner, 1992). An enforced prescriptive measure could lead to an erroneous calibration. This could happen if a modeller adjusts aquifer properties to ensure a better match of simulated heads with field observations, when in fact the field data might be wrong. A model will in reality have a variable calibration performance, perhaps "good" in places, perhaps "poor" in places.

Quantitative calibration performance measures generally relate to the calculation of potentiometric head residuals (the difference between measured and modelled heads) and associated statistics at known monitoring locations. It is not possible to draw absolute quantitative comparisons in regard to groundwater level contours, because contours are the result of interpolations between data points, and are subjective, at least in part (subjective choices are made even when selecting parameters or methods of generating contours through software packages). It is more appropriate to shift the modelled data to times at which measured data were collected, than to interpolate measured data to simulation times.

## Prediction Uncertainty

Verification (also called validation) is a test of whether the model can be used as a predictive tool, by demonstrating that the calibrated model is an adequate representation of the physical system. The main purpose of groundwater flow modelling would most usually be to carry out resource management predictions for specified future periods, which often range into tens to hundreds of years. For low complexity models, it may be acceptable to undertake predictions even though the model may not have been calibrated.

Predictions are undertaken by running the model with the adopted (calibrated) parameters, and imposing hydrological stresses to represent the expected future climatic conditions, and the expected future groundwater management scenarios. Whereas a model could be used to assess scenarios in relative terms with little uncertainty, there is considerable uncertainty associated with absolute predictions.

Prediction uncertainty arises mainly from the uncertain confidence in the (calibrated) model as a predictive tool, and uncertainties in predicting the magnitude and timing of future climatic and management stresses. Much greater value can be obtained from modelling studies by undertaking a staged programme of prediction scenarios. The first stage could comprise the simulation of a base case, against which other predictions may be compared. For a medium complexity numerical model with few parameters, sensitivity coefficients should be determined for at least the best case and worst case extremes of each parameter.

Numerical groundwater models can provide a quantitative basis for estimating recharge, but long-term averages are sensitive to the length of the averaging period and the start date for the averaging. By calculating all possible averages for a chosen minimum averaging period, and using random sequencing of the periods to avoid bias, the average recharge estimates can be ranked and assigned probabilities. The resulting cumulative distribution function allows managers to quantify the risk in setting a value for sustainable yield, and provides users with a measure of confidence in the surety of groundwater supply. The main limitations of this approach are the appropriateness of the conceptual model, the robustness of the numerical model, and the representativeness of the simulation time period. Nevertheless, the method is an advance on current practice, which provides a single deterministic estimate with no hint of uncertainty. Where the purpose of a high complexity numerical model is the assessment of average annual recharge or sustainable yield, post-processing of model water budgets should be done to produce a probability distribution for total recharge.

## Model Reviews

Model reviews may be made at any of four levels: model appraisal, peer review, model audit, and post-audit. The nominal difference between a peer review and an audit is that a peer review would usually involve a detailed review of a modelling study report, while an audit would also require an in-depth review of the model data files, simulations and outputs.

A peer review or a model audit should only be done by an experienced groundwater modeller, different from the person who has developed the model. The conceptualisation stage of model development should be reviewed only by a competent hydrogeologist with local knowledge. A complete set of data files for at least one representative simulation should be provided to the auditor, so that he/she can verify that the model structure is as reported and runs successfully without numerical errors or mass imbalances. For medium and high complexity models, an internal model audit should be carried out progressively as part of an in-house quality control programme.

A post-audit describes the process of revisiting the modelling study several years after it is completed, to assess the accuracy of model predictions. This will preferably include distinct hydrological conditions compared to the data set used for model calibration, which will also allow assessment of the model non-uniqueness issue. The fundamental lesson from reported post-audits is that a valid and complete conceptual model is essential for making accurate predictions.

## Concluding Remarks

There are very few published and accepted guidelines on groundwater flow modelling. The notable example is the suite of Standard Guides from the American Society for Testing and Materials (ASTM), which are reasonably well-accepted standard practice guidelines.

The fundamental reason that most models are developed is to assess alternative resource management plans or predict the impacts of proposed projects. Therefore, the modelling report needs to provide sufficient documentation to support decision-making by various parties in relation to the project. In particular, the report needs to clearly communicate the current system understanding, the range of alternative management scenarios considered, and the predicted system responses.

## References

1.  American Society for Testing and Materials (1993). Standard Guide for Application of a Ground-Water Flow Model to a Site-Specific Problem. ASTM Standard D 5447-93, West Conshohocken, PA.

2.  American Society for Testing and Materials (1993). Standard Guide for Comparing Ground-Water Flow Model Simulations to Site-Specific Information. ASTM Standard D 5490-93, West Conshohocken, PA.

3.  American Society for Testing and Materials (1994). Standard Guide for Defining Boundary Conditions in Ground-Water Flow Modeling. ASTM Standard D 5609-94, West Conshohocken, PA.

4.  American Society for Testing and Materials (1995). Standard Guide for Subsurface Flow and Transport Modeling. ASTM Standard D 5880-95, West Conshohocken, PA.

5.  American Society for Testing and Materials (1996). Standard Guide for Calibrating a Ground-Water Flow Model Application, ASTM Standard D5981-96, West Conshohocken, PA.

6.    Anderson, M.P. and Woessner, W.W. (1992). Applied Groundwater Modeling. Academic Press, Inc., San Diego, CA., 381 p.

7.    Bear, J., and Verruijt, A. (1987). Modeling Groundwater Flow and Pollution. D. Reidel Publishing Company, 414 p.

8.    Franke, O.L., Bennett, G.D., Reilly, T.E., Laney, R.L., Buxton, H.T., and Sun, R.J. (1991), Concepts and Modeling in Ground-Water Hydrology -- A Self-Paced Training Course. U.S. Geological Survey Open-File Report 90-707.

9.    McDonald, M.G. and Harbaugh, A.W. (1988). A Modular Three-Dimensional Finite-Difference Ground-Water Flow Model, USGS TWRI Chapter 6-A1, 586 p.

10.   Murray-Darling Basin Commission (2000). Groundwater Flow Modelling Guideline, Aquaterra Consulting Pvt Ltd, South Perth 6151, Australia.

11.   Spitz K. and Moreno, J. (1996). A Practical Guide to Groundwater and Solute Transport Modeling, Wiley, New York, 461 p.

12.   Wang, H.F. and Anderson, M.P. (1982). Introduction to Groundwater Modeling. W.H. Freeman and Company, San Francisco, CA, 237 p.

# 9-Conceptualisation and Model Design

## Introduction

Groundwater systems are complicated beyond our capability to evaluate them comprehensively in detail. Comprehensive analysis means that need to take into account all the characteristics of the system, and predict the effects of hydrological and land use stresses. There are usually insufficient data to completely characterise the groundwater system under investigation, and assumptions and simplifications are required to obtain a quantitative solution for a given problem. We use groundwater models to integrate our hydrogeological understanding with the available data, to develop a predictive tool for evaluating groundwater systems, subject to assumptions and limitations.

A groundwater model is a computer-based representation of the essential features of a natural hydrogeological system that uses the laws of science and mathematics. Its two key components are a conceptual model and a mathematical model. The conceptual model is an idealised representation (usually graphical) of our hydrogeological understanding of the essential flow processes of the system. A mathematical model is a set of equations, which, subject to certain assumptions, quantifies the physical processes active in the aquifer system(s) being modelled.

While the model itself obviously lacks the detailed reality of the groundwater system, the behaviour of a valid model approximates that of the aquifer(s). A groundwater model provides a scientific means to synthesise the available data into a numerical characterisation of a groundwater system. The model represents the groundwater system to an adequate level of detail, and provides a predictive tool to quantify the effects on the system of specified hydrological, pumping or irrigation stresses.

Typical model purposes include improving hydrogeological understanding (synthesis of data); aquifer simulation (evaluation of aquifer behaviour); designing practical solutions to meet specified goals (engineering design); optimising designs for economic efficiency and account for environmental effects (optimisation); evaluating recharge, discharge and aquifer storage processes (water resources assessment); predicting impacts of alternative hydrological or development scenarios (to assist decision-making); quantifying the sustainable yield (economically and environmentally sound allocation policies); resource management (assessment of alternative policies); sensitivity and uncertainty analysis (to guide data collection and risk-based decision-making); and visualisation (to communicate aquifer behaviour).

There is no such thing as a perfect model (Spitz and Moreno, 1996). The application of numerical simulation models to groundwater problems involves both an art and a science (Anderson and Woessner, 1992). Understanding the science is critical, and requires a sound knowledge of aspects of geology, hydrogeology, groundwater hydraulics, and engineering. The art is no less critical, and is gained from experience of applying numerical models to practical problems, working in a multi-disciplinary team, with on-going review by experienced modelling and hydrogeology specialists.

A generic modelling methodology should be applied to any modelling study. The tasks of the methodology include conceptualisation, calibration, prediction, uncertainty analysis, reporting and model reviews. Completing a successful modelling project is not primarily a question of understanding the numerical techniques, but of understanding the best application of a modelling approach to a particular site or catchment for the stated objectives.

## Hydrogeological Interpretation

The definition of the purpose and objective is the critical first step in any groundwater modelling study. The purpose and objective are also closely related to the concept of model complexity, which must also be specified at the outset. The modelling study objective and purpose must be clearly stated in specific and measurable terms, along with the resource management objectives that the model will be required to address. It is also important that the overall management constraints are outlined in terms of budget, schedule, staged development, and eventual ownership and use of the model. These issues are important because the ability of the data acquired or available to provide an increasingly accurate representation (model) of the groundwater system increases with time, money, and the technical expertise applied (USACE, 1999).

The data requirements for a groundwater model comprise hydrogeological framework data and hydrological stress data. The framework data describe the physical system (aquifer geometry, hydrological interaction processes), and parameters that do not change with time. The stress data describe the dynamic hydrological stresses on the system (initial conditions, time-varying data and the translation of management strategies into modelling scenarios). Compiling the data is not a simple task, and many diverse data sources need to be accessed.

It is necessary to collate and analyse the available data in order to develop an understanding of the important aspects of the physical system, and of the hydrological processes that control or impact the groundwater flow system. This leads to the development of the conceptual model. The data requirements can be quite onerous, especially for high complexity applications, or where complex surface-groundwater interaction flow processes apply. If there is a lack of data, it is also likely that there is a lack of hydrogeological understanding of the important flow processes, and it will not be possible to immediately develop a high complexity model. Data acquisition and interpretation are ongoing activities in the staged development of a high complexity model, with the model study findings often being used to help guide field programmes to obtain critical data that can improve the model accuracy and reliability. The data review may redefine the study objectives or initiate data collection networks that are tailored and specific to the management issue and the required modelling study.

The selected modelling approach should be consistent with the available data sets and the current conceptual model of the groundwater system. The model should also be flexible enough to be expanded or refined into a more complex model if more data becomes available and the model is to be used for long term management of the catchment or aquifer system. Following the initial hydrogeological interpretation, a computer model may be developed by inputting data to groundwater modelling software, which is essentially a complex, three-dimensional, interactive database, with time variability.

The process of collating data into a form for input to a model usually identifies data gaps, which can be used to guide monitoring efforts, or to modify the modelling objectives to establish achievable outcomes for the available level of data and understanding. The associated data analysis often provides an improved understanding of the groundwater flow system, and is often overlooked and under-resourced in the rush to develop a model.

To be able to be used in a modelling study, all data must be specified in consistent space and time units (metres and days are accepted standard modelling practice), and to a consistent horizontal and vertical datum. In addition, it is necessary for aquifer head measurements to be reduced to a common datum density (e.g. fresh water), and to a standard temperature of 25°C (if temperature differences are significant), so that resulting heads can be contoured meaningfully. The need for consistent units and datum usually means that surveying of bores and other features is required before data compilation can be completed. Any modelling study, but especially one involving a site close to the coast, needs to confirm that the elevation datum for all data is tied to consistent datum. This is particularly important for tide data, which usually forms a boundary condition for coastal models.

## Development of Conceptual Model

Development of a valid conceptual model is the most important step in a computer modelling study. A conceptual model is a simplified representation of the essential features of the physical hydrogeological system, and its hydrological behaviour, to an adequate degree of detail. The conceptual model is usually presented graphically as a cross-section or block diagram, with supporting documentation outlining in descriptive and quantitative terms the essential system features. It forms the foundation upon which the interactive, site-specific model is built, and is itself based on an initial literature review, data collation and hydrogeological interpretation.

While the conceptual model is an idealised summary of the current understanding of catchment conditions, and the key aspects of how the flow system works, it is subject to some simplifying assumptions. The assumptions are required partly because a complete reconstruction of the field system is not feasible, and partly because there is rarely sufficient data to completely describe the system in comprehensive detail. However, the conceptual model should be developed using the principle of simplicity (or parsimony), such that the model is as simple as possible, while retaining sufficient complexity to adequately represent the physical elements of the system, and to reproduce system behaviour.

In developing an adequate (parsimonious) conceptual model, however, sufficient degrees of freedom must be incorporated to the model features to allow simulation of a broad range of responses. It must be possible for the model to predict system responses ranging from desired to undesired outcomes. In other words, the model must not be configured or constrained such that it artificially produces a restricted range of prediction outcomes.

Each of the conceptual model features needs to be described to a level of detail commensurate with the ability of the data to represent the system, and with our ability to understand the system (given the current data). Our understanding of the system improves as more data are gathered and analysed, and with assessment of how the system responds to stresses induced by climatic, hydrologic and human-induced changes. The conceptual model will evolve with new data, and the first conceptual model developed will not be the last (Spitz and Moreno, 1996), with improvements achieved by augmenting the data base.

When post-audits are undertaken some years after modelling studies are completed, one of the key causes of inaccurate predictions has been found to be that the conceptual model is invalid or incomplete (Anderson and Woessner, 1992). The other key element is inaccuracies in the assumed future stresses (pumping rates, rainfall recharge, etc.). If a conceptual model inadequately configures certain features (i.e. is invalid), or does not incorporate important features due to lack of data or understanding (i.e. is incomplete), then the model application will not make accurate predictions. Such failures, however unavoidable at the time or understandable in hindsight, are generally not due to numerical or theoretical deficiencies in the model itself, provided the model selection is valid for the particular problem. Such failures are usually attributable to errors in the conceptual model, or the assumed stresses, which highlights the importance of these aspects of modelling.

Often, models are developed in "crisis" mode to answer pressing questions so that management decisions can be made (Anderson and Woessner, 1992). This particularly applies to low and medium complexity models, often where the data available may be quite limited, and such an approach may be appropriate for those studies. However, high complexity models should be developed in "management" mode, allowing successive improvements to be made to the model, based on augmenting the data base. With refinements to the conceptual model and improved calibration to a wider range of hydrological conditions over an extended simulation timeframe, the accuracy and reliability of model predictions will improve.

The definition of a water budget and associated boundary conditions for the model domain are integral components of the conceptual model. The water budget is a description of the inflows and outflows across the model domain boundaries, along with any internal consumptive uses, in descriptive and quantitative terms where possible. The model domain covers the entire area of interest, including areas of potential future impact, although its size should be minimised to reduce computational effort. Model boundaries are the interface between the model calculation domain and the surrounding environment (Spitz and Moreno, 1996), and occur notably on the edges of the domain. Other ("internal") boundary conditions reflect influences from other environmental factors (such as rivers, wells, etc.) that are manifest inside the domain. The external boundaries of the model domain should take advantage of natural or physical groundwater boundaries (e.g. aquifer extents, coastlines, rivers, lakes).

## Selection of Modelling Code

The modelling code is the computer programme that contains algorithms to numerically solve the mathematical model. Most modelling codes in common use today also have a graphical user interface for the pre- and post-processing of modelling data. The mathematical model is the basic hydraulic equation that governs the flow of groundwater in the saturated zone. It is a partial differential equation in time and three-dimensional space. The conceptual model and the hydrogeological framework data together help to define the boundary conditions for the solution of the mathematical model. The hydrogeological stresses complete the boundary condition definition, and provide the temporal and spatial data for the solution of the hydraulic equation.

A modelling code can be thought of as a very complex, three-dimensional, interactive database, with time variability, because it incorporates: (a) the means to input data to describe the model domain and hydrologic stresses in space and time; (b) the numerical algorithms to solve the mathematical model (hydraulic equation of groundwater flow): and (c) the means to output the results of the simulation. Most mathematical models use analytical (simple) or numerical (complex) means to solve the governing differential equations. There are other methods, such as the boundary integral method (a combination of analytical and numerical methods), and the analytical element method (an elaborate semi-analytical method), but they are not in common use.

Analytical models are equations representing exact solutions to the hydraulic equation for one- or two- dimensional flow problems under broad simplifying assumptions, usually including aquifer homogeneity. They can be solved by hand, or by simple computer programs but do not allow for spatial variability and often do not allow for temporal variability. They are useful to provide rough approximations for many applications with little effort, as they usually do not involve calibration (site- specific monitoring data is often not available for these simple problems). This approach can suit most simple, low-complexity modelling studies. There is no systematic approach for simplifying a given groundwater problem and for selecting the appropriate analytical solution. In fact, it depends entirely on the capability of the model user to visualise the problem and to apply professional judgement to select a valid analytical approach.

In numerical models, the continuous differential terms in the governing equations are replaced by finite quantities. The computational power of the computer is used to solve the resulting algebraic equations by matrix arithmetic. In this way, problems with complex geometry, dynamic response effects and spatial and temporal variability may be solved accurately. This approach must be used in cases where the essential aquifer features form a complex system, and where surface-groundwater interaction is an important component. To facilitate the data input, flow simulation and results output, most computer modelling codes in common use provide a graphical user interface (GUI), based on the Microsoft Windows system.

The selection of an appropriate modelling code and GUI for a particular study is a matter of ensuring that the code has the capability to adequately represent the essential features and flow processes of the groundwater system being studied. It is also important to ensure that the selected code has been verified and benchmarked against standard test problems, to confirm that the code accurately solves the equations that comprise the mathematical model. Most of the commercially available codes have been verified, but it is still worthwhile for the modeller to run test problems to confirm that the modelling software installation on their PC can reproduce the test problem simulation results accurately. Public domain modelling codes are often preferred to proprietary codes.

Public domain codes are defined as commercially available and widely distributed, relatively inexpensive, and generally accepted models with features that can be and have been used to simulate a wide range of hydrogeological conditions. Public domain codes have received extensive peer review, and case histories documenting their general applicability, as well as their limitations, have been published in the scientific literature. Many were originally developed (and are continually being refined) by US government agencies (e.g. US Geological Survey and EPA, with substantial assistance from specialist consultants. Proprietary codes are those developed in-house by certain companies, and while they may share many attributes with public domain codes, the source code program is not available, the purchase price is usually much more expensive, and the peer review is often limited.

Some of the frequently used ground water modelling codes include MODFLOW, MT3D, MOC3D, AQUIFEM, FEMWATER, SUTRA, FEFLOW, PMWIN, Groundwater Vistas, Visual Modflow, Groundwater Modelling System etc. It is generally accepted that Modflow, originally developed by the US Geological Survey (McDonald and Harbaugh, 1988), is the industry-leading public domain numerical flow model, although it is not necessarily suitable for every modelling study. There are a number of graphical user interfaces for Modflow. There are also several text books that detail modelling methodologies with Modflow (notably Anderson and Woessner, 1992).

The main reasons that Modflow enjoys a good reputation are that the code has been verified against a range of analytical solutions, it has been used to successfully simulate a wide range of hydrogeological systems across the world, the source code is in the public domain and there are several relatively cheap GUIs available (some good GUIs are available for free - e.g. Processing Modflow for Windows PMWIN4 can be downloaded from www.uovs.ac.za/igs/index.htm). Another great strength is that the Modflow code was developed with a modular structure (e.g. modules for certain hydrological processes may be turned on or off), and new modules for flow processes or improved numerical methods are being continually produced and integrated seamlessly. Modflow is known to have the best range of stream-aquifer interaction modules.

**Model Study Plan**

A model study plan should be completed and reviewed at the end of the conceptualisation stage with a report that includes details of the (a) study purpose, objectives, model complexity, and resources required to complete the study; (b) initial hydrogeological interpretation and conceptual model, data summary, boundary conditions and preliminary water budget; (c) selected modelling code and limitations/uncertainties in the modelling approach; and (d) model design and configuration specifics including details on the boundaries; grid; layers; aquifer units and parameters; recharge, discharge and water balance; surface-groundwater interaction; calibration and prediction timeframes and accuracy targets; steady state or transient calibration and/or prediction runs; and data available and required to complete the study.

**Boundary Conditions**

Boundary conditions are constraints imposed on the model grid to represent the interface between the model calculation domain and the surrounding environment. There are three major types of boundary conditions, all of which may vary with time. The type of boundary selected should be consistent with the conceptual model and the water budget, and should be located and oriented consistent with the physical features it represents. In particular, model domain boundaries should be set far from the area of interest (e.g. a water supply borefield) so that imposed stresses on the grid interior do not reach the boundaries. Alternatively, the boundary needs to be configured such that the simulated boundary effect is realistic (e.g. using a head-dependent flow boundary at a groundwater divide or a surface-groundwater interaction feature).

Boundary conditions should be designed to take advantage of physical or hydraulic boundaries. Physical boundaries usually relate to the physical presence of an impermeable geological formation or a large body of surface water. An impermeable boundary typically forms the lower and/or lateral boundaries of modelled systems, and may be justified provided there is at least a two order of magnitude contrast in hydraulic conductivity between the two units (Anderson and Woessner, 1992). Hydraulic boundaries form as a result of hydrologic conditions, notably at groundwater divides and streamlines, although these features are not permanent, and may shift their location or magnitude (of flux or head). Care must be taken in specifying hydraulic boundary conditions, whereas physical boundaries are more easily handled.

## Model Grids

Grids define the spatial area of the numerical model domain in terms of finite differences or finite elements, (grids are not required for analytical models). Finite differences divide the aquifer into a rectangular grid of nodes that define the corners or the centres of model cells. Most finite difference grids are "block-centred", where nodes lie in the centre of cells, but they can be mesh-centred, where the nodes lie at the intersections of the grid lines and define the corners of the cell. For block-centred grids, flux boundaries need to fall on the edge of cells, and head boundaries need to fall on the node in the centre of the cell. Finite elements divide the aquifer into a mesh of node points that form polygonal (usually triangular) cells. In a finite element mesh (or in mesh-centred finite difference grids), however, head and flux boundaries need to be aligned with the nodes. The boundary condition location must be consistent with the adopted grid design.

The spatial discretisation of the grid should be fine in areas of interest or areas of stress, but may be coarse away from these areas, or where data are sparse. The size of the nodal spacing is dependent on the expected curvature of the water table or potentiometric surface, with fine spacing required to accurately define highly curved surfaces (e.g. around pumping wells or near rivers, etc.) or steep hydraulic gradients in the horizontal or vertical directions. Nodes may be regularly spaced, or the spacing may be increased as the grid is expanded towards regional boundaries. For finite difference grids, the grid expansion factor (ratio of larger to smaller adjacent nodal spacings) should not exceed 1.5. The aspect ratio (ratio of maximum to minimum cell dimensions) should ideally be close to unity, and should not exceed 10 for finite difference grids, or a value of 5.0 for finite element meshes (Anderson and Woessner, 1992). The external boundaries of the model domain should be oriented parallel to the primary groundwater flow direction, if possible. Often, particularly for regional models with variable flow direction, it is more convenient to align a model grid with cardinal directions.

## Layers

Layers are used in models to represent hydrostratigraphic units, which comprise geological units with similar aquifer properties. Several geological formations may be combined into one hydrostratigraphic unit (or model layer), or a geological formation may be subdivided into aquifer and confining units (or several layers). Quasi-3D (multi-layer) models usually simulate horizontal flow in each of the stacked aquifer layers, and vertical leakage through confining units between layers. Aquifer head and storage is often not simulated in the confining unit. This is considered an acceptable approximation when there is more than two orders of magnitude contrast in hydraulic conductivity between the aquifer and confining units (Anderson and Woessner, 1992). Otherwise, a profile model or a fully-3D model may be preferred. However, fully-3D models have more onerous data requirements, and expert advice is required.

The number of layers in a model will depend on the conceptualisation of the aquifer system and on all of the factors which define a model's complexity. More layers are needed when vertical head gradients are significant, as the head in each model layer is effectively averaged over the thickness of that layer.

Layer elevation data (top and bottom surfaces or layer thickness) are a key data requirement for models, to define aquifer thickness, and therefore help define aquifer transmissivity (product of hydraulic conductivity and thickness) and storage volumes. Geometry is also used when the top elevation of a layer is compared in the mathematical model to the simulated water level to decide whether the aquifer at that point is confined or unconfined, while the bottom elevation is used to identify when a cell is drained ("goes dry"), or should be "re-wet". If it is known beforehand that the aquifer is confined, and the saturated thickness will not vary significantly through simulations, then substantial data processing savings can be made by not specifying layer elevations, and modelling the confined aquifer using the transmissivity parameter. For unconfined aquifers, any dewatering-type simulations, or where the saturated thickness varies significantly, it is critical to define the layer geometry as accurately as possible.

## Profile Models

Profile or slice models consist of two-dimensional models oriented vertically, and are used when vertical flows or vertical hydraulic gradients are important, but a fully-3D model may not be warranted. A profile model is usually a vertical slice of unit width of aquifer, which must be oriented along a flow-line to remain consistent with the assumption of conservation of mass. As more layers can be accommodated in a profile model (than an areal model) for the same level of computational demand, profile models are suitable for situations in which detailed simulation of vertical flow components is essential. Point sinks and sources (e.g. wells) cannot be accurately simulated in a profile model, as they represent radial flow features, and a profile model cannot account for components of flow outside the cross-section. Line sinks or sources (i.e. a long linear feature such as a river or drain aligned transverse to the profile) can be simulated in a standard profile model. An axi-symmetric profile model (in simple terms, a wedge-shaped slice of aquifer) can be used to simulate point sinks or sources, although this requires special computer codes and techniques. Layers may correspond with undulating geological units, or they may be horizontal slices in which the lithological variation is handled by variable aquifer properties along the layer.

## Aquifer Parameters

Parameter values need to be assigned to the appropriate models cells, based on the extent of the corresponding hydrostratigraphic units, and the associated field measurements or literature estimates for the aquifer parameters. This is where Graphical User Interface (GUI) software packages have improved modelling productivity by making data processing more manageable. Typically, parameter data are sparse, and some form of interpolation is required to represent the overall spatial variability of aquifer parameters over the model domain based on a few point measurements. Geostatistical methods are sometimes used, and automated calibration techniques (notably using PEST software: Doherty, 1994) are becoming much more popular methods for accounting for spatial parameter variability, while reducing the model calibration effort.

## Water Budget

The preliminary water budget should be outlined in terms of the major components of natural recharge and discharge, and human-induced stresses (abstraction and seepage). The locations where these inputs and outputs are manifest must be detailed, along with known or expected changes with time (due to climatic variations, pumping, seepage and surface-groundwater interaction). The water budget needs to be viewed in conjunction with the conceptual model to estimate the overall throughputs of water through the groundwater system. The initial estimates should be cross-checked with subsequent modelling estimates of long term and short term water budget components.

## Surface Water-Groundwater Interaction

Surface water-groundwater interaction can form a critical component of the water budget, as well as an essential feature of the conceptual model, and often forms the most complex, sensitive and uncertain parts of a model. This is particularly so because the relevant flow processes commonly involve consideration of unsaturated flow, recharge and evapotranspiration processes. There are analytical solutions covering a range of surface water-groundwater interaction conditions, although rigid boundary conditions and simplifying assumptions mean that these analytical models can usually only be applied to simple one-dimensional problems. Dynamic simulation of surface water-groundwater interaction requires mathematical description of transient effects within complex surface and groundwater flow systems, such as (a) rainfall-runoff processes, surface water flow routing, and evapotranspiration; (b) fluctuations in the stage (elevation) and volume of a surface water body, and leakage to groundwater; (c) infiltration through the ground surface and flow in the unsaturated (vadose) zone; (d) flow in the saturated (aquifer zone) and discharge to the ground surface.

From a groundwater perspective, it is commonly assumed that direct simulation of the unsaturated flow system is not critical, and that leakage from surface water to groundwater occurs instantaneously. As most groundwater models utilise a one month stress period, this approximation is usually valid. In some cases, however, these assumptions will not be valid, and specialised treatment of surface water-groundwater interaction process will be required, using specialised computer code, and modelling expertise. For most groundwater modelling projects, however, the treatment of surface water-groundwater interaction is effected by utilising the major boundary conditions types, based on appropriate simplifying assumptions, which will vary for the site-specific conditions involved. The method adopted for any particular case should be as simple as possible or as complex as necessary, appropriate for the study purpose, complexity and data available.

The simplest approaches involve the use of specified head and specified flow boundaries, while more complex approaches involve head-dependent flow boundaries, which also sometimes account for surface flow volumes. With specified head boundaries, the aquifer head at the surface water location is commonly specified as the elevation of the water surface, and the model computes the flow across the boundary, dependent on the gradient to or from the adjacent model cells. With specified flow boundaries, the flow rate across the boundary is specified as a "known" value, and the model computes the corresponding head value at the boundary. Both of these approaches can be used to simulate a range of surface water-groundwater interaction processes, including rivers, drains, springs, lakes, coastlines, evapotranspiration, etc. Both of these approaches, however, suffer from the limitations that the hydraulic conductivity across the interface is not limited (e.g. by a low permeability silt layer in a stream bed), the applicable hydraulic gradient can be over-estimated, and the flow rate for specified head boundaries is potentially unlimited.

When a head-dependent flow boundary is used, flow is computed at the surface water-groundwater interface as a function of the relative water levels at any time and a conductance term at the boundary interface, with the conductance term becoming a calibration parameter. For leakage from a stream, the head difference can and should be limited to the sum of the depth of water in the stream, and the streambed thickness. This ensures that leakage to groundwater occurs at its maximum potential (unsaturated flow) rate when the water table drops below the river bed. Depending on the actual model in use, the amount of surface water that can leak into the aquifer can be potentially unlimited, or (preferably) can be limited to the flow volume specified in the stream. This latter approach ensures that flow interchange volumes are physically realistic. The best-known examples of this approach are the various "streamflow-routing" packages of Modflow, which is one of the major strengths relating to the widespread use of Modflow. The amount of groundwater flow that can discharge to the stream (or ground surface) is dependent on the aquifer storage available (accounted for in the model), the hydraulic gradient between the aquifer and the stream, and the interface conductance term.

Other surface water-groundwater interaction processes that can be considered head-dependent flow conditions are rising water tables that intersect the ground surface, evapotranspiration and interactions between lakes or reservoirs and groundwater. When water tables intersect the ground surface, surface flow occurs as spring or drain flow. This is a special case of stream-aquifer interaction, where the discharge only ever occurs from groundwater to surface water.

Special evapotranspiration features are available in some codes that simulate the discharge of groundwater at a rate that increases with increasing water table level, up to a maximum limit (usually free water evaporation that applies at the ground surface). This approach can be used to simulate evapotranspiration from vegetation, as well as from lake features. It can also be used to simulate otherwise very complex surface water-groundwater interaction processes, such as "rejected recharge" and "interflow" surface discharge during the wet season from groundwater systems where the water table is very close to the ground surface. In this case, there is no conductance term to limit the flow rate, although the evapotranspiration flux is usually limited to some factor of the pan evaporation rate.

A number of lake and reservoir simulation algorithms have recently become available, which take account of the water balance interactions for these mini-systems. The new, complex algorithms for treating lake-type features give quite consistent results compared to the more traditional method (e.g. Chung and Anderson, 1998). The traditional method is to specify lake-cells in models with a storage value of 1.0, and a very high permeability (say, 10,000 m/d), and to also allow for other surface water-groundwater interaction fluxes (e.g. evapotranspiration and groundwater inflow/outflow).

**Time Frames**

Two types of time interval are used in flow models: (i) stress periods during which boundary conditions and stresses (e.g. hydrologic conditions and pumping) are constant, and between which boundary conditions and stresses can vary; and (ii) time steps during which model calculations are made to simulate the effect of stresses on the system.

The definition of suitable stress periods is complicated by the fact that individual hydrological conditions (e.g. rainfall, stream flow and pumping regimes) can vary quite independently, and yet stress periods must be defined when each of the stresses may be considered constant. This can result in a very complex data processing operation to prepare model input and process model output files, and appropriate study resources need to be allowed for this purpose. It is common for monthly stress periods (e.g. 30.44 days) to be adopted, with individual stress rates being averaged, although this can cause problems when checking model calibration against measured data, which reflects the dynamic response of the system to the actual (potentially short-lived) stress occurring. Some codes allow separate stress periods for different types of stress.

Ideally, models should use small time steps to obtain accurate iterative solutions, but this can cause inefficiencies due to long run times. Time steps may be constant or increasing during a stress period. The model solution is sensitive to rapidly fluctuating water levels caused by introducing or changing stresses, and a number of small time steps should be used to capture the early response, even if one is only interested in the solution at later times.

To guide the selection of time steps, Anderson and Woessner (1992) recommend a minimum critical time step ($T_c = Sa^2/4T$, where representative values are inserted for storage S, transmissivity T and cell/element size a). This estimate, however, is very conservative as it is pertinent only to explicit solution methods. In practice, iterative methods of solution can tolerate a much larger minimum time step, in the order of 100 times the critical value. A further rule of thumb suggests that the solution should proceed through at least 3 to 5 time steps, with no significant stress or boundary condition changes, before the solution is considered accurate. In any case, the sensitivity of the solution to time step changes may need to be evaluated.

## Concluding Remarks

There are very few published and accepted guidelines on groundwater flow modelling. The notable example is the suite of Standard Guides from the American Society for Testing and Materials (ASTM), which are reasonably well-accepted standard practice guidelines. The groundwater modelling guideline documents are quite consistent in regard to the accepted general approach to groundwater modelling (with greater or lesser emphasis on certain aspects, depending on the application of the guideline) which may be summarised as:

* Define purpose of study and objectives
* Develop conceptual model
* Select model approach/code (analytical or numerical and software package)
* Develop and calibrate model
* Assess parameter sensitivity and calibration uncertainty
* Complete prediction scenarios and sensitivity/uncertainty analysis
* Report
* Post-audit (at some time in future)

## References

1. Anderson, M.P. and Woessner, W.W. (1992). Applied Groundwater Modeling. Academic Press, Inc., San Diego, CA., 381 p.

2. American Society for Testing and Materials (1993). Standard Guide for Application of a Ground-Water Flow Model to a Site-Specific Problem. ASTM Standard D 5447-93, West Conshohocken, PA.

3. American Society for Testing and Materials (1993). Standard Guide for Comparing Ground-Water Flow Model Simulations to Site-Specific Information. ASTM Standard D 5490-93, West Conshohocken, PA.

4. American Society for Testing and Materials (1994). Standard Guide for Defining Boundary Conditions in Ground-Water Flow Modeling. ASTM Standard D 5609-94, West Conshohocken, PA.

5. American Society for Testing and Materials (1995). Standard Guide for Subsurface Flow and Transport Modeling. ASTM Standard D 5880-95, West Conshohocken, PA.

6. American Society for Testing and Materials (1996). Standard Guide for Calibrating a Ground-Water Flow Model Application, ASTM Standard D5981-96, West Conshohocken, PA.

7. Bear, J., and Verruijt, A. (1987). Modeling Groundwater Flow and Pollution. D. Reidel Publishing Company, 414 p.

8.    Chung, K-P, and Anderson M.P. (1998). Simulation of Lake Levels: Comparison between Modflow's Lake Package and the Use of High Conductivity Zones. Proceedings, ModelCare98 Conference, Denver.

9.    Doherty, J. (1994). PEST: A Unique Computer Program for Model-independent Parameter Optimisation. Proceedings, Water Down Under 94, Adelaide, 21-25 November 1994, Vol. 1, 551-554.

10.   Groundwater Modeling Program (2000). Land and Water Management, Hydrologic Studies Unit, Department of Environmental Quality, State of Michigan.

11.   McDonald, M.G. and A.W. Harbaugh, 1988, A Modular Three-Dimensional Finite-Difference Ground-Water Flow Model, USGS TWRI Chapter 6-A1, 586 p.

12.   Spitz, Karlheinz and Moreno, Joanna (1996). A Practical Guide to Groundwater and Solute Transport Modeling. John Wiley & Sons, Inc.

13.   USACE (1999). Engineering and Design - Groundwater Hydrology. U.S. Army Corps of Engineers Manual No. 1110-2-1421.

14.   Wang, H.F. and Anderson, M.P. (1982). Introduction to Groundwater Modeling. W.H. Freeman and Company, San Francisco, CA, 237 p.

# 10-Guidelines for Groundwater Modelling

## Introduction

Groundwater models provide a scientific and predictive tool for determining appropriate solutions to water allocation, surface water - groundwater interaction, landscape management or impact of new development scenarios. However, if the modelling studies are not well designed from the outset, or the model doesn't adequately represent the natural system being modelled, the modelling effort may be largely wasted, or decisions may be based on flawed model results, and long term adverse consequences may result.

A groundwater model is a computer-based representation of the essential features of a natural hydrogeological system that uses the laws of science and mathematics. Its two key components are a conceptual model and a mathematical model. The conceptual model is an idealised representation of our hydrogeological understanding of the key flow processes of the system. A mathematical model is a set of equations, which, subject to certain assumptions, quantifies the physical processes active in the aquifer system(s) being modelled. While the model itself obviously lacks the detailed reality of the groundwater system, the behaviour of a valid model approximates that of the aquifer(s).

A groundwater model provides a scientific means to draw together the available data into a numerical characterisation of a groundwater system. The model represents the groundwater system to an adequate level of detail, and provides a predictive scientific tool to quantify the impacts on the system of specified hydrological, pumping or irrigation stresses. Typical model purposes include improving hydrogeological understanding (synthesis of data); aquifer simulation (evaluation of aquifer behaviour); designing practical solutions to meet specified goals (engineering design); optimising designs for economic efficiency and account for environmental effects (optimisation); evaluating recharge, discharge and aquifer storage processes (water resources assessment); predicting impacts of alternative hydrological or development scenarios (to assist decision-making); quantifying the sustainable yield (economically and environmentally sound allocation policies); resource management (assessment of alternative policies); sensitivity and uncertainty analysis (to guide data collection and risk-based decision-making); visualisation (to communicate aquifer behaviour).

Groundwater investigations, and modelling studies in particular, involve both a science and an art. The scientific basis is important, and requires a sound knowledge of geology, hydrogeology, groundwater hydraulics, hydrology, surface-groundwater interaction and engineering, as well as sufficient spatial and time series data to describe the system. The art is manifest in the creative processes required for developing a groundwater model as a simple computer-based representation of a complex natural system. There is also an art in applying experienced judgement where data are lacking to sufficiently rationalise natural processes, and in effectively communicating the modelling study results.

Best practice modelling is not primarily a question of understanding and implementing the appropriate mathematical techniques, but of understanding and implementing an appropriate modelling approach. That is, the approach must be appropriate for the particular site conditions and the stated study objectives. The end-users of model studies generally do not have (or need to have) this understanding and capability, which is the mark of a competent modeller. In other words, modellers are specialist service providers to end-users. There is normally a lack of consistency in approaches, communication and understanding among and between modellers and end-users, which often results in considerable uncertainty for decision-making. Therefore, proper guidelines are needed to promote best practice modelling methodologies. They should provide the means by which end-users can plan, initiate and manage modelling studies, and assess outcomes with reduced uncertainty.

Significant changes in catchments across the country have occurred in the last hundred years. Many of these changes have been induced by changes in the hydrology and hydrogeology of catchments, and are today reflected in stressed rivers and groundwater systems. For groundwater systems, these stresses can either be water level rises or water level declines (and associated issues such as water quality impacts) which are impacting the productivity and environmental sustainability of catchments.

Groundwater models provide a relevant and useful scientific and predictive tool for predicting impacts and developing management plans. Groundwater models should be seen as an integral part of the water resource management process. This is so because models are increasingly being used to demonstrate the effects of proposed developments and alternative policies for the purposes of gaining consensus on improved allocation distributions and management plans.

**Groundwater Modelling Tool Box**

It is not possible to see into the sub-surface, and observe the geological structure and the groundwater flow processes. The best we can do is to construct bores, use them for pumping and monitoring, and measure the effects on water levels and other physical aspects of the system. It is for this reason that groundwater flow models have been, and will continue to be, used to investigate the important features of groundwater systems, and to predict their behaviour under particular conditions.

Models also form an integral part of decision support systems in the process of managing water resources, salinity and drainage, and should not be regarded as just an end point in themselves. The development and evaluation of resource management strategies for sustainable water allocation, and for control of land and water resource degradation, are heavily dependent on groundwater model predictions. Regional scale groundwater flow modelling studies are commonly used for water resource evaluation and to help quantify sustainable yields and allocations to end-users.

Groundwater modelling is only one management tool available to catchment managers for developing solutions to complex catchment issues. Models provide one of the best tools for determining the most appropriate land/water management options or strategies to adopt. They are rarely the only component of a large catchment or resource availability study, and are often linked to other socio-economic models and extensive community consultation initiatives. Modelling can be a very powerful tool when used in the right circumstances and when models are properly constructed. Accurate and reliable data must be available, together with a modelling team with proven skills in the local hydrogeology/hydrology and the designated modelling package.

An inventory of available data is required prior to the commissioning of any modelling study. This data should also be briefly evaluated prior to preparing any modelling brief to determine its suitability for the management problem and the type of modelling study required. This data review may redefine the study objectives or initiate data collection networks that are tailored and specific to the management issue and the required modelling study. The selected modelling approach should be consistent with the available data sets and the current conceptual model of the groundwater system. The model should also be flexible enough to be expanded into a more complex model if more data becomes available and the model is to be used for long term management of the catchment or aquifer system.

Model complexity is defined as the degree to which a model application resembles, or is designed to resemble the physical hydrogeological system. There are three main classifications of model complexity (in order of increasing complexity):

*Basic model* - a simple model suitable for preliminary assessment (rough calculations), not requiring substantial resources to develop, but not suitable for complex conditions or detailed resource assessment (less than three weeks work)

*Impact Assessment model* - a moderate complexity model, requiring more data and a better understanding of the groundwater system dynamics, and suitable for predicting the impacts of proposed developments or management policies (one to six months work)

*Aquifer Simulator* - a high complexity model, suitable for predicting responses to arbitrary changes in hydrological conditions, and for developing sustainable resource management policies for aquifer systems under stress (more than six months work).

In simple terms, model complexity can be described by the "quick-cheap-good" paradox. The end-user can readily obtain a model with one or two of these three attributes, but not all three. If a model is required to be done quickly, it can also be done cheaply, but the results may not be good enough on which to base important resource development or management decisions. Such a simple model may be good enough for rough calculations to guide a field program, or to assess the broad impacts of a certain proposal, but would usually not be sufficient for project approval or licensing purposes. Alternatively, if a good, reliable model is required, then it is not likely to be able to be developed quickly or cheaply.

In less simple terms, the "quick-cheap-good" attributes are better defined in terms of model complexity. The level of model complexity needs to discussed and agreed between the end-user and the modeller, to ensure that it suits the study purpose, objectives and resources available for each study. This involves consideration of the complexities of the hydrogeological system and the design of an appropriate modelling approach. The data requirements and the level of modelling effort also need to be considered in relation to the resources available and the overall project objective. Long term requirements may involve the staged development of a complex model from a simple application (with ongoing data acquisition and interpretation), and eventual transfer to an end-user as a predictive scientific tool for resource management.

**Basic Guidelines**

The following sections enumerate systematic procedures for modelling study best practice.

*(a)    Modelling Approach*

1.    Clearly state, at the outset, the model study objectives and the model complexity required.
2.    Adopt a level of complexity that is high enough to meet the objective, but low enough to allow conservatism where needed.
3.    Develop a conceptual model that is consistent with available information and the project objective. Document the assumptions involved.
4.    If possible, a suitably experienced hydrogeologist/modeller should undertake a site visit at the conceptualisation stage.
5.    Address the non-uniqueness problem by using measured hydraulic properties, and calibrating to data sets collected from multiple distinct hydrologic conditions.
6.    Perform an assessment of the model uncertainty by undertaking application verification, and sensitivity or uncertainty analysis of calibration and prediction simulations.
7.    Provide adequate documentation of the model development and predictions.
8.    Undertake peer review of the model at various stages throughout its development, and to a level of detail appropriate for the model study scope and objectives.
9.    Maintain effective communication between all parties involved in the modelling study through regular progress reporting (technical issues and project management) and review.

*(b)    Development of Conceptual Model*

1.    The modelling study objective and purpose must be clearly stated in specific and measurable terms, along with the resource management objectives that the model will be required to address.
2.    The overall management constraints should be outlined in terms of budget, schedule, staged development and long term maintenance, and eventual ownership and use of the model.
3.    The model complexity must be assessed and defined to suit the study purpose, objectives and resources available for each model study
4.    The available reports on the study area should be collated and listed by the project manager and a broad description of the essential features of the hydrogeological system outlined in the study brief. The brief should also identify and list data sources, types and quality, and known issues that may affect the selection of an appropriate model complexity and the setting of calibration accuracy targets.
5.    The modelling study should be initiated with a literature review and data analysis in order to develop an understanding of the important aspects of the physical system, data reliability, and of the hydrological processes that control

or impact the groundwater flow system. The data analysis should identify data gaps that may affect the model development, and recommend field programmes necessary for additional data acquisition. The initial literature review and data analysis step needs to be adequately resourced for the purposes of the modelling study.

6. The available data to be used in model input or in calibration assessment should be collated into a database (spreadsheet format as a minimum).

7. Spatial coordinate and elevation data must be specified to a consistent standard datum. Head measurements should be reduced to a common density (freshwater is suggested) and common temperature (25°C is suggested) datum.

8. Data with a length component should be specified in units of metres. Data with a volume component should be specified in units of cubic metres. Data with a time component should be specified in units of days. Database compilations must explicitly state the units of the data.

9. A conceptual model must be developed, presented and reviewed prior to undertaking model construction, calibration and prediction. Assumptions must be documented.

10. The conceptual model should be based on an initial literature review, data collation and hydrogeological interpretation. It should be developed by making use of the principle of simplicity/parsimony to ensure the model is not too complex for the purposes of the study.

11. The conceptual model should present in descriptive and quantitative terms the essential system features - geological framework and boundaries, and the hydrological behaviour (natural and human-induced stresses), including a preliminary water balance.

12. The conceptual model must have sufficient degrees of freedom to allow a broad range of prediction responses spanning the criteria of acceptable or unacceptable impacts.

13. The conceptual model features must be described to an adequate level of detail commensurate with the ability of the data to represent the system, and with the collective ability to understand the system, given the current data and likely future data acquisition.

14. The conceptual model should be reviewed and revised as the database is augmented.

*(c)*    *Model Calibration and Prediction*

1. Any assumptions or modifications required to refine the conceptual hydrogeological understanding during its transformation into a mathematical model should be fully documented.

2. Medium to high complexity models should be calibrated to measured data before they are used for prediction simulations, and the calibration performance should be presented in qualitative and quantitative terms in comparison to agreed target criteria.

3. A calibration sensitivity analysis should be undertaken and a journal of the calibration process should be kept.

4. Since an objective function is used to compare how the model simulation matches the historical groundwater system behaviour, the formulation of the

objective function is a critical step in automated model calibration and should be discussed and justified. The objective function should be sensitive to deviations from calibration targets.

5. Automated model calibration should be preceded by a manually-instigated calibration effort to check that the mathematical model is in fact performing correctly in terms of data accuracy and conceptual functionality.

6. An inverse model (e.g. PEST, UCODE or MODFLOWP) should be run for one iteration initially to identify the correlated parameters and insensitive parameters. One of the correlated parameters and the insensitive parameters should be fixed before the automated model calibration is to proceed.

7. It is highly preferable that a model is calibrated to a range of distinct hydrological conditions (e.g. prolonged or short term dry or wet periods, and ranges of induced stresses), and that calibration is achieved with hydraulic conductivity and other parameters that are consistent with measured values, as this helps address the non-uniqueness problem of model calibration.

8. For medium to high complexity models where early time simulation output is critical, the initial head data for transient simulations should be consistent with (i.e. dynamically calibrated to) the initially specified boundary conditions and parameters, and should closely match the measured conditions at the start of the simulation period. The modeller should provide justification for the initial conditions adopted.

9. Model calibration acceptability should be judged in relation to water balance, residual error, and qualitative performance measures and criteria, and to selected reasonable quantitative performance measures.

10. Model calibration acceptability should be judged in relation to selected lumped quantitative performance measures, the value of which should be minimised. Listings of measured and modelled head values should be reported, along with relevant calibration performance measures for selected calibration data sets.

11. Plots of measured and modelled heads, residuals and/or error statistics should also be presented to indicate the spatial distribution of errors (e.g. scattergrams or contour plots of modelled heads with measured spot heights or other error plots).

12. Calibrated models should ideally be verified by running the model in predictive mode to check whether the simulation reasonably matches the observations of a reserved data set, deliberately excluded from consideration during calibration. Sensitivity analysis should also be completed.

13. The initial set of prediction scenarios to be addressed following model calibration and verification should be limited in range, and outlined in terms of:

• number of prediction simulations required and the types of prediction runs required (e.g. pumping rate ranges and timing, climatic variations, etc.)

• prediction run timeframe and hydrological data set to be used (e.g. a repeat of the historical record, or the development of a synthetic data set for prediction)

• type of sensitivity and/or uncertainty assessment.

14.    For subsequent programmes of model predictions, the scope of model prediction scenarios and uncertainty/sensitivity analysis should be based on the findings of previous programmes. It should be possible for these subsequent scenarios to be undertaken on a lump sum basis per scenario.

*(d)    Uncertainty Analysis*

1.    The modeller should outline the uncertainty assessment methodology at the outset, indicating how outcomes will be presented in terms that are meaningful in relation to the study objectives.
2.    For all models, some form of assessment of the underlying inaccuracy, sensitivity and/or limitation of the modelling approach and results needs to be explained.
3.    For low complexity models, perform either a complete sensitivity analysis or a review; for medium complexity models, perform at least a partial sensitivity analysis, taking into account best case and worst case parameter extremes; for medium and high complexity models, a partial sensitivity analysis is recommended during trial- and-error calibration to enhance modeller understanding and accelerate calibration; for high complexity numerical models, perform only a limited sensitivity analysis (not violating the calibration conditions) after calibration is completed, in order to indicate qualitatively the impact of key parameters in critical areas.
4.    Where the purpose of a high complexity numerical model is the assessment of average annual recharge or sustainable yield, post-processing of model water budgets should be done to produce a probability distribution for total recharge.
5.    For short periods of prediction (say, less than 10 years), a comprehensive scenario analysis is required as a minimum; where it is important to quantify the risk in prediction over short periods of time (say, less than 10 years), a stochastic approach is warranted.
6.    For long periods of prediction (say, more than 10 years), a steady state prediction should be performed for at least three situations representing expected, dry and wet conditions; each situation should have an agreed probability of exceedance indicated by cumulative probability distributions for each stress.   Alternatively, transient prediction approaches would also be acceptable, especially if it is important to also predict the time taken to achieve a new equilibrium ("steady state").
7.    For low complexity models, a stochastic (e.g. Monte Carlo) analysis may be performed in order to assess the uncertainty in model outcomes due to uncertain aquifer property values; for medium complexity models, either a worst case combination of parameters should be adopted, or a stochastic (e.g. Monte Carlo) analysis may be performed.

*(e)    Model Reporting and Reviews*

1.    Reports should be submitted at specified stages throughout a modelling study to enable review of the technical and contractual progress achieved, and decisions to be taken on whether and how to progress the study. A minimum

recommended reporting schedule comprises reports at the completion of the stages of Conceptualisation, Calibration and Prediction.

2. The extent and detail of the model report structure and composition should be consistent with the model study purpose and complexity. It is critical that all assumptions are clearly documented.

3. Model archive documentation should be maintained, consistent with the procedures of the organisation undertaking the work. Commonly, an archive would comprise a combination of modelling journals, documents on pre- and post-processing data analysis, and modelling data and software program files. The objective is to document the modelling effort sufficiently that such that the model could be re-generated for review and/or further refinement at some time in the future.

4. For medium and high complexity models, an internal model audit should be carried out progressively as part of an in-house quality control programme.

5. For medium and high complexity models, a post-audit should be carried out several years after original development, as part of the ongoing use of the model as a management tool. Reviews of and adjustments to the conceptual model and the model calibration may be required, which relies on the model archive produced at the end of the original study.

## Model Limitations

Limitations and uncertainties exist in any modelling study in regard to our hydrogeological understanding, the conceptual model design, and model calibration and prediction simulations, as well as recharge and evapotranspiration estimation and simulation. There are also limitations associated with the capabilities of the existing groundwater modelling software packages to adequately represent the complexities of any given hydrogeological system, and particularly in regard to surface-groundwater interaction. These limitations are best addressed by careful scoping of proposed modelling approaches at the outset, and review at various stages throughout the project. It is also important that modellers properly document model limitations at the proposal stage and in technical reports, as well as outlining possible methods of resolving them by subsequent work programmes of data acquisition and analysis and/or modelling. In some cases, the limitations may be so severe that there may be little value in putting the effort into a modelling study until more data and hydrogeological understanding is obtained, or until new technical methods are developed.

## References

1. Anderson, M.P. and Woessner, W.W. (1992). Applied Groundwater Modeling. Academic Press, Inc., San Diego, CA., 381 p.

2. American Society for Testing and Materials (1993). Standard Guide for Application of a Ground-Water Flow Model to a Site-Specific Problem. ASTM Standard D 5447-93, West Conshohocken, PA.

3. American Society for Testing and Materials (1993). Standard Guide for Comparing Ground-Water Flow Model Simulations to Site-Specific Information. ASTM Standard D 5490-93, West Conshohocken, PA.

4.   American Society for Testing and Materials (1994). Standard Guide for Defining Boundary Conditions in Ground-Water Flow Modeling. ASTM Standard D 5609-94, West Conshohocken, PA.

5.   American Society for Testing and Materials (1995). Standard Guide for Subsurface Flow and Transport Modeling. ASTM Standard D 5880-95, West Conshohocken, PA.

6.   American Society for Testing and Materials (1996). Standard Guide for Calibrating a Ground-Water Flow Model Application, ASTM Standard D5981-96, West Conshohocken, PA.

7.   Bear, J., and Verruijt, A. (1987). Modeling Groundwater Flow and Pollution. D. Reidel Publishing Company, 414 p.

8.   Groundwater Modeling Program (2000). Land and Water Management, Hydrologic Studies Unit, Department of Environmental Quality, State of Michigan.

9.   Spitz, Karlheinz and Moreno, Joanna (1996). A Practical Guide to Groundwater and Solute Transport Modeling. John Wiley & Sons, Inc.

10.  Wang, H.F. and Anderson, M.P. (1982). Introduction to Groundwater Modeling. W.H. Freeman and Company, San Francisco, CA, 237 p.

# 11-Data Requirements for Groundwater Modelling

## Introduction

A groundwater model is any computational method that represents an approximation of an underground water system (modified after Anderson and Woessner 1992). While groundwater models are, by definition, a simplification of a more complex reality, they have proven to be useful tools over several decades for addressing a range of groundwater problems and supporting the decision-making process.

Groundwater systems are affected by natural processes and human activity, and require targeted and ongoing management to maintain the condition of groundwater resources within acceptable limits, while providing desired economic and social benefits. Groundwater management and policy decisions must be based on knowledge of the past and present behaviour of the groundwater system, the likely response to future changes and the understanding of the uncertainty in those responses.

The location, timing and magnitude of hydrologic responses to natural or human-induced events depend on a wide range of factors - for example, the nature and duration of the event that is impacting groundwater, the subsurface properties and the connection with surface water features such as rivers and oceans. Through observation of these characteristics a conceptual understanding of the system can be developed, but often observational data is scarce (both in space and time), so our understanding of the system remains limited and uncertain.

Groundwater models provide additional insight into the complex system behaviour and (when appropriately designed) can assist in developing conceptual understanding. Furthermore, once they have been demonstrated to reasonably reproduce past behaviour, they can forecast the outcome of future groundwater behaviour, support decision-making and allow the exploration of alternative management approaches. However, there should be no expectation of a single 'true' model, and model outputs will always be uncertain. As such, all model outputs presented to decision-makers benefit from the inclusion of some estimate of how good or uncertain the modeller considers the results.

A *groundwater flow model* simulates hydraulic heads (and watertable elevations in the case of unconfined aquifers) and groundwater flow rates within and across the boundaries of the system under consideration. It can provide estimates of water balance and travel times along flow paths. A *solute transport model* simulates the concentrations of substances dissolved in groundwater. These models can simulate the migration of solutes (or heat) through the subsurface and the boundaries of the system. Groundwater models can be used to calculate water and solute fluxes between the groundwater system under consideration and connected source and sink features such as surface water bodies (rivers, lakes), pumping bores and adjacent groundwater reservoirs.

## Groundwater Modelling Process

The groundwater modelling process has a number of stages. As a result, the modelling team needs to have a combination of skills and at least a broad or general knowledge of: hydrogeology; the processes of groundwater flow; the mathematical equations that describe groundwater flow and solute movement; analytical and numerical techniques for solving these equations; and the methods for checking and testing the reliability of models.

The modeller's task is to make use of these skills, provide advice on the appropriate modelling approach and to blend each discipline into a product that makes the best use of the available data, time and budget. In practice, the adequacy of a groundwater model is best judged by the ability of the model to meet the agreed modelling objectives with the required level of confidence. The modelling process can be subdivided into seven stages (shown schematically in Figure 1) with three hold points where outputs are documented and reviewed.

The process starts with *planning*, which focuses on gaining clarity on the intended use of the model, the questions at hand, the modelling objectives and the type of model needed to meet the project objectives. The next stage involves using all available data and knowledge of the region of interest to develop the conceptual model *(conceptualisation)*, which is a description of the known physical features and the groundwater flow processes within the area of interest. The next stage is *design*, which is the process of deciding how to best represent the conceptual model in a mathematical model. It is recommended to produce a report at this point in the process and have it reviewed. *Model construction* is the implementation of model design by defining the inputs for the selected modelling tool.

The *calibration and sensitivity analysis* of the model occurs through a process of matching model outputs to a historical record of observed data. It is recommended that a calibration and sensitivity analysis report be prepared and reviewed at this point in the process.

*Predictions* comprise those model simulations that provide the outputs to address the questions defined in the modelling objectives. The predictive analysis is followed by an analysis of the implications of the *uncertainty* associated with the modelling outputs.

Clear communication of the model development and quality of outputs through *model reporting and review* allows stakeholders and reviewers to follow the process and assess whether the model is fit for its purpose, that is, meets the modelling objectives.

The process is one of continual iteration and review through a series of stages. For example, there is often a need to revisit the conceptual model during the subsequent stages in the process. There might also be a need to revisit the modelling objectives and more particularly reconsider the type of model that is desired once calibration has been completed. Any number of iterations may be required before the stated modelling objectives are met. Accordingly, it is judicious at the planning stage to confirm the iterative nature of the modelling process so that clients and key stakeholders are receptive to and accepting of the approach.

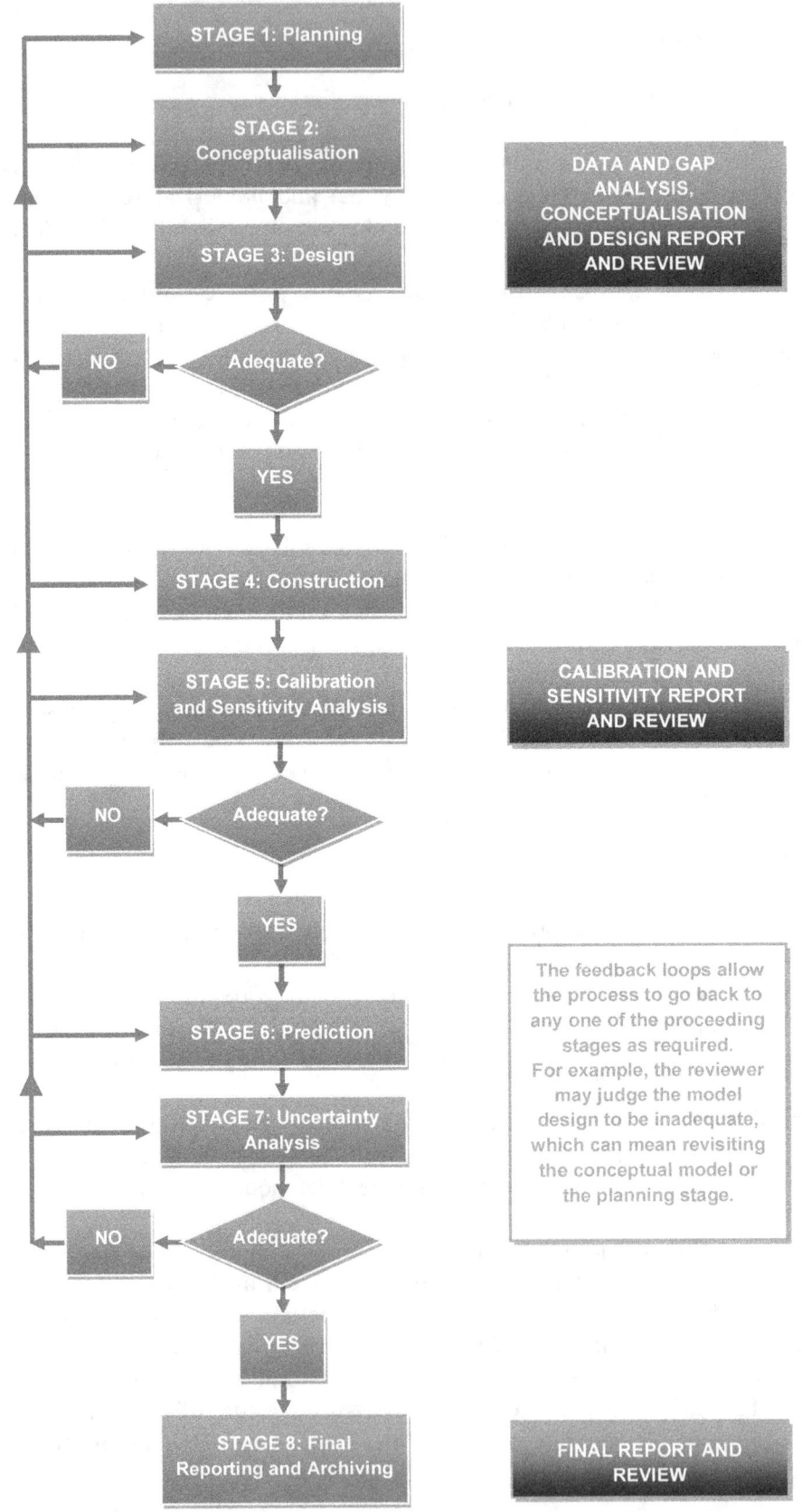

Figure 1:       Groundwater Modelling Process
*(modified after MDBC, 2001 and Yan et al., 2010)*

# Data Requirements for Groundwater Modelling

The first phase of any groundwater study consists of collecting all existing geological and hydrological data on the groundwater basin in question. This will include information on surface and subsurface geology, water tables, precipitation, evapotranspiration, pumped abstractions, stream flows, soils, land use, vegetation, irrigation, aquifer characteristics and boundaries, and groundwater quality. If such data do not exist or are very scanty, a program of field work must first be undertaken, for no model whatsoever makes any hydrological sense if it is not based on a rational hydrogeological conception of the basin. All the old and newly-found information is then used to develop a conceptual model of the basin, with its various inflow and outflow components.

A conceptual model is based on a number of assumptions that must be verified in a later phase of the study. In an early phase, however, it should provide an answer to the important question: does the groundwater basin consist of one single aquifer (or any lateral combination of aquifers) bounded below by an impermeable base? If the answer is yes, one can then proceed to the next phase: developing the numerical model. This model is first used to synthesize the various data and then to test the assumptions made in the conceptual model. Developing and testing the numerical model requires a set of quantitative hydrogeological data that fall into two categories:

- Data that define the physical framework of the groundwater basin
- Data that describe its hydrological stress

These two sets of data are then used to assess a groundwater balance of the basin. The separate items of each set are listed below.

| *Physical framework* | *Hydrological stress* |
|---|---|
| 1. Topography | 1. Water table elevation |
| 2. Geology | 2. Type and extent of recharge areas |
| 3. Types of aquifers | 3. Rate of recharge |
| 4. Aquifer thickness and lateral extent | 4. Type and extent of discharge areas |
| 5. Aquifer boundaries | 5. Rate of discharge |
| 6. Lithological variations within the aquifer | |
| 7. Aquifer characteristics | |

It is common practice to present the results of hydrogeological investigations in the form of maps, geological sections and tables - a procedure that is also followed when developing the numerical model. The only difference is that for the model, a specific set of maps must be prepared. These are:

- Contour maps of the aquifer's upper and lower boundaries
- Maps of the aquifer characteristics
- Maps of the aquifer's s net recharge
- Water table contour maps

Some of these maps cannot be prepared without first making a number of auxiliary maps. A map of the net recharge, for instance, can only be made after topographical, geological, soil, land use, cropping pattern, rainfall, and evaporation maps have been made.

The data needed in general for a groundwater flow modelling study can be grouped into two categories: (a) Physical framework and (b) Hydrogeologic framework (Moore, 1979). The data required under physical framework are:

1. Geologic map and cross section or fence diagram showing the areal and vertical extent and boundaries of the system.
2. Topographic map at a suitable scale showing all surface water bodies and divides. Details of surface drainage system, springs, wetlands and swamps should also be available on map.
3. Land use maps showing agricultural areas, recreational areas etc.
4. Contour maps showing the elevation of the base of the aquifers and confining beds.
5. Isopach maps showing the thickness of aquifers and confining beds.
6. Maps showing the extent and thickness of stream and lake sediments.

These data are used for defining the geometry of the groundwater domain under investigation, including the thickness and areal extent of each hydrostratigraphic unit.

Under the hydrogeologic framework, the data requirements are:

1. Water table and potentiometric maps for all aquifers.
2. Hydrographs of groundwater head and surface water levels and discharge rates.
3. Maps and cross sections showing the hydraulic conductivity and/or transmissivity distribution.
4. Maps and cross sections showing the storage properties of the aquifers and confining beds.
5. Hydraulic conductivity values and their distribution for stream and lake sediments.
6. Spatial and temporal distribution of rates of evaporation, groundwater recharge, surface water - groundwater interaction, groundwater pumping, and natural groundwater discharge.

The data collection and analysis stage of the modelling process involves:

- confirming the location and availability of the required data
- assessing the spatial distribution, richness and validity of the data
- data analysis commensurate with the level of confidence required. Detailed assessment could include complex statistical analysis together with an analysis of errors that can be used in later uncertainty analysis.
- developing a model project database. The data used to develop the conceptualisation should be organised into a database, and a data inventory should be developed, which includes data source lists and references.
- evaluating the distribution of all parameters/observations so that model calibration can proceed with parameters that are within agreed and realistic limits. Parameter distributions for the conceptual model are sometimes best represented as statistical distributions.

- justification of the initial parameter value estimates for all hydrogeological units
- quantification of any flow processes or stresses (e.g. recharge, abstraction).

Some of the compiled information will be used not only during the conceptualisation, but also during the design and calibration of the model. This includes the data about the model layers and hydraulic parameters as well as observations of hydraulic head, watertable elevation, and fluxes.

The conceptualisation stage may involve the development of maps that show the hydraulic heads in each of the aquifers within the study area. These maps help illustrate the direction of groundwater flow within the aquifers, and may infer the direction of vertical flow between aquifers.

The data used to produce maps of groundwater head is ideally obtained from water levels measured in dedicated observation wells that have their screens installed in the aquifers of interest. More often than not, however, such data is scarce or unavailable and the data is sourced from, or complemented by, water levels from production bores. These may have long well screens that intersect multiple aquifers, and be influenced by preceding or coincident pumping. The accuracy of this data is much less than that obtained from dedicated observation wells. The data can be further supplemented by information about surface expressions of groundwater such as springs, wetlands and groundwater-connected streams. It provides only an indication of the minimum elevation of the watertable (i.e. the land surface) in areas where a stream is gaining and local maximum elevation in areas where a stream is losing. As such, this data has a low accuracy, but can be very valuable nonetheless.

*The hydrogeological domain*

The hydrogeological domain should be conceptualised to be large enough to cover the location of the key stresses on the groundwater system (both the current locations and those in the foreseeable future) and the area influenced or impacted by those stresses. It should also be large enough to adequately capture the processes controlling groundwater behaviour in the study area.

All hydrogeological systems are 'open' and it is debatable whether the complete area of influence of the hydrogeological system can be covered. As such, some form of compromise is inevitable in defining the hydrogeological domain.

The hydrogeological domain comprises the architecture of the hydrogeologic units (aquifers and aquitards) relevant to the location and scale of the problem, the hydraulic properties of the hydrogeological units, the boundaries and the stresses.

One of the difficult decisions early on in developing a conceptual model relates to the limits of the hydrogeological domain. This is best done so that all present and potential impacts on the groundwater system can be adequately accounted for in the model itself. The extent of the conceptual model can follow natural boundaries such as those formed by the topography, the geology or surface water features. It should also account for the extent of the potential impact of a given stress, for example pumping or injection. It is important that the extent of the hydrogeological domain is larger than the model domain developed during the model design stage.

Defining the hydrogeological domain involves:

- describing the components of the system with regard to their relevance to the problem at hand, such as *the hydrostratigraphy* and the *aquifer properties*
- describing the relationships between the components within the system, and between the system components and the broader environment outside of the hydrogeological domain
- defining the specific processes that cause the water to move from recharge areas to discharge areas through the aquifer materials
- defining the *spatial scale* (local or regional) and *time scale* (steady-state or transient on a daily, seasonal or annual basis) of the various processes that are thought to influence the water balance of the specific area of interest
- in the specific case of solute transport models, *defining the distribution of solute concentration* in the hydrogeological materials (both permeable and less permeable) and the processes that control the presence and movement of that solute
- making simplifying assumptions that reduce the complexity of the system to the appropriate level so that the system can be simulated quantitatively. These assumptions will need to be presented in a report of the conceptualisation process, with their justifications.

*Hydrostratigraphy*

The layout and nature of the various hydrogeological units present within the system will guide the definition of the distribution of various units in the conceptual model. Generally, where a numerical simulation model is developed, the distribution of hydrogeologic layers typically provides the model layer structure. In this regard, the conceptualisation of the units should involve consideration of both the lateral and vertical distribution of materials of similar hydraulic properties.

Typical information sources for this data are from geological information such as geological maps and reports, drill-hole data and geophysical surveys and profiles. Where the data is to be used to define layers in numerical models, surface elevation data (usually from digital elevation models) is required.

A hydrostratigraphic description of the system will consist of:

- stratigraphy, structural and geomorphologic discontinuities (e.g. faults, fractures, karst areas)
- the lateral extent and thickness of hydrostratigraphic units
- classification of the hydrostratigraphic units as aquifers (confined or unconfined) or as aquitards
- maps of aquifer/aquitard extent and thickness (including structure contours of the elevation of the top and bottom of each layer)

*Aquifer properties*

The aquifer and aquitard properties control water flow, storage and the transport of solutes, including salt, through the hydrogeological domain. Quantified aquifer properties are critical to the success of the model calibration. It is also well understood that aquifer properties vary spatially and are almost unknowable at the detailed scale. As such, quantification of aquifer properties is one area where simplification is often applied, unless probabilistic parameterisation methods are applied for uncertainty assessment. Hydraulic properties that should be characterised include hydraulic conductivity (or transmissivity), specific storage (or storativity) and specific yield.

*Conceptual boundaries*

The conceptualisation process establishes where the boundaries to the groundwater flow system exist based on an understanding of groundwater flow processes. The conceptualisation should also consider the boundaries to the groundwater flow system in the light of future stresses being imposed (whether real or via simulations). These boundaries include the impermeable base to the model, which may be based on known or inferred geological contacts that define a thick aquitard or impermeable rock.

Assumptions relative to the boundary conditions of the studied area should consider:

- where groundwater and solutes enter and leave the groundwater system
- the geometry of the boundary; that is, its spatial extent
- what process(es) is(are) taking place at the boundary, that is, recharge or discharge
- the magnitude and temporal variability of the processes taking place at the boundary. Are the processes cyclic and, if so, what is the frequency of the cycle?

*Stresses*

The most obvious anthropogenic stress is groundwater extraction via pumping. Stresses can also be those imposed by climate through changes in processes such as evapotranspiration and recharge. Description and quantification of the stresses applied to the groundwater system in the conceptual domain, whether already existing or future, should consider:

- if the stresses are constant or changing in time; are they cyclic across the hydrogeological domain?
- what are their volumetric flow rates and mass loadings?
- if they are localised or widespread (i.e. point-based or areally distributed).

Fundamental to a conceptual groundwater model is the identification of recharge and discharge processes and how groundwater flows between recharge and discharge locations. As for many features of a groundwater model, the level of detail required is dependent on the purpose of the model. The importance attached to individual features such as recharge and discharge features in any given study area should be discussed among the project team.

Representation of surface water - groundwater interaction is required in increasing detail in modelling studies. An interaction assessment should outline the type of interaction between surface water and groundwater systems in terms of their connectedness and whether they are gaining or losing systems. Techniques such as hydraulic measurements, tracer tests, temperature measurements and mapping, hydrogeochemistry and isotopic methods may be used. The need to account for spatial and temporal variability, for example during flood events, in describing interaction between surface water and groundwater should also be assessed.

*Physical processes*

The processes affecting groundwater flow and/or transport of solutes in the aquifer will need to be understood and adequately documented in the model reporting process. Description of the actual processes, as opposed to the simplified model representation of processes, is required to facilitate third-party scrutiny of the assumptions used in the model development. Flow processes within the hydrogeological domain need to be described, including the following:

- the equilibrium condition of the aquifer, that is, whether it is in steady state or in a transient state. This is established by investigating the historical records in the form of water-level hydrographs, groundwater-elevation surfaces made at different times, or readings from piezometers.
- the main flow direction(s). Is groundwater flowing in one direction predominantly? Is horizontal flow more significant than vertical flow?
- water properties such as density. Are they homogeneous throughout the aquifer? What are the effects of dissolved solutes and/or temperature? Can the flow field be assumed to be driven by hydraulic gradients only?

Additional tasks related to describing the flow processes include:

- creating flow nets from groundwater elevation contours. These will describe the directions of flow, and can be used in a semi-quantitative manner to derive flow volumes.
- quantifying the components of recharge and discharge to the hydrogeological domain, including all those related to point and diffuse recharge and discharge.
- undertaking analysis of the interactions between surface water and groundwater in the hydrogeological domain where it has been highlighted as a significant process.

A database (e.g. GIS-based) will capture all the data that has been collated, whether or not it has been used to develop the conceptual model, with data sources listed and references to previous studies.

*Boundary conditions*

Groundwater flow models require information about the head and/or head gradient at the boundaries of the model domain. There are three types of boundary conditions:

- *Type 1, Dirichlet or specified head boundary condition*: The head of a boundary cell or node is specified. When the head is specified along a section of the model boundary, the flow across this model boundary section is calculated.

- *Type 2, Neumann or specified head-gradient boundary condition*: The gradient of the hydraulic head is specified at the boundary, which implies that the flow rate across the boundary is specified.
- *Type 3, Cauchy or specified head and gradient boundary condition*: Both the head and the head gradient are specified. In flow models this type of boundary condition is implemented in an indirect manner by specifying a head and a hydraulic conductance or resistance. Both represent effects of features that are located outside the model domain. For example, if a confined aquifer underlies a lake, the flow between the aquifer and the lake can be represented by a Type 3 boundary condition in which the specified head represents the lake level, and the conductance is that of the aquitard that separates the aquifer from the lake.

All three types of model boundary conditions can be assigned as either constant or variable with time. For example, rivers can be modelled as Type 3 Cauchy boundary conditions with time-varying river stages obtained from water-level records.

Groundwater stresses are defined as those processes that lead to the removal or addition of water from or to a groundwater domain. Stresses are typically separated into those associated with the climate (rainfall infiltration and evapotranspiration) and those associated with human activity (such as groundwater extraction). Groundwater stresses are often considered or treated as boundary conditions both by modellers and model GUIs alike. Technically they are 'sink and source' terms that are included in the equations that describe water movement and storage in the model.

Most groundwater model codes and GUIs allow the modeller to implement boundary conditions and stresses that are tailored to represent typical near-surface groundwater phenomena such as rainfall-derived recharge, interaction with rivers or lakes and evapotranspiration fluxes from shallow or outcropping groundwater.

*Initial conditions*

Initial conditions in a transient simulation should be obtained, wherever possible, from a previous model run (e.g. a steady state solution) to avoid spurious results at early times in the transient model run.

Initial conditions define the groundwater conditions present at the start of the model run. In practice, the modeller must define initial heads in all model cells. The choice of initial conditions for a steady state model does not influence the model outcome, but the steady state solution is obtained more rapidly when initial conditions are defined that are reasonably close to the final solution.

For a transient groundwater model, the initial conditions are part of the mathematical problem statement and will influence the model outcomes during the subsequent time steps. It is therefore important that the models are chosen so that they are consistent with the boundary conditions and stresses. When field data is used to define the initial conditions there is a risk that the assigned heads (and solute concentrations) are not in equilibrium with the boundary conditions and stresses applied to the model. Remedies to this problem include:

- allowing for an initial model equilibration time. After a certain amount of time the influence of the initial heads on the calculated heads becomes negligible
- using the results of a steady state model with the boundary conditions and stresses, as they are believed to be at the start of the transient simulation. This approach is only strictly valid if the system can be assumed to be in a steady state at some point in time. In practice, however, it can provide a useful initial condition that is both stable and close to the correct starting condition for a transient model
- using the results of another variant of the model. This is appropriate, for example, when the model is used for predictive simulations; the calculated heads from the (calibrated) model are used to define the initial heads of the predictive model.

## *Initial estimates of model parameters*

All available information should be used to guide the parameterisation and model calibration. All parameters should initially be considered to be uncertain. Before a model can be run it is necessary to assign initial values to all model parameters.

Parameter values representing hydrogeological properties are normally chosen based on aquifer tests undertaken in the area of interest or through simple calculations that use observed groundwater behaviour to indicate key parameter values. Where parameter values have not been calculated they are typically estimated from values reported in the literature for the hydrostratigraphic units being modelled or from text books that provide more generic ranges of values for the type of sediments or rocks included in the model.

Even when aquifer tests provide values for hydraulic conductivity and storage parameters for some of the hydrogeological units being modelled, these parameters are typically variable within an individual unit. As a result the initial values of hydrogeological parameters should be considered as approximate guides only and subsequent adjustment or modification of these parameters during the calibration process is expected.

## *Solute transport data*

All available solute concentration data should be used during conceptualisation to determine the spatial distribution of solutes, identify source zones and migration pathways, and to determine appropriate boundary conditions. An assessment of the relative importance of advection, diffusion and dispersion should be made during the conceptualisation stage, and a decision should be made on which processes are to be included in the solute transport model. The importance of variable-density flow should be assessed with a quantitative analysis using all available head and concentration data. Conceptualisation for the purpose of solute transport involves:

- collection of solute concentration data, and solute conditions at the start of transient simulations
- identification of solute transport processes
- delineation of the area of interest (which may be different from that of the flow model) and an assessment of the relevant time scales
- identification of solute concentration boundary conditions and sources and sinks of solutes

- assessment of the spatial variability (i.e. heterogeneities) in the aquifer's geological properties
- quantification of solute transport parameters to be used in simulations.

All available solute concentration data should be used during conceptualisation to determine the spatial distribution of solutes, identify source zones and migration pathways, and to determine appropriate boundary conditions.

Measurements of the spatial distribution and temporal variations of solute concentrations are essential elements of the conceptualisation process. Solute concentration data is also required to determine the initial conditions for transient solute transport models and during the calibration stage. Solute concentrations should be obtained from all available sources within the study area, including pumping bores, injection wells, monitoring wells, surface water bodies and rainfall. If insufficient solute concentration data is available for an adequate site characterisation, new data collection efforts should be undertaken.

*Solute concentration boundary conditions*

Similar to flow models, boundary conditions must be defined for transport models, and similar considerations apply in the selection of their location, that is, preferably coinciding with physical features and sufficiently far away from the area of interest. There are three types of solute concentration boundary conditions.

- *Type 1, Dirichlet or specified concentration boundary condition.* The concentration of a boundary cell or node is specified. Solute mass can be added or removed through Dirichlet boundaries by advection and/or diffusion and dispersion.
- *Type 2, Neumann or specified concentration gradient boundary condition.* The gradient of the solute concentration is specified at the boundary, which implies that the diffusive/dispersive flux across the boundary is specified.
- *Type 3, Cauchy or specified concentration and gradient boundary condition.* Both the concentration and the gradient are specified.

The specified mass flux boundary condition can be implemented as either a Type 2 or a Type 3 boundary condition, depending on which transport process dominates. If dispersive and diffusive mass transport across the boundary is small, which is often a defensible assumption, the concentration gradient across the boundary can be set to zero. The specified mass flux is the product of the specified flow rate and the solute concentration of groundwater entering the system.

Specified boundary concentrations and fluxes can be constant during the entire duration of the simulation or vary as a function of time. The type of boundary condition may even change during a simulation, which could occur where surface water features are variable in extent, or where tidal fluctuations occur on a sloping beach face.

*Sources and sinks*

Sources and sinks either add water to or remove water from the model domain, and the water entering or leaving the model has an associated solute concentration that must be known or approximated. Sources can be injection wells, rivers, lakes or recharge. Abstraction wells are one example of a sink, and the concentration of the water leaving the model domain in this way is typically considered to be equal to that of the groundwater immediately adjacent to the well. Evapotranspiration represents a sink of water, but not of solutes, and causes an increase in solute concentrations. This is typically encountered in the simulation of groundwater discharge in riparian zones or salt lakes.

In coastal aquifers, the source of saline groundwater may not always be modern seawater, but may reflect other sources such as rock dissolution, connate water entrapped in marine deposits, paleo-seawater that intruded during land surface inundations, and/or anthropogenic contaminants. Also, tidal creeks, rivers and estuaries may also be sources of salt water in coastal aquifers, and knowledge of their tidal limits and the annual salinity variations along their lengths is usually required. Failing to account for these factors may result in a flawed conceptual understanding of the system, leading to erroneous model outcomes. The data collection effort during the conceptualisation stage must therefore allow for various hypotheses to be evaluated, for example, by collecting information on various hydrochemical and isotope tracer techniques that can identify solute origins.

*Heterogeneity*

Groundwater flow conceptualisation usually involves identification and delineation of the primary hydrostratigraphic units, and the heterogeneities in hydraulic conductivity and porosity within geological strata are often neglected or implicitly incorporated (e.g. through an anisotropic hydraulic conductivity field). While this is usually a reasonable approach for determining the distribution of aquifer heads and for estimating average groundwater flows, aquifer heterogeneities within geological units have a more profound influence on solute transport. Therefore, solute transport models generally require a higher resolution of geological information, in particular in the vertical direction.

An assessment must be made of the extent to which solute concentration patterns are influenced by heterogeneities, by considering the existence of preferential flow pathways, aquitard windows, dual-porosity effects, and the degree of the variability of porosity and permeability within aquifers. Heterogeneities are usually characterised from various data sources, such as geological maps, borehole logs, geophysical surveys, solute concentration distributions, aquifer tests and slug tests, and knowledge about the depositional environment or fracture density, connectivity and aperture. The depositional environments of some unconsolidated aquifers can result in heterogeneities that impose considerable effects on concentration distributions. These include unconsolidated aquifers comprising fluvial sediments, where permeable sand and/or gravel may alternate with relatively impermeable clay layers over short distances.

*Solute transport parameters*

Solute transport models require input parameters that describe the combined effect of advection, dispersion and diffusion. This typically involves quantification of the following parameters:

- effective porosity

- longitudinal and transverse dispersivity
- diffusion coefficient
- an equation(s) of state (for variable density problems).

Solute transport models require the *effective porosity* and spatial variations thereof to be specified. The porosity has a dual role in solute transport models: it determines the advective flow rate, and it determines the volume of water in the model for storage of solute mass. Total porosity values are relatively easy to quantify when undisturbed cores are available. If this is not the case, values can sometimes be obtained from geophysical logs or estimated from the literature. A range of values exist for different lithological units, but the variability of this parameter is not as large as the hydraulic conductivity variability.

The processes associated with the spreading of solute plumes are challenging to reproduce explicitly (i.e. in a process-based way) because of the small scale of many dispersive factors. The associated transport parameters (such as *dispersivity*) are equally difficult to quantify, especially under field conditions, and the approach to solute transport parameterisation is usually one where transport parameters are modified so that field observations are optimally reproduced by the transport model. The transverse dispersivity is usually much lower than the longitudinal dispersivity, and the sparse data that exists suggest that (i) the horizontal transverse dispersivity is about one order of magnitude lower than the longitudinal dispersivity and (ii) the vertical transverse dispersivity is one or two orders of magnitude smaller than the horizontal transverse dispersivity (Zheng and Bennett 2002).

*Diffusion* can be an important transport process in solute transport problems (i) at the local (i.e. metres or less) scale; (ii) in low-permeability units (e.g. shale, clay); or (iii) at long timescales (i.e. centuries or more) in stagnant groundwater systems. Unless these problems are being considered, the value of the diffusion coefficient has little effect on the simulation outcomes. The parameterisation of diffusion depends on the solute of interest. The value of the diffusion coefficient is dependent on temperature, and varies for different solute species. However, the diffusion coefficient of chloride, which only ranges between $10^{-9}$ and $2 \times 10^{-9}$ $m^2$/s in pure water, can be used as a good approximation under most circumstances for solutes like major ions, or in a simulation that considers an aggregate solute concentration, like total dissolved solids, or salinity. Specialised application could require the use of different diffusion coefficients for individual ions, for example, with long-term transport processes in clay layers (e.g. safety assessment of nuclear waste repositories).

Variable-density problems further require an *equation of state* that relates the water density to concentration, temperature, and/or pressure. The equation of state couples the groundwater flow equation to the advection–dispersion equation. The flow is affected by the density, and the flow affects the concentrations, and through this, the density. Equations of state are typically linear or exponential functions, and their parameters are readily available in the literature and the supporting documentation of model codes. The parameter values depend on the chemical composition of the groundwater, and the modeller needs to evaluate which relationships are appropriate for the system under consideration.

## Modelling Software

Groundwater modelling sometimes requires the use of a number of software types. These include:

- the model code that solves the equations for groundwater flow and/or solute transport, sometimes called simulation software or the computational engine
- a GUI that facilitates preparation of data files for the model code, runs the model code and allows visualisation and analysis of results (model predictions)
- software for processing spatial data, such as a geographic information system (GIS), and software for representing hydrogeological conceptual models
- software that supports model calibration, sensitivity analysis and uncertainty analysis
- programming and scripting software that allows additional calculations to be performed outside or in parallel with any of the above types of software.

Some software is public domain and open source (freely available and able to be modified by the user) and some is commercial and closed (only available in an executable form that cannot be modified by the end user).

Some software fits several of the above categories, for example, a model code may be supplied with its own GUI, or a GIS may be supplied with a scripting language. Some GUIs support one model code while others support many. Software packages are increasingly being coupled to other software packages, either tightly or loosely. Table 1 lists some examples of modelling software commonly used.

## Concluding Remarks

It is generally agreed that modelling and model calibration should utilise and take into account all available information. In the context of groundwater flow modelling, available information includes:

- observations of watertable elevations and piezometric heads (at depth)
- prior estimates of hydrogeological properties obtained following aquifer tests, slug tests and even permeameter tests on cores
- geophysical data, including seismic and ground-based or airborne electromagnetic data used to define stratigraphy
- downhole geophysics leading to understanding of fracture density and orientation
- records of pumping abstraction and irrigation rates
- estimates of recharge and evapotranspiration
- measurements of streamflow or water quality in losing and gaining streams
- concentrations of solutes and tracers that could provide insights about flow directions and/or groundwater age.

Some of this data are measurements of state variables (e.g. head or concentration), some are observations of quantities derived from state variables (e.g. flux of water or solute) and some are observations of hydrogeological properties or boundary conditions represented by model parameters.

# Table 1:     Groundwater Modelling Software Commonly Used

| Name of Software | Type of Software | Description |
|---|---|---|
| MODFLOW | Simulation of saturated flow | Open source software developed by the USGS, based on a block-centred finite difference algorithm. Relies on a large number of modular packages that add specific capabilities. Most packages are also open source and can therefore be modified by end users. Can be coupled to MT3DMS and other codes to simulate solute transport, as well as MIKE 11 for flow in river and stream networks. |
| FEFLOW | Simulation of saturated and unsaturated flow, transport of mass (multiple solutes) and heat, with integrated GUI | Commercial software based on the finite element method. Several versions with different capabilities. Extendable using plug-ins that can be developed by end users to expand the capabilities, during or after computations. Can be coupled to MIKE 11 to simulate flow in river and stream networks. |
| SUTRA | Simulation of saturated and unsaturated flow, transport of mass and heat | Open source software based on the finite element method, designed for density-coupled flow and transport. |
| MT3DMS | Simulation of transport of multiple reactive solutes in groundwater | Open source software that can be coupled with MODFLOW to compute coupled flow and transport. |
| SEAWAT | Simulation of saturated flow and transport of multiple solutes and heat | Open source software combining MODFLOW and MT3DMS for density-coupled flow and transport. |
| MIKE SHE | Integrated catchment modelling, with integrated GUI | Commercial software that uses the finite difference method for saturated groundwater flow, several representations of unsaturated flow, including the 1D Richards equation, MIKE 11 for flow in river and stream networks and the 2D diffusive-wave approach for overland flow. |
| Visual MODFLOW | GUI | Commercial software. Supports MODFLOW (with many packages), MODPATH, SEAWAT, MT3DMS, MT3D99, RT3D, PHT3D, MGO, PEST, MODFLOW-SURFACT, MIKE 11. |
| Groundwater Vistas | GUI | Commercial software. Supports MODFLOW (with many packages), MODPATH, SEAWAT, MT3DMS, PEST, MODFLOW-SURFACT. |
| GMS | GUI | Commercial software. Supports MODFLOW (with many packages), MODPATH, MODAEM, SEAWAT, MT3DMS, RT3D, SEAM2D, PEST, SEEP2D, FEMWATER. |
| PMWIN | GUI | Commercial software. Supports MODFLOW (with many packages), MODPATH, SEAWAT, MT3DMS, PHT3D, PEST. |
| ArcGIS | GIS | Commercial software to manage spatial data. Capabilities can be extended using ArcPy, an implementation of the Python scripting language. |
| Surfer | Gridding and contouring | Commercial software to manage and plot spatial data. |
| Hydro GeoAnalyst | Management of hydrogeological data | Visualisation of bore logs, fence diagrams. Creation of hydrostratigraphic layers. Incorporates elements of ArcGIS. |
| RockWorks | Management of hydrogeological data | Visualisation of bore logs, fence diagrams. Creation of hydrostratigraphic layers. Can be linked to ArcGIS. |
| ArcHydro Groundwater | Management of hydrogeological data | Visualisation of bore logs, fence diagrams. Creation of hydrostratigraphic layers. Tightly linked with ArcGIS. |
| PEST | Parameter estimation and uncertainty analysis | Open-source software designed to allow parameter estimation for any model. Available in many implementations to support specific groundwater models and GUIs. |

Historical measurements may reflect the behaviour of a groundwater system subject only to natural stresses, and with head gradients and flows that are much smaller than after development of the project (e.g. a water supply borefield, an irrigation scheme or a mine). The changes in levels of stress on an aquifer mean that the future behaviour of the groundwater-flow model depends on different model parameters. Calibration may lead to good estimates of some model parameters that have little influence on the accuracy of predictions and such estimates will not improve the level of confidence in predictions.

## References

1.  Anderson, M. P. and Woessner, W. W. (1992). Applied Groundwater Modeling, Simulation of Flow and Advective Transport. Academic Press, San Diego, USA.

2.  Boonstra, J. and de Ridder, N. A. (1981). Numerical Modelling of Groundwater Basins. A User-oriented Manual, International Institute for Land Reclamation and Improvement, The Netherlands, 226 p.

3.  Kumar, C. P. (2014), Groundwater Data Requirement and Analysis. Scientific Essay, ISBN (eBook): 978-3-656-75554-8, ISBN (Book): 978-3-656-75553-1, GRIN Publishing GmbH, Munich, Germany.

4.  Merz, Sinclair Knight and National Centre for Groundwater Research and Training (2012). Australian Groundwater Modelling Guidelines. Waterlines Report Series No. 82, 191 p.

5.  Moore, J. E. (1979). Contributions of Groundwater Modelling to Planning. Journal of Hydrology, Vol. 43, pp. 121-128.

6.  Murray–Darling Basin Commission (MDBC) (2001). Groundwater Flow Modelling Guideline. Report prepared by Aquaterra.

7.  Yan, W., Alcoe, D., Morgan, L, Li C and Howles, S, (2010). Protocol for Development of Numerical Groundwater Model. Version 1, Report prepared for the Government of South Australia, Department for Water.

8.  Zheng, C. and Bennett, G. D. (2002). Applied Contaminant Transport Modelling. 2nd edition, John Wiley and Sons Inc., New York.

# 12-Capabilities of Groundwater Modelling Software

## Introduction

Groundwater modelling has become an important methodology in support of the planning and decision-making processes involved in groundwater management. Groundwater models provide an analytical framework for obtaining an understanding of the mechanisms and controls of groundwater systems and the processes that influence their quality, especially those caused by human intervention in such systems. Increasingly, models are an integral part of water resources assessment, protection and restoration studies, and provide essential and cost-effective support for planning and screening of alternative policies, regulations, and engineering designs affecting ground water.

There are many different groundwater modelling codes available, each with their own capabilities, operational characteristics, and limitations. If modelling is considered for a project, it is important to determine if a particular code is appropriate for that project, or if a code exists that can perform the simulations required in the project.

In practice, it is often difficult to determine the capabilities, operational characteristics, and limitations of a particular groundwater modelling code from the documentation, or even impossible without actual running the code for situations relevant to the project for which a code is to be selected due to incompleteness, poor organization, or incorrectness of a code documentation. Systematic and comprehensive description of a code features based on an informative classification provides the necessary basis for efficient selection of a groundwater modelling code for a particular project or for the determination that no such code exists.

## ASTM Guides

*Standard Guide for Describing the Functionality of a Ground-Water Modelling Code* (ASTM D6033 – 96, 2002)

This guide is intended to encourage correctness, consistency, and completeness in the description of the functions, capabilities, and limitations of an existing groundwater modelling code through the formulation of a code classification system and the presentation of code description guidelines.

This guide presents a systematic approach to the classification and description of computer codes used in groundwater modelling. Due to the complex nature of fluid flow and biotic and chemical transport in the subsurface, many different types of groundwater modelling codes exist, each having specific capabilities and limitations. Determining the most appropriate code for a particular application requires a thorough analysis of the problem at hand and the required and available resources, as well as a detailed description of the functionality of potentially applicable codes.

Typically, groundwater modelling codes are non-parameterized mathematical descriptions of the causal relationships among selected components of the aqueous subsurface and the chemical and biological processes taking place in these systems. Many of these codes focus on the presence and movement of water, dissolved chemical species and biota, either under fully or partially saturated conditions, or a combination of these conditions. Other codes handle the joint movement of water and other fluids, either as a gas or a non-aqueous phase liquid, or both, and the complex phase transfers that might take place between them. Some codes handle interactions between the aqueous subsurface (for example, a groundwater system) and other components of the hydrologic system or with non-aqueous components of the environment.

The classification protocol is based on an analysis of the major function groups present in groundwater modelling codes. Additional code functions and features may be identified in determining the functionality of a code. A complete description of a code's functionality contains the details necessary to understand the capabilities and potential use of a groundwater modelling code. Tables are provided with explanations and examples of functions and function groups for selected types of codes. Consistent use of the descriptions provided in the classification protocol and elaborate functionality analysis form the basis for efficient code selection.

Although groundwater modelling codes exist for simulation of many different groundwater systems, one may encounter situations in which no existing code is applicable. In those cases, the systematic description of modelling needs may be based on the methodology presented in this guide. This guide is one of a series of guides on groundwater modelling codes and their applications, such as Guides D 5447, D 5490, D 5609, D 5610, D 5611, and D 5718.

Complete adherence to this guide may not be feasible. For example, research developments may result in new types of codes not yet described in this guide. In any case, code documentation should contain a section containing a complete description of a code's functions, features, and capabilities.

This guide offers an organized collection of information or a series of options and does not recommend a specific course of action. This document cannot replace education or experience and should be used in conjunction with professional judgment. Not all aspects of this guide may be applicable in all circumstances. This ASTM standard is not intended to represent or replace the standard of care by which the adequacy of a given professional service must be judged, nor should this document be applied without consideration of a project's many unique aspects. The word "Standard" in the title of this document means only that the document has been approved through the ASTM consensus process.

*Other ASTM Guides*

| | |
|---|---|
| D5447 | Guide for Application of a Ground-Water Flow Model to a Site-Specific Problem |
| D5490 | Guide for Comparing Ground-Water Flow Model Simulations to Site-Specific Information |
| D5609 | Guide for Defining Boundary Conditions in Ground-Water Flow Modelling |
| D5610 | Guide for Defining Initial Conditions in Ground-Water Flow Modelling |
| D5611 | Guide for Conducting a Sensitivity Analysis for a Ground-Water Flow Model Application |

D5718      Guide for Documenting a Ground-Water Flow Model Application
D653       Terminology Relating to Soil, Rock, and Contained Fluids

## Selection of Groundwater Modelling Software

The model code should be selected provisionally at the end of the scoping study and definitely by the end of the conceptual modelling. Two fundamental areas need to be considered in selecting a model code; is it appropriate?, is it useable?

*Is it appropriate?*

During the scoping study the current ideas about how the aquifer is behaving will be gathered and throughout the conceptual modelling these hypotheses will be challenged and developed. We need to consider the modelling objectives (the questions which the model is being asked to address) and the key flow mechanisms identified during conceptual modelling and ask:

1. Can the code represent the key flow mechanisms? For example, a two dimensional single layer model will be inappropriate if vertical groundwater head gradients are known to be significant.

2. What simplifications are required to represent these key mechanisms?

3. Has the code been tested for similar problems to ours?

4. Can the code perform the kind of predictive runs that the modelling objectives will require?

5. Can the model be readily updated as our conceptual understanding grows?

*Is it useable?*

Since a regional groundwater model represents a significant capital investment and is likely to be used and updated for many years, it will probably be used by several in-house staff in addition to the people who develop it. Therefore, we should also ask:

1. How much effort is required to become familiar with the code?

2. Do we have access to the source code so that the way in which the calculations are performed can be investigated?

3. How good is the user manual? (Does it adequately explain how the code works? Are the input instructions correct? Are there examples data sets?)

4. How good is the user support (speed of response and technical content)?

5. Is there a good and well-tested user interface?

6. How good is the quality control of the code and its user interface?

7. Does the user interface allow digital data to be imported or exported in variety of common formats?

*Graphical user interface (GUI) selection*

A number of the codes have a graphical user interface (GUI) which aids in the creation of input files for the model code to read and for visualizing the model output. The AQUA, ICMM and MIKE-SHE codes are supplied with their own proprietary GUIs while for MODFLOW there are a number of different interfaces available, including Processing MODFLOW, Visual MODFLOW and Groundwater Vistas (GV).

Experience shows that although user interfaces are helpful in creating the input files containing the spatial distributions of aquifer geometry and hydraulic parameters, they have limitations, which must be borne in mind. These include a lack of flexibility in creating time variant data (e.g. recharge and abstractions) for which spreadsheets or utilities written in FORTRAN or Visual Basic are often more powerful. In addition, it is not uncommon for a model code to be able to perform functions which the interface does not support.

Finally, the rapid development of these user interfaces has too frequently been at the expense of rigorous quality control. Consequently, the necessity for a model user to be familiar with and constantly check the ASCII data input and output files remains as important as ever.

MODFLOW GUIs are generally supplied with a compiled (executable) version of MODFLOW-96 which may contain minor enhancements (e.g. MODFLOWwin32 with Groundwater Vistas GUI)

The modular finite-difference groundwater flow model (MODFLOW) was developed by the U.S. Geological Survey (USGS) and first released in 1988 (McDonald & Harbaugh, 1988). It has evolved continually and is now the most widely used program in the world for representing groundwater flow. It reads ASCII files containing the input data (recharge, abstraction, aquifer parameters, boundary conditions etc.) and performs calculations in order to output heads and flows.

MODFLOW's success as a modelling tool owes much to its original design. It was written in the early 1980s as a modular code to replace the 500 or so pieces of groundwater modelling software which were scattered over the USGS's mainframes. It had to be accurate, easy to understand, easy to enhance and modify and computationally efficient. MODFLOW is well documented, logically programmed but has never pretended to do everything. As people have wanted to simulate processes of which MODFLOW was incapable, they have written their own modifications. MODFLOW is continuously under development. Details of the latest version can be downloaded from http://water.usgs.gov/nrp/gwsoftware/modflow.html.

A number of proprietary enhancements of the basic public domain MODFLOW code are available. These codes have been written to make significant changes to the original public domain code. In contrast to the public domain versions, the source code for these proprietary codes is not publicly available.

## Classification of Groundwater Software

Frequently used groundwater software in various categories have been listed below. Some of the software belong to more than one category. Considering the large variability and the quick development of groundwater models, a new, more sophisticated model can often replace a previously applied model. Additionally, the reconsideration of the conceptual model and the regeneration of the mesh may need a new allocation of the parameters. Therefore, it is important that model data (information) are stored independently from a given model, with a preference for GIS-based databases. Considerable development in the field of user-friendly GIS and data base servers makes the set-up and the modification of models easier and more time-effective. One such model is FEFLOW which incorporates mathematical modelling with GIS-based data exchange interfaces.

The input data for a groundwater model include natural and artificial stress, and parameters, dimensions, and physico-chemical properties of all aquifers considered in the model. A finer level of detail of the numerical approximation (solution) greatly increases the data requirements. Input data for aquifers are common values such as transmissivities, aquitard resistances, abstraction rates, groundwater recharges, surface water levels etc. The most common output data are groundwater levels, fluxes, velocities and changes in these parameters due to stress put into the model.

### Aquifer and Pump Testing

*Aquiferwin32* - Pumping test analysis, slug test analysis and step test analysis plus analytical groundwater flow modelling and pumping test simulations

*AquiferTest Pro* - Provides graphical analysis and reporting of pumping test and slug data for Confined Aquifers, Unconfined Aquifers, Leaky Aquifers, and Fractured Rock Aquifers

*Infinite Extent* - Pump test analysis software

### Groundwater Databases

*ChemGraph* - ChemGraph provides a very efficient system for entering, storing, and reporting environmental and groundwater monitoring data.

*Enviro-Base Pro* - EnviroBase Pro is a user-customizable database of soil and chemical properties, drinking water standards, adsorption, dispersivity, and chemical half-life information.

*EQuIS Geology/EQuIS Chemistry* - EQuIS Geology/EQuIS Chemistry provide advanced database management system for environmental geology and chemistry data.

*HydroGeo Analyst* - HydroGeo Analyst is the complete solution for groundwater and borehole data management and visualization technology.

## Geochemical Software

*AquaChem* - AquaChem is a powerful aqueous geochemistry package for graphical and numerical analysis and modelling of water quality data, featuring fully customizable database of physical and chemical parameters.

*Hydrogeochem* - Hydrogeochem is a coupled model of hydrologic transport and geochemical reaction in saturated-unsaturated media.

*MINETQA2* for Windows is a Windows version of the popular EPA geochemical equilibrium speciation model capable of computing equilibria among the dissolved, adsorbed, solid, and gas phases

## Public Domain Software

*BALANCE* - BALANCE is a USGS computer program for calculating mass transfer for geochemical reactions in ground water.

*GEOPACK* - A comprehensive U.S. EPA geostatistical software system for conducting analysis of the spatial variability of one or more random functions.

*HST3D* - A Computer Code for Simulation of Heat and Solute Transport in Three-Dimensional Ground-Water Flow Systems

*MOC* - MOC simulates 2D solute transport in flowing ground water

*MOCDENSE* - MOCDENSE is a variable density groundwater flow and solute-transport model. This version of MOCDENSE simulates the flow in a cross-sectional plane rather than in an areal plane.

*MODFLOW* - USGS Modular Three-Dimensional Ground-Water Flow Model

*PEST* - non-linear parameter estimation software for any numerical model.

*PESTAN* - Pesticide Transport is a U.S. EPA program for evaluating the transport of organic solutes through the vadose zone to ground water.

*PHREEQE* - PHREEQE is a USGS computer program designed to model geochemical reactions.

*PLOTCHEM* - A Water-quality data plotting software package

*RECESS* - RECESS is a group of six computer programs (RECESS, RORA, PART, TRANS, CURV and STREAM) for describing the Recession of Ground-Water Discharge and for Estimating Mean Groundwater Recharge and Discharge from Streamflow Records.

*RETC* - Retention Curve Program for Unsaturated Soils

*VLEACH* - VLEACH, a One-Dimensional Finite-Difference Vadose Zone Leaching Model, is a U.S. EPA program which describes the movement of an organic contaminant within and between three phases: (1) as a solute dissolved in water, (2) as a gas in the vapor phase, and (3) as an adsorbed compound in the solid phase.

*VS2DT* - VS2DT is a USGS program for flow and solute transport in variably-saturated, single-phase flow in porous media. A finite-difference approximation is used in VS2DT to solve the advection-dispersion equation.

**USGS Ground Water Software**

*AQTESTSS* (Win) Version 1.2, 2004/07/02
Several spreadsheets for the analysis of aquifer-test and slug-test data

*GSFLOW* (Linux/Win) Version 1.1.6, 2013/03/28
Coupled Groundwater and Surface-water FLOW model based on the USGS Precipitation-Runoff Modeling System (PRMS) and Modular Groundwater Flow Model (MODFLOW-2005)

*GW_Chart* (Windows) Version 1.27.0.0, 2014/03/06
A program for creating specialized graphs used in groundwater studies.

*HST3D* (Unix/Linux/Win) Version 2.2.16, 2010/10/21
Three-dimensional flow, heat, and solute transport model

*HYDROTHERM* (Linux/Win) Version 3.1.0, 2008/03/01
Three-dimensional finite-difference model to simulate multiphase groundwater flow and heat transport in the temperature range of 0 to 1,200 degrees Celsius

*INFIL3.0* (Win) Version 1.0, 2008/06/19
A grid-based, distributed-parameter watershed model to estimate net infiltration below the root zone

*MF2K-VSF* (Win) Version 1.01, 2006/07/05
Three-Dimentional finite-difference groundwater model (MODFLOW) - 2000 version -with variably saturated flow

*Model Viewer* (Win) Version 1.6, 2012/09/24
A computer program that displays the results of three-dimensional groundwater models

*ModelMuse* (Windows) Version 3.5.0.0, 2015/02/27
A Graphical User Interface for MODFLOW-2005, MODFLOW-LGR, MODFLOW-LGR2, MODFLOW-NWT, MODFLOW-CFP, MODFLOW-OWHM, MT3DMS, SUTRA, PHAST, MODPATH, and ZONEBUDGET

*MODFLOW-2005*, v. 1.11.00
Three-dimensional finite-difference groundwater model. Codes included on this link include MODPATH, RADMOD, and ZONEBUDGET.

*MODFLOW-GUI* (Win) Version 4.35.0.0, 2013/11/29
Graphical Pre- and post-processor for use with the MODFLOW, MOC3D, MF2K-GWT, MT3DMS, MODPATH, and ZONEBUDGET models, for use within Argus ONE

*MODFLOW-NWT* (WIN) Version 1.0.9, 2014/06/24
A Newton-Raphson formulation for MODFLOW-2005 to improve solution of unconfined groundwater-flow problems

*MODFLOW-OWHM* (WIN) Version 1.00.00, 2014/11/05
MODFLOW-based integrated hydrologic flow model for the analysis of human and natural water movement within a supply-and-demand framework

*MODFLOW-USG* (WIN) Version 1.2.0, 2014/03/21
Unstructured grid version of MODFLOW for simulating groundwater flow and tightly coupled processes

*MODOPTIM* (Win) Version 1.0, 2005/05/05
A General Optimization Program for Groundwater Flow Model Calibration and Groundwater Management in MODFLOW

*PHAST* (Linux/Win) Version 2.4.6-8538, 2014/03/24
A program for simlulating ground-water flow, solute transport, and multicomponent geochemical reactions

*PULSE* (Win), 2007/01/29
Model-Estimated Groundwater Recharge and Hydrograph of Groundwater Discharge to a Stream

*SEAWAT* Version 4.00.05, 2012/10/19
A computer program for simulation of three-dimensional variable-density groundwater flow and transport

*SHARP* (DOS/DG/SGI/Sun) Version 1.1, 2004/09/03
A quasi-three-dimensional, numerical finite-difference model to simulate freshwater and saltwater flow separated by a sharp interface in layered coastal aquifer systems

*SUTRA* Version 2.2 (Sept 2010)
2D or 3D saturated-unsaturated, variable-density groundwater flow with solute or energy transport

*Utility_PIEs* (Windows) Version 1.10.1.0, 2014/03/06
Programs for simplifying the analysis of geographic information in U.S. Geological Survey ground-water models for use within Argus Open Numerical Environments (Argus ONE)

*VS2DT* (Sun/Win) Version 3.2, 2004/10/18
Model for simulating water flow and solute transport in variably saturated porous media

*WTAQ* (Win) Version 2.1, 2012/06/29
A computer program for calculating drawdowns and estimating hydraulic properties for confined and water-table aquifers

## MODFLOW-based Software

*GMS (Groundwater Modelling System)* - GMS supports both finite-difference and finite-element groundwater models in 2D and 3D including MODFLOW 2000, MODPATH, MT3DMS/RT3D, SEAM3D, ART3D, UTCHEM, FEMWATER, PEST, UTEXAS, MODAEM and SEEP2D.

*Groundwater Vistas* - Complete groundwater model design and analysis for MODFLOW, MODFLOW-SURFACT, MODPATH, MT3D and PEST. Now available with GW3D visualization and Remote Model Launch.

*MODFLOW SURFACT* - MODFLOW-based groundwater flow and contaminant transport model

*MODPATH* - A Particle Tracking Post-Processing Package for MODFLOW, the USGS 3-D Finite-Difference Ground-Water Flow Model

*MODTECH* - ModTech is a new 3D groundwater flow and mass transport software that enables you to model groundwater flow and contaminant transport within a real-world geographic (GIS) environment.

*PMWIN (Processing MODFLOW for Windows)* - PMWIN is a graphical interface for MODFLOW, MODPATH, PMPATH, MT3D, RT3D, MOC3D, PEST and UCODE

*Visual MODFLOW* - Visual MODFLOW Premium is the proven standard for professional 3D groundwater flow and contaminant transport modelling using MODFLOW-2000, MODPATH, MT3DMS and RT3D.

## Density-Dependent Flow

*FEFLOW* - 2D/3D finite element subsurface flow system - model for density dependent groundwater flow, heat flow and contaminant transport

*GMS (Groundwater Modelling System)* - GMS supports both finite-difference and finite-element groundwater models in 2D and 3D including MODFLOW 2000, MODPATH, MT3DMS/RT3D, SEAM3D, ART3D, UTCHEM, FEMWATER, PEST, UTEXAS, MODAEM and SEEP2D.

*MOCDENSE* – It is a variable density groundwater flow and solute-transport model. This version of MOCDENSE simulates the flow in a cross-sectional plane rather than in an areal plane.

*SEAWAT* - A computer program for simulation of three-dimensional variable-density ground water flow

*SUTRA* – It is a 2D/3D ground water saturated-unsaturated transport model, a complete saltwater intrusion and energy transport model. SUTRA is integrated with Argus ONE for graphical pre and post processing.

## Unsaturated Flow & Transport

*BIOF&T* - BIOF&T models biodegradation and bioremediation, flow and transport in the sat/unsat zones in 2 or 3 dimensions in heterogeneous, anisotropic porous media or fractured media

*ChemFlux* - ChemFlux is a contaminant transport software modelling package for modelling of mass transport, contaminant concentrations and plume migration.

*FEFLOW* - 2D/3D finite element subsurface flow system - model for density dependent groundwater flow, heat flow and contaminant transport

*POLLUTE* - Layered contaminant transport analysis from sources such as landfills, buried waste, spills, and disposal ponds.

*SUTRA* - SUTRA is a 2D/3D ground water saturated-unsaturated transport model, a complete saltwater intrusion and energy transport model. SUTRA is integrated with Argus ONE for graphical pre and post processing.

*SVFlux 2D/3D* - Finite element seepage analysis software. Perfomr 2D and 3D flow analysis in unsaturated or saturated soil.

*Visual HELP for Windows* - The most complete modelling environment for the HELP model for evaluating and optimizing landfill designs!

*WHI UnSat Suite* - A compilation of the most popular 1-D groundwater unsaturated zone models including VLEACH, VS2DT, PESTAN and HELP.

## Fracture Flow Software

*BIOF&T* - BIOF&T models biodegradation and bioremediation, flow and transport in the sat/unsat zones in 2 or 3 dimensions in heterogeneous, anisotropic porous media or fractured media

*FEFLOW* - 2D/3D finite element subsurface flow system - model for density dependent groundwater flow, heat flow and contaminant transport

*FRAC3DVS* - Three-dimensional, steady-state or transient, saturated or unsaturated flow in fractured or unfractured porous media.

*SWIFT* - Three-dimensional transient flow in fractured or unfractured, anisotropic, heterogeneous porous media. Viscosity dependency as a function of temperature and brine concentrations.

**Review of Selected Groundwater Software**

*AQTESOLV*

AQTESOLV by HydroSOLVE Inc. is software designed to calculate hydraulic conductivity, storativity and other aquifer properties from data sets collected during slug and aquifer (pumping) tests.

AQTESOLV is user friendly. It can be mastered using the tutorial and help file. Most students pick up the software quickly. The Slug Test and Pumping Tests Wizards do a good job of walking the user through the input of data. A useful feature is the basic pictures on each input screen that show what each variable represents.

AQTESOLV can import text files generated by commonly used pressure transducers. Also, data can be manually entered or pasted from a spreadsheet. It is easy to change the input values once you have entered them, and to switch between English and SI units. After importing, the raw data can be manipulated using mathematical functions. For example, hydraulic head data can be converted to drawdown data.

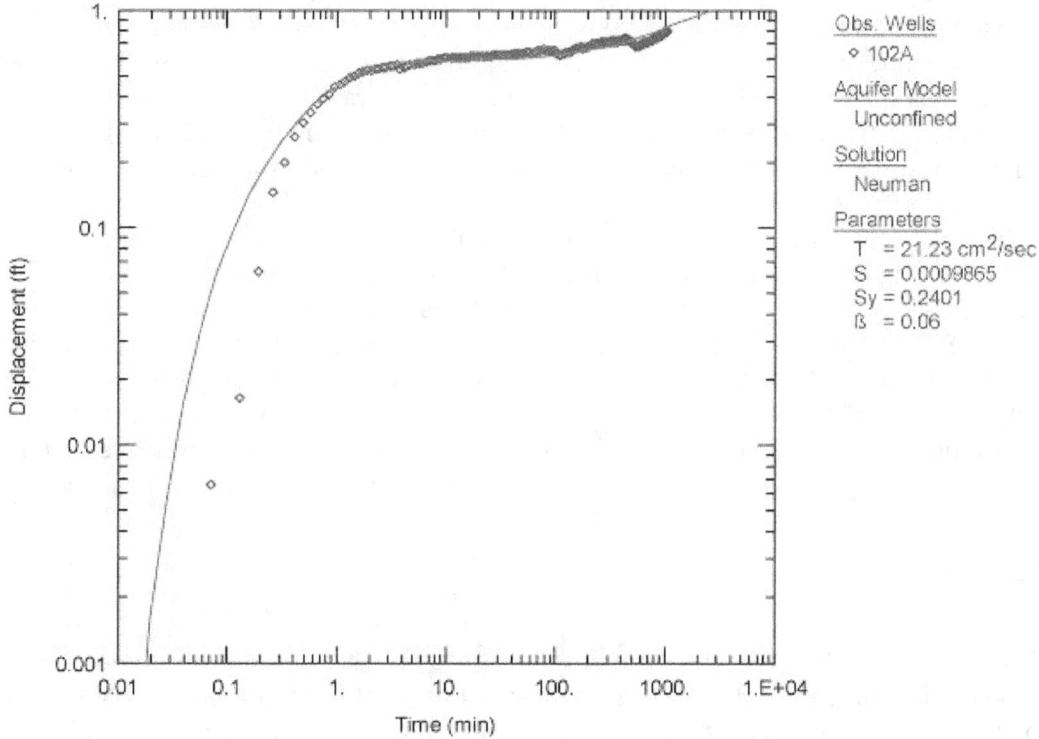

Once the data are entered, the software offers a variety of solutions. This is where user-knowledge is important. AQTESOLV gives little guidance on selecting the appropriate solution for the data and hydrogeologic setting and the user is referred to the relevant literature for details on each solution. The user must also know how to correctly display the data (e.g. on linear or log scales), and how to transform the raw data into the form used by each solution. The software provides an automated matching feature, but the automated match is usually poor, and manual fitting of the solution lines to the data is recommended.

Incorporating output from AQTESOLV into a presentation or report is functional, but not fancy. A hard copy of the graphed data, your best fit line, and calculated parameters can be sent to a printer, or, exported as a Windows metafile. Overall, the software is easy to use, offers a variety of solutions, creates presentation-quality output, and saves time when compared to performing the analysis by hand.

AQTESOLV can be downloaded at (http://www.aqtesolv.com/). Depending on choices such as single-user or site license, standard or professional version, and commercial or academic application, prices vary from $500-1500.

## AQUACHEM

AquaChem manages water quality data, allowing the user to plot data, perform common statistical analyses, model water chemistry with PHREEQC, and create simple reports. AquaChem is compatible with Microsoft (MS) Office; the database created by AquaChem can be opened and modified in MS Access and data can be imported to AquaChem from text formatted and MS Excel files.

The most useful feature is the ability to plot data quickly. Map plots, displaying water quality data in diagrams using X and Y coordinates for station location, can be created and AutoCAD or common graphics files can be used as base map. Also, chemistry of water from different wells, or groups of wells, can be compared in various diagrams such as ternary, Piper, and Stiff. Saturation indices can be calculated with a built-in basic version of PHREEQC and data files can be constructed in AquaChem for input to PHREEQC for more rigorous modelling.

Statistical tools include trend analysis (Mann-Kendall, linear), normality tests, and evaluation of correlation between parameters. Although, database management and report preparation may be as easily performed in Excel and Access, the statistical tools in AquaChem and the ease and speed of creating diagrams used in aquatic chemistry allows the user to quickly analyze and present data. AquaChem is not a substitute for more advanced geochemical modelling programs but is a useful tool for interpreting data and discovering trends in the data. Commercial Standalone License for AquaChem is available at http://www.swstechnology.com/groundwater-software/water-quality-analysis/aquachem for $1595.

## GFLOW

GFLOW is an analytic element code that belongs in the arsenal of commonly used groundwater modelling tools. It solves groundwater flow problems by superimposing analytic solutions under Dupuit-Forchheimer assumptions. In its most simple dress the code facilitates rapid construction and solution of a steady-state single-layer groundwater model containing multiple recharge and transmissivity zones with complicated internal boundary conditions. However, one of its most innovative aspects is its ability to perform conjunctive surface water/groundwater modelling.

GFLOW uses stream networks to route baseflow so that available streamflow responds to stresses such as pumping, an option that is particularly useful in determining the effect of wells on the headwaters of streams. It has several ways of simulating the interaction of groundwater with lakes, including the ability to perform detailed water balances in the presence of through-flowing streams, a capacity not yet available in other analytic element codes or MODFLOW. Together these options make it particularly useful in modelling unconfined flow in terrains (e.g. glacial) with closely spaced surface-water bodies.

A second innovation of the code is to facilitate use of the "stepwise" modelling approach. Often Dupuit-Forchheimer assumptions are more valid for regional flow problems than for site-specific problems where vertical head gradients are important. To allow the speed of analytic elements to be mated with the flexibility of finite-differences, GFLOW supports automatic conversion of part of its domain to a local MODFLOW model with flux or constant head boundary conditions around the inset grid. Once the conversion is made, the user can add more complexity to geometry and properties inside the MODFLOW model.

Another advanced feature of the code is particle tracking that takes advantage of mass balance considerations to simulate flow lines from the water table below a stream to a well. There are "tricks" that are needed to most effectively perform analytic element modelling; the ample HELP material introduces the user to the technique's special vocabulary (e.g., "linesinks") and also provides a general guide to modelling, which makes the code a first-rate teaching tool. The GUI, compatible with all versions of Windowsâ, is workmanlike with excellent visualization tools related to surface water and calibration. A lot of thought has gone into converting readily available maps such as USGS coverages into base maps. GFLOW is available for $475 from http://www.haitjema.com/ .

*GMS (Groundwater Modelling System)*

GMS (Groundwater Modelling System) is a comprehensive software package for developing computer simulations of groundwater problems (flow and contaminant transport). The Environmental Modelling Research Laboratory at Brigham Young University in Utah oversees the continued development of GMS, but GMS is distributed commercially through a variety of vendors.

GMS provides tools for site characterization, model development, post-processing, calibration, and visualization. It supports TINs, solids, borehole data, 2D and 3D geostatistics, and both finite element and finite difference models in 2D and 3D, including MODFLOW, MODPATH, MT3D, RT3D, FEMWATER, SEEP2D, ART3D, MODAEM, SEAM3D, and UTCHEM. Parameter estimation is supported through the processes included with MODFLOW, PEST, and UCODE.

GMS's design, allows the user to select "modules" in custom combinations. These modules are related to pre- and post-processing, model selection, and calibration routines. Discussion of all modules is beyond the scope of this review, but many commercial vendors have websites with detailed descriptions of each module and how they can be used. New users can select specific modules for a current project and add modules as needs expand or change. Use of GMS saves time by having a modelling "package" that is the same for a variety of programs. The initial learning curve is steep. However, GMS uses the same "conceptual model" approach (discussed below) for alternative modelling programs (e.g., MODFLOW and FEMWATER).

A "conceptual model" approach is built into the map/GIS module. Here, a user defines the model properties (i.e., boundary conditions, hydraulic conductivity values, 2D and 3D model domain, etc.) in a GIS interface that is independent of the simulation codes. With a click of a button, model properties are transferred from the GIS interface to the appropriate grid cells or mesh elements. The advantage of this approach is the ability to quickly change a conceptual model, transfer these conditions to a new simulation, and evaluate the results. In addition, GMS allows more advanced users to easily see and edit the grid by grid values, without going to the final text file (although a quick check of the text files is always a good idea). GMS 6.0 can be fully integrated with ArcGIS if the user also has an ArcGIS license, allowing for a seamless transfer in creating conceptual models and producing final figures of modelling results.

In summary, GMS is a useful all around groundwater modelling package that offers the advantages of modular purchases, multiple model support, linkages to ArcGIS, conceptual model development, and integrated inversion routines. As with many software programs, new releases of GMS should be used cautiously with the understanding that some of the latest features may require revision before the full range of capabilities are functional.

Latest version GMS 10.0 can be purchased at http://www.aquaveo.com/software/gms-pricing. There are four commercial versions available - MODFLOW I for $1800, MODFLOW II for $3950, MODFLOW III for $6150 and Premium for $11350.

*HYDRUS*

The HYDRUS software package (including the much anticipated three-dimensional capability) is a major upgrade and extension of the HYDRUS-2D/MESHGEN-2D software package (originally developed and released by U.S. Salinity Laboratory, PC-Progress and the International Groundwater Modelling Center). HYDRUS is a Microsoft Windows based modelling environment for analysis of water flow, solute, and heat transport in variably saturated porous media.

For flow, the code solves the mixed form of the Richards' equation with many functions for simulating hydraulic conductivity versus water content (or pressure head) relationships, including hysteresis. It allows root water uptake with compensation; spatial root distribution functions; and includes new soil hydraulic property models. The new code allows for dynamic, system-dependent boundary conditions (e.g., switching between pressure heads, seepage face, zero flux or atmospheric boundary depending on the position of the water level).

For solute transport, the code solves the advection dispersion equation, but with many processes not usually include in unsaturated zone codes. For example, in addition to the typical linear partitioning between soil, water, and gas phases, the code also simulates the following processes: non-linear and non-equilibrium partitioning between phases; diffusion in the gas phase; zero- and first-order degradation kinetics; including decay chains (e.g., for nitrates and radionuclides); adjective flow in a dual-porosity system allowing for preferential flow in fractures or macropores while storing water and dissolved chemicals in the matrix; transport of viruses, colloids, and/or bacteria using an attachment/detachment model;

filtration theory; and blocking functions; and flowing particles in two-dimensional applications. Hydrus also includes a new constructed wetland module (only in 2D).

HYDRUS still allows optimization in 1D and 2D. Unfortunately, optimization is not provided for 3D applications. Another useful new feature is better print management (printing at regular time intervals or after a constant number of time steps).

The GUI is much improved. It includes many new functions improving the user-friendliness, such as drag-and-drop, context sensitive pop-up menus after clicking on objects, selection and editing of multiple objects in the same dialog window, and allowing multiple projects and views to be opened at the same time in the HYDRUS main window. One of the best new features is that time-varying and cumulative fluxes can be calculated and displayed across internal mesh lines.

The only negatives associated with the program are that the User's Guide could be more complete. Not all features are explained well enough for a modeler without HYDRUS experience to easily follow. However, the on-line discussion forum is very helpful for those who take advantage of it. The forum can be found at http://www.pc-progress.cz/_Forum/default.asp

HYDRUS pricing is based on HYDRUS Editions (sometimes also called Levels), that reflect the complexity of the computational domain. Currently there are five levels: 2D-Lite, 2D-Standard, 3D-Lite, 3D-Standard and 3D-Professional. Higher levels always cover all functionalities of lower levels. Price for Single License of 2D-Lite (two-dimensional applications for simple rectangular 2D geometries) is about $650 while price for for Single License of 3D-Professional (three-dimensional applications for general 3D geometries) is about $2650.

The original, public-domain version of HYDRUS1D is included in the HYDRUS package and may still be downloaded for free from the web site http://www.pc-progress.com/en/Default.aspx?H1d-downloads. HYDRUS may be purchased by visiting http://www.pc-progress.com/en/Default.aspx.

*MIKE SHE*

The MIKE SHE code is a powerful, physically-based, distributed parameter, fully integrated code for three-dimensional simulation of hydrologic systems. It has been successfully applied at multiple scales, using spatially distributed, continuous climate data to simulate a broad range of integrated hydrologic, hydraulic and transport problems in humid as well as more arid areas. For example, in the U.S., it has been used extensively in South Florida on Everglades Restoration projects, in Colorado, at Rocky Flats (a former DOE nuclear manufacturing facility) and in the Black Mesa basin of northeastern Arizona.

MIKE SHE was developed by DHI Water and Environment (http://www.dhi.dk/). It was originally developed as 'SHE' along with the British Institute of Hydrology and SOGREAH (France). In the 2003 version, all major hydrologic flow processes are dynamically coupled, including 2-D overland flow, 1-D channel flow, 3-D saturated zone flow, 1-D (Richard's based) unsaturated zone flow, snowmelt and evapotranspiration. Overland flow utilizes

DEMs, while channel flow is simulated through the MIKE 11 code, which has extensive capabilities including, user-defined regulating structures, water quality and sediment transport, and a morphological module. It is also capable of simulating integrated advective-dispersive transport, sorption, biodegradation, geochemistry (including PHREEQC), and macropore flow, generally applicable for most hydrologic, water resources and contaminant transport applications. In contrast to similar codes, MIKE SHE utilizes rigorous physical flow equations for all major flow processes, but also permits more simplified descriptions.

Though sophisticated and flexible, its ability to simulate evapotranspiration and stream-aquifer interactions could be improved. Verification emphasized that development and calibration requires a systematic approach and careful consideration of scale-dependent integrated processes. Developing more complex, larger-scale integrated models requires specialized skills and a substantial investment in time.

The graphical interface is significantly improved, offering a dynamic navigation tree, dynamic dialogs, and limited on-line documentation, and notably improved output animation capabilities. It seamlessly links with Arcview shape files and has well-organized spreadsheet-graphical functionality for ease in editing spatial and temporal input.

DHI offers the full version *MIKE SHE Enterprise* for about \$17,800 though simpler version (*MIKE SHE Studio*) is less expensive (about \$10500) and specialized add-on modules increase the price.

*MODFLOW*

MODFLOW-2005 is the most recent version of the U.S. Geological Survey (USGS) public domain MODFLOW software for simulating flow in saturated porous media in three dimensions with a finite difference grid. Its popularity has continued, in part due to the modularity of the program and the resulting ability of USGS and others to add capabilities.

MODFLOW includes observation, sensitivity and parameter-estimation options with a convenient "Parameter" approach that facilitates model setup. A wide range of conditions can be represented using "Packages," including many recently-released new packages: The Hydrogeologic-Unit Flow (HUF) Package facilitates connection with hydrogeologic framework models such as EarthVision by populating the model grid with hydraulic parameters using defined hydrogeologic units. The Multi-Node Well (MNW) Package distributes flow in wells that intersect multiple model layers (for horizontal or angled wells, multiple nodes) and accounts for reduced pumpage caused by drawdown at the pumping well. The Geometric Multi-Grid (GMG) Package is efficient for solving large problems and is the first part of MODFLOW to be written in C rather than FORTRAN. The StreamFlow-Routing (SFR, replaces STR) Package routes surface water to streams and lakes, adjusting the river stage as stream discharge changes, such that flux to and from the groundwater system adjusts to the changing stage. The Subsidence and Aquifer-System Compaction (SUB3) Package simulates elastic compaction and expansion, and inelastic compaction of compressible fine-grained beds within the aquifer. Unsaturated Flow Packages are due out soon.

Some simple pre- and post-processing programs for MODFLOW are available from the USGS. These include GW_CHART, a program for evaluating model fit to observations and sensitivity analysis; and ModelViewer, a program to contour simulated heads. More powerful capabilities are available in a variety of interfaces that require purchase of commercial products. These can be found by searching for "MODFLOW interface" using a Web search engine. These programs facilitate construction of input files and assessment of results, but they do not support the latest MODFLOW capabilities, and errors in user understanding or bugs in the interfaces can yield misleading results.

Familiarity with the MODFLOW text files is important for trouble shooting, for using new MODFLOW features, and for dealing with unusual modelling situations. The new online guide of the USGS MODFLOW Web page provides a quick way to look up options and input formats. Download documentation, source code, and executables of MF2K and associated programs at no charge.

*MT3DMS*

MT3DMS is a public domain solute transport model for saturated porous media. In its current version it handles three-dimensional advective-dispersive transport of multiple chemical species and a range of simple chemical reactions. Moreover it serves also as the "engine" to compute advective-dispersive transport in other models that have more advanced reaction capabilities and for the variable density simulator SEAWAT. Notable features of MT3DMS that distinguish it from many other solute transport models include the availability of several different numerical solution techniques for advective transport and the variety of boundary conditions and features that are needed to build realistic models. In particular the particle-tracking based solution schemes can produce results that are essentially free of numerical dispersion in the absence of excessive grid refinement - a common problem among many other models.

Since its initial release as the single species simulator in 1990 MT3D, it has undoubtedly become the workhorse of computational contaminant hydrology. Of course, the most obvious reason for this is its full compatibility with MODFLOW. Other important reasons for its popularity, among practitioners and researchers, are its robustness, which allows new users to quickly come up with accurate results, and its very good documentation which includes the availability of its source code. The source code of MT3DMS is written in Fortran using a MODFLOW-type easy to follow modular structure.

With MT3DMS, the transport of multiple species can be carried out simultaneously, whereby each species can be subject to sorption, either using linear, Freundlich or Langmuir isotherms, and/or first-order degradation reactions. This limits the reaction capabilities to cases where the reaction rates of each species can be assumed not to depend significantly on concentrations of other chemical species. That feature can be used, for example, to calibrate a (nonreactive) transport model (see Figures) by simultaneously carrying out simulations for several nonreactive species (or species that can in a particular case study be considered as being inert). MT3DMS capabilities also include a dual domain transport formulation that might be used for simulations of fractured or highly heterogeneous aquifers. MT3DMS, like MODFLOW, does not come with its own graphical user interface and thus its ease of use ultimately depends on the GUI selected for processing data input/output. However, as MT3DMS is supported by essentially all commonly used MODFLOW GUIs, users have a good choice among several products. MT3DMS can be downloaded for free from http://hydro.geo.ua.edu/mt3d/index.htm. The page contains also links to several GUI suppliers.

## ROCKWORKS

RockWorks 16, the latest version of subsurface visualization software from the Golden, Colorado-based RockWare, Inc., provides a large array of tools useful for site characterization in many earth science disciplines. For geologists, hydrogeologists, and geophysicists involved in water-resources investigations, the software package is particularly applicable for developing hydrostratigraphic models that facilitate construction of groundwater flow and transport models.

RockWorks provides centralized borehole database management in the Borehole Data Manager window. From this interface, borehole data, including lithology, stratigraphic contacts,geophysical data, geochemical measurements, fracture information, and groundwater levels can be used to create maps, cross sections, fence diagrams, single or multi-log plots, as well as three-dimensional surface and solid models. Other capabilities useful to water resources investigations are available in the RockWare Utilities window. This interface facilitates creation of contour maps of water levels and drawdowns, as well as Piper and Stiff diagrams. Utilities include statistical analyses andcontrol of display options.

In addition to visualization of the subsurface for conceptual model development, RockWorks provides useful ways to create grids that can be used for modelling. For example, solids models created in RockWare can be exported as ASCII XYZ files, which can then be imported intovarious ground water modelling interfaces for defining model units and layers. Waterresource investigators will like the new Well Construction interface. This addition allows for detailed well construction information to be stored in adatabase format. With this data, RockWorks can be used to plot well construction diagrams alongside 2-D and 3-D strip-logs, allowing the user to identify the formation or lithology accessed by the screened intervals. Well features such as screen depth and length, casing length and construction materials can be included in the well construction diagram.

Several enhancements have been made in RockWorks. One of the major enhancements is the Access database format now implemented for storage and management of borehole information. This new format increases the user's ability to query borehole information, and link data from various tables within the database. Checks on data integrity within the database have been improved (for example, checks on consistent layer elevations). The Access format in RockWorks does not require Microsoft Access to be installed for operation of the software. The option to import and manipulate data in a spreadsheet format is still available. The tool layout for creating borehole strip-logs has been improved and the edit capability allows for editing of all graphics created in RockWorks. In addition to the export capabilities for grids, RockWorks now has more user-friendly options for exporting data to AutoCad and ArcGIS programs. RockWorks comes with an extensive help menu and several tutorials that help introduce the new user to the major features available with the software. The RockWare website (http://www.rockware.com/) provides further options for technical support. The software can be purchased online for $700 at www.rockware.com, or via telephone order at (800) 775-6745.

# References

1.  Anderson, M.P. and W.W. Woessner (1992). Applied Groundwater Modelling. Academic Press, Inc., San Diego, CA., 381 p.

2.  American Society for Testing and Materials (1993). Standard Guide for Application of a Ground-Water Flow Model to a Site-Specific Problem. ASTM Standard D 5447-93, West Conshohocken, PA, 6 p.

3.  American Society for Testing and Materials (1995). Standard Guide for Subsurface Flow and Transport Modelling. ASTM Standard D 5880-95, West Conshohocken, PA, 6 p.

4.  Bear, J., and A. Verruijt (1987). Modelling Groundwater Flow and Pollution. D. Reidel Publishing Company, 414 p.

5.  Franke, O.L., Bennett, G.D., Reilly, T.E., Laney, R.L., Buxton, H.T., and Sun, R.J. (1991). Concepts and Modelling in Ground-Water Hydrology -- A Self-Paced Training Course. U.S. Geological Survey Open-File Report 90-707.

6.  Kashyap, Deepak (1989). Mathematical Modelling for Groundwater Management – Status in India. Indo-French Seminar on Management of Water Resources, 22-24 September, 1989, Festival of France-1989, Jaipur, pp. IV-59 to IV-75.

7.  Kinzelbach, W. (1986). Groundwater Modelling: An Introduction with Sample Programs in BASIC. Elsevier, New York, 333 p.

8.  Kumar, C. P. (1992). Groundwater Modelling – In. Hydrological Developments in India Since Independence. A Contribution to Hydrological Sciences, National Institute of Hydrology, Roorkee, pp. 235-261.

9.  Kumar, C. P. (2001). Common Ground Water Modelling Errors and Remediation. Journal of Indian Water Resources Society, Volume 21, Number 4, October 2001, pp. 149-156.

10. McDonald, M.G. and A.W. Harbaugh (1988). A Modular Three-Dimensional Finite-Difference Ground-Water Flow Model, USGS TWRI Chapter 6-A1, 586 p.

11. Pinder, G.F., and J.D. Bredehoeft (1968). Application of the Digital Computer for Aquifer Evaluation, Water Resources Research, Vol. 4, pp. 1069-1093.

12. Wang, H.F. and M.P. Anderson (1982). Introduction to Groundwater Modelling. W.H. Freeman and Company, San Francisco, CA, 237 p.

16. Website - Kumar Links to Hydrology Resources:
    http://www.angelfire.com/nh/cpkumar/hydrology.html

17.   Website - Scientific Software Group:
      http://www.scisoftware.com/

18.   Website - USGS Ground-Water Software:
      http://water.usgs.gov/nrp/gwsoftware/

# 13-Overview of Groundwater Models

## Introduction

The use of groundwater models is prevalent in the field of environmental science. Models have been applied to investigate a wide variety of hydrogeologic conditions. More recently, groundwater models are being applied to predict the transport of contaminants for risk evaluation.

In general, models are conceptual descriptions or approximations that describe physical systems using mathematical equations; they are not exact descriptions of physical systems or processes. By mathematically representing a simplified version of a hydrogeological system, reasonable alternative scenarios can be predicted, tested, and compared. The applicability or usefulness of a model depends on how closely the mathematical equations approximate the physical system being modeled. In order to evaluate the applicability or usefulness of a model, it is necessary to have a thorough understanding of the physical system and the assumptions embedded in the derivation of the mathematical equations.

Groundwater models describe the groundwater flow and transport processes using mathematical equations based on certain simplifying assumptions. These assumptions typically involve the direction of flow, geometry of the aquifer, the heterogeneity or anisotropy of sediments or bedrock within the aquifer, the contaminant transport mechanisms and chemical reactions. Because of the simplifying assumptions embedded in the mathematical equations and the many uncertainties in the values of data required by the model, a model must be viewed as an approximation and not an exact duplication of field conditions. Groundwater models, however, even as approximations, are a useful investigation tool that groundwater hydrologists may use for a number of applications.

Application of existing groundwater models include water balance (in terms of water quantity), gaining knowledge about the quantitative aspects of the unsaturated zone, simulating of water flow and chemical migration in the saturated zone including river-groundwater relations, assessing the impact of changes of the groundwater regime on the environment, setting up/optimising monitoring networks, and setting up groundwater protection zones.

The modelling studies in India have so far been confined to academic and research organisations. The practising professionals mostly still prefer to employ lumped models for planning of groundwater development and recharge. Such models completely ignore the distributed character of the groundwater regime. Thus, they are based upon rather conservative concepts like safe yields and are incapable of accounting for the stream-aquifer interaction and the dependence of lateral recharge on the water table pattern. Consequently, permissible mining (i.e. withdrawals in excess of vertical recharge) and perennial yield can not be arrived at. The objectives of modelling studies in India have been mainly (i) groundwater recharge, (ii) dynamic behaviour of the water table, (iii) stream-aquifer interaction, and (iv) sea-water intrusion etc. It is important to understand general aspects of both groundwater flow and transport models so that application or evaluation of these models may be performed correctly.

# Groundwater Models

Groundwater flow modelling studies are required to resolve groundwater and catchment management issues. A model, no matter how sophisticated, will never describe the groundwater system under investigaion without deviation of model simulations from the actual physical processes. As a consequence, in applying a numerical model to a field study, the model user should always understand the implications of simplifying assumptions. Salient features of a number of unsaturated and groundwater models have been presented below.

## 1.   3DFEMFAT

(3-D Finite-Element Model of Flow and Transport through Saturated-Unsaturated Media)

3DFEMFAT is a 3-Dimensional Finite-Element Model of Flow And Transport through Saturated-Unsaturated Media. Typical applications are infiltration, wellhead protection, agriculture pesticides, sanitary landfill, radionuclide disposal sites, hazardous waste disposal sites, density-induced flow and transport, saltwater intrusion, etc. 3DFEMFAT can do simulations of flow only, transport only, combined sequential flow and transport, or coupled density-dependent flow and transport. In comparison to conventional finite-element or finite-difference models, the transport module of 3DFEMFAT offers several advantages: (1) it completely eliminates numerical oscillation due to advection terms, (2) it can be applied to mesh Peclet numbers ranging from 0 to infinity, (3) it can use a very large time step size to greatly reduce numerical diffusion, and (4) the hybrid Lagrangian-Eulerian finite-element approach is always superior to and will never be worse than its corresponding upstream finite-element or finite-difference method. Because of these advantages, 3DFEMFAT is suitable for applications to large field problems. It is flexibile and versatile in modeling a wide range of real-world problems.

## 2.   AQUA3D

(3-D Groundwater Flow and Contaminant Transport Model)

AQUA3D is a program developed to solve three-dimensional groundwater flow and transport problems using the Galerkin finite-element method. AQUA3D solves transient groundwater flow with inhomogeneous and anisotropic flow conditions. Boundary conditions may be prescribed nodal head and prescribed flow as a function of time or head-dependent flow. AQUA3D also solves transient transport of contaminants and heat with convection, decay, adsorption and velocity-dependent dispersion. Boundary conditions may be either prescribed nodal concentration (temperature) or prescribed dispersive mass (heat) flux.

## 3.   AT123D

(Analytical Groundwater Transport Model for Long-Term Pollutant Fate and Migration)

AT123D, analytical, transient One-, Two-, and Three-Dimensional Model, is an analytical groundwater transport model. AT123D computes the spatial-temporal concentration distribution of wastes in the aquifer system and predicts the transient spread of a contaminant plume through a groundwater aquifer. The fate and transport processes accounted for in AT123D are advection, dispersion, adsorption, and decay. AT123D estimates all the above components on a monthly basis for up to 99 years of simulation time. AT123D can be used as an assessment tool to help the user estimate the dissolved concentration of a chemical in three dimensions in groundwater resulting from a mass release over a source area. AT123D can handle: two kinds of source releases – instantaneous, continuous with a constant loading or time-varying releases; three types of waste–radioactive, chemicals, heat; four types of source configurations–a point source, a line source parallel to the x-, y-, z-axis, an area source perpendicular to the z-axis, a volume source; four variations of the aquifer dimensions–finite depth and finite width, finite depth and infinite width, infinite depth and finite width, infinite depth and infinite width.

## 4. BIOF&T 2-D/3-D

(Biodegradation, Flow and Transport in the Saturated/Unsaturated Zones)

BIOF&T 3-D models biodegradation, flow and transport in the saturated and unsaturated zones in two or three dimensions in heterogeneous, anisotropic porous media or fractured media. BIOF&T allows real world modeling not available in similar packages. Model convection, dispersion, diffusion, adsorption, desorption, and microbial processes based on oxygen-limited, anaerobic, first-order, or Monod-type biodegradation kinetics as well as anaerobic or first-order sequential degradation involving multiple daughter species.

## 5. Chemflo

(Simulates Water and Chemical Movement in Unsaturated Soils)

Chemflo is an interactive software system for simulating one-dimensional water and chemical movement in unsaturated soils. Chemflo was developed to enable decision-makers, regulators, policy-makers, scientists, consultants, and students to simulate the movement of water and chemicals in unsaturated soils. Water movement is modeled using Richards equation. Chemical transport is modeled by means of the convection-dispersion equation. These equations are solved numerically for one-dimensional flow and transport using finite differences. Results of Chemflo can be displayed in the form of graphs and tables.

## 6. ChemFlux

(Finite Element Mass Transport Model)

ChemFlux is a stable finite element contaminant transport modeling software. It is a finite element software package characterized by automatic mesh generation, automatic mesh refinement and automatic time-step refinement. The solver offers speed and reduction in convergence problems. Results of benchmark tests run against MT3D confirm the effectiveness of the solver. ChemFlux is able to provide the same level of accuracy as MT3D in solutions dominated by advection while implementing the irregular geometry benefits of the finite element method. ChemFlux can also import groundwater gradients from the SVFlux groundwater modeling package. Predicting the movement of contaminant plumes through the processes of advection, diffusion, adsorption and decay is possible. The

ChemFlux design module provides an elegant and simple user interface. Problem geometry and groundwater gradients may be imported from the SVFlux software.

## 7.    FEFLOW

(Finite Element Subsurface Flow System)

FEFLOW is a finite-element package for simulating 3D and 2D fluid density-coupled flow, contaminant mass (salinity) and heat transport in the subsurface. It is capable of computing:

- Groundwater systems with and without free surfaces (phreatic aquifers, perched water tables, moving meshes);
- Problems in saturated-unsaturated zones;
- Both salinity-dependent and temperature-dependent transport phenomena (thermohaline flows);
- Complex geometric and parametric situations.

The package is fully graphics-based and interactive. Pre-, main- and post-processing are integrated. There is a data interface to GIS (Geographic Information System) and a programming interface. The implemented numerical features allow the solution of large problems. Adaptive techniques are incorporated.

## 8.    FLONET/TRANS

(2-D cross-sectional groundwater flow and contaminant transport modeling)

FLONET/TRANS is a software package for 2-D cross-sectional groundwater flow and contaminant transport modeling. The modeling environment offers all the advantages of finite-element modeling (numerical stability and flexible geometry) together with a logical and intuitive graphical interface that makes finite-element modeling fast and easy. It uses the dual formulation of hydraulic potentials and streamlines to solve the saturated groundwater flow equation and create accurate flownet diagrams for any two-dimensional, saturated groundwater flow system. In addition, it also simulates advective-dispersive contaminant transport problems with spatially-variable retardation and multiple source terms.

## 9.    FLOWPATH

(2-D Groundwater Flow, Remediation, and Wellhead Protection Model)

FLOWPATH for Windows is a popular model for groundwater flow, remediation, and wellhead protection. It is a comprehensive modeling environment specifically designed for simulating 2-D groundwater flow and contaminant transport in unconfined, confined and leaky aquifers with heterogeneous properties, multiple pumping wells and complex boundary conditions. Some typical applications of FLOWPATH include:

- Determining remediation well capture zones
- Delineating wellhead protection areas
- Designing and optimizing pumping well locations for dewatering projects

- Determining contaminant fate and exposure pathways for risk assessment

## 10.  GFLOW

(Analytic Element Model with Conjunctive Surface Water and Groundwater Flow and a MODFLOW Model Extract Feature)

GFLOW is an efficient stepwise groundwater flow modeling system. It is a Windows program based on the analytic element method. It models steady-state flow in a single heterogeneous aquifer using the Dupuit-Forchheimer assumption. While GFLOW supports some local transient and three-dimensional flow modeling, it is particularly suitable for modeling regional horizontal flow. To facilitate detailed local flow modeling, it supports a MODFLOW-extract option to automatically generate MODFLOW files in a user-defined area with aquifer properties and boundary conditions provided by the GFLOW analytic element model. GFLOW also supports conjunctive surface water and groundwater modeling using stream networks with calculated baseflow.

## 11.  GMS

(Groundwater Modeling Environment for MODFLOW, MODPATH, MT3D, RT3D, FEMWATER, SEAM3D, SEEP2D, PEST, UTCHEM, and UCODE)

GMS is a sophisticated and comprehensive groundwater modeling software. It provides tools for every phase of a groundwater simulation including site characterization, model development, calibration, post-processing, and visualization. GMS supports both finite-difference and finite-element models in 2D and 3D including MODFLOW 2000, MODPATH, MT3DMS/RT3D, SEAM3D, ART3D, UTCHEM, FEMWATER and SEEP2D. The program's modular design enables the user to select modules in custom combinations, allowing the user to choose only those groundwater modeling capabilities that are required.

## 12.  Groundwater Vistas

(Model Design and Analysis for MODFLOW, MODPATH, MT3D, RT3D, PEST and UCODE)

Groundwater Vistas (GV) is a sophisticated windows graphical user interface for 3-D groundwater flow and transport modeling. It couples a model design system with comprehensive graphical analysis tools. GV is a model-independent graphical design system for MODFLOW MODPATH (both steady-state and transient versions), MT3DMS, MODFLOWT, MODFLOW-SURFACT, MODFLOW2000, GFLOW, RT3D, PATH3D, SEAWAT and PEST, the model-independent calibration software. The combination of PEST and GV's automatic sensitivity analysis make GV a good calibration tool. The advanced version of Groundwater Vistas provides the ideal groundwater risk assessment tool. Groundwater Vistas is a modeling environment for the MODFLOW family of models that allows for the quantification of uncertainty. Stochastic (Advanced) Groundwater Vistas includes, Monte Carlo versions of MODFLOW, MODPATH and MT3D, Geostatistical Simulators SWIFT support advanced output options and more. GV displays the model design in both plan and cross-sectional views using a split window (both views are visible at the same time). Model results are presented using contours, shaded contours, velocity vectors, and detailed analysis of mass balance.

## 13.  HST3D

(3-D Heat and Solute Transport Model)

HST3D is a powerful user-friendly interface for HST3D integrated within the Argus Open Numerical Environments (Argus ONE) modeling environment. HST3D allows the user to enter all spatial data, graphically run HST3D, and visualize the results. Argus ONE integrates CAD, GIS, Database, Conceptual Modeling, Geostatistics, Automatic Grid and Mesh Generation, and Scientific Visualization within one comprehensive graphical user interface (GUI). The Heat and Solute Transport Model HST3D simulates ground-water flow and associated heat and solute transport in three dimensions. The HST3D model may be used for analysis of problems such as those related to subsurface-waste injection, landfill leaching, saltwater intrusion, freshwater recharge and recovery, radioactive waste disposal, water geothermal systems, and subsurface energy storage. The Argus ONE GIS and Grid Modules are required to run HST3D.

## 14.  MicroFEM

(Finite-Element Program for Multiple-Aquifer Steady-State and Transient Groundwater Flow Modeling)

The Windows version of MicroFEM is a new program, based on the DOS package Micro-Fem. It takes you through the whole process of ground-water modeling, from the generation of a finite-element grid through the stages of preprocessing, calculation, postprocessing, graphical interpretation and plotting. Confined, semi-confined, phreatic, stratified and leaky multi-aquifer systems can be simulated with a maximum of 20 aquifers. Irregular grids, as typically used by finite-element programs, have several advantages compared to the more or less regular grids used by finite-difference codes. A model with a well-designed irregular grid will show more accurate results with fewer nodes, so less computer memory is required while calculations are faster. MicroFEM offers extensive possibilities as to the ease of creating such irregular grids. Other MicroFEM features include the ease of data preparation and the presentation and analysis of modeling results. A flexible way of zone-selection and formula-assignment is used for all parameters: transmissivities, aquitard resistances, well discharges and boundary conditions for each layer. Depending on the type of model, this can be extended with layer thicknesses, storativities, spatially varying anisotropy, topsystem and user-defined parameters. To inspect and interpret model results, maps and profiles can be used to visualize contours, heads, 3D-flowlines, flow vectors, etc. Time-drawdown curves and water balances can be selected with just a few keystrokes or mouse clicks.

## 15.  MOC

(2-D Solute Transport and Dispersion in Groundwater)

MOC simulates 2-D solute transport in flowing groundwater. MOC is both general and flexible in that it can be applied to a wide range of problem types. MOC is applicable for one-

or two-dimensional problems involving steady-state or transient flow. MOC computes changes in concentration over time caused by the processes of convective transport, hydrodynamic dispersion, and mixing (or dilution) from fluid sources. MOC assumes that gradients of fluid density, viscosity and temperature do not affect the velocity distribution. However, the aquifer may be heterogeneous and/or anisotropic. MOC is based on a rectangular, block-centered, finite-difference grid. It allows the specification of injection or withdrawal wells and of spatially-varying diffuse recharge or discharge, saturated thickness, transmissivity, boundary conditions and initial heads and concentrations. MOC incorporates first-order irreversible rate-reaction; reversible equilibrium controlled sorption with linear, Freundlich, or Langmuir isotherms; and reversible equilibrium-controlled ion exchange for monovalent or divalent ions.

## 16.    MOCDENSE

(Two-Constituent Solute Transport Model for Groundwater Having Variable Density)

MOCDENSE is a modified version of the ground-water flow and solute-transport model of Konikow and Bredehoeft which was designed to simulate the transport and dispersion of a single solute that does not affect the fluid density. This modified version of MOCDENSE simulates the flow in a cross-sectional plane rather than in an areal plane. Because the problem of interest involves variable density, the modified model solves for fluid pressure rather than hydraulic head in the flow equation; the solution to the flow equation is still obtained using a finite-difference method. Solute transport is simulated in MOCDENSE with the method of characteristics as in the original model. Density is considered to be a function of the concentration of one of the constituents.

## 17.    MODFLOW

(Three-Dimensional Finite-Difference Ground-Water Flow Model)

MODFLOW is the name that has been given the USGS Modular Three-Dimensional Ground-Water Flow Model. Because of its ability to simulate a wide variety of systems, its extensive publicly available documentation, and its rigorous USGS peer review, MODFLOW has become the worldwide standard ground-water flow model. MODFLOW is used to simulate systems for water supply, containment remediation and mine dewatering. When properly applied, MODFLOW is the recognized standard model.

The main objectives in designing MODFLOW were to produce a program that can be readily modified, is simple to use and maintain, can be executed on a variety of computers with minimal changes, and has the ability to manage the large data sets required when running large problems. The MODFLOW report includes detailed explanations of physical and mathematical concepts on which the model is based and an explanation of how those concepts were incorporated in the modular structure of the computer program. The modular structure of MODFLOW consists of a Main Program and a series of highly-independent subroutines called modules. The modules are grouped in packages. Each package deals with a specific feature of the hydrologic system which is to be simulated such as flow from rivers or flow into drains or with a specific method of solving linear equations which describe the flow system such as the Strongly Implicit Procedure or Preconditioned Conjugate Gradient. The division of MODFLOW into modules permits the user to examine specific hydrologic features of the model independently. This also facilitates development of additional capabilities because new modules or packages can be added to the program without modifying the existing ones. The input/output system of MODFLOW was designed for optimal flexibility.

Ground-water flow within the aquifer is simulated in MODFLOW using a block-centered finite-difference approach. Layers can be simulated as confined, unconfined, or a combination of both. Flows from external stresses such as flow to wells, areal recharge, evapotranspiration, flow to drains, and flow through riverbeds can also be simulated.

## 18.    MODFLOW SURFACT

(MODFLOW-Based Ground-Water Flow and Contaminant Transport Model)

A new flow and transport model, MODFLOW SURFACT, is based on the USGS MODFLOW code, the most widely-used ground-water flow code in the world. MODFLOW, however, has certain limitations in simulating complex field problems. Additional computational modules have been incorporated to enhance the simulation capabilities and robustness. MODFLOW SURFACT is a seamless integration of flow and transport modules.

## 19.    MODFLOWT

(An Enhanced Version of MODFLOW for Simulating 3-D Contaminant Transport)

MODFLOWT is an enhanced version of the USGS MODFLOW model which includes packages to simulate advective-dispersive contaminant transport. Fully three-dimensional, MODFLOWT simulates transport of one or more miscible species subject to adsorption and decay through advection and dispersion. MODFLOWT performs groundwater simulations utilizing transient transport with steady-state flow, transient flow, or successive periods of steady-state flow. Groundwater flow data sets created for the original MODFLOW model function without alteration in MODFLOWT; thus extension of modeling projects to simulate contaminant transport is very easy using MODFLOWT. It is thoroughly tested and has been bench-marked against other transport codes including MT3D, SWIFT and FTWORK. A comprehensive and pragmatic approach to contaminant transport has been incorporated into MODFLOWT which allows for three distinct directional dispersivity values, multiple chemicals and a rigorous treatment of the hydrodynamic dispersion tensor.

## 20.    MODFLOWwin32

(MODFLOW for Windows)

MODFLOWwin32 has all the features of other MODFLOW versions including the newest packages added over the years since MODFLOW's original release by the USGS. These new packages include the Stream Routing Package, Aquifer Compaction Package, Horizontal Flow Barrier Package, BCF2 and BCF3 Packages, and the new PCG2 solver. In addition, MODFLOWwin32 will create files for use with MODPATH (particle-tracking model for MODFLOW) and MT3D (solute transport model). MODFLOWwin32, as its name implies, is a 32-bit program designed to address all the memory available to Windows. MODFLOWwin32 will run in all versions of Windows.

## 21.  MODPATH

(3-D Particle Tracking Program for MODFLOW)

MODPATH, "A Particle Tracking Post-Processing Package for MODFLOW, the USGS 3-D Finite-Difference Ground-Water Flow Model (MODFLOW)," is a widely-used particle-tracking program.

## 22.  MOFAT

(Multiphase (Water, Oil, Gas) Flow and Multicomponent Transport Model)

MOFAT for Windows includes a graphical preprocessor, mesh editor and postprocessor with on-line help. Simulate multiphase (water, oil and gas) flow and transport of up to five non-inert chemical species in MOFAT. Model flow of light or dense organic liquids in three fluid phase systems. Simulate dynamic or passive gas as a full three-phase flow problem. Model water flow only, oil-water flow, or water-oil-gas flow in variably-saturated porous media. MOFAT achieves a high degree of computational efficiency by solving flow equations at each node (on the finite-element mesh) only for phases that are undergoing changes in pressures and saturations above specified tolerances using a new adaptive solution domain method. Therefore, if NAPL is absent or exists at a residual saturation, MOFAT will locally eliminate those flow equations. MOFAT analyzes convective-dispersive transport in water, NAPL, and gas phases by assuming local equilibrium or nonequilibrium partitioning among the fluid and solid phases. MOFAT considers interphase mass transfer and compositional dependence of phase densities. A concise but accurate description of soil capillary pressure relations is used which assures natural continuity between single-phase, two-phase and three-phase conditions.

## 23.  MS-VMS

(MODFLOW-Based Visual Modeling System with Comprehensive Flow and Transport Capability)

MS-VMS is a comprehensive MODFLOW-based ground-water flow and contaminant transport modeling system. The USGS modular ground-water flow model, MODFLOW, is the most widely-used ground-water flow model in the world. But, in its original form, MODFLOW has certain limitations and cannot be used to simulate some complex problems encountered regularly by modelers, hydrogeologists, and engineers in the field. MS-VMS overcomes these limitations.

## 24. MT3D

(A Modular 3D Solute Transport Model)

MT3D is a comprehensive three-dimensional numerical model for simulating solute transport in complex hydrogeologic settings. MT3D has a modular design that permits simulation of transport processes independently or jointly. MT3D is capable of modeling advection in complex steady-state and transient flow fields, anisotropic dispersion, first-order decay and production reactions, and linear and nonlinear sorption. It can also handle bioplume-type reactions, monad reactions, and daughter products. This enables MT3D to do multi-species reactions and simulate or assess natural attenuation within a contaminant plume. MT3D is linked with the USGS groundwater flow simulator, MODFLOW, and is designed specifically to handle advectively-dominated transport problems without the need to construct refined models specifically for solute transport.

## 25. PEST

(Parameter Estimation for Any Model)

PEST is a nonlinear parameter estimation and optimization package. It can be used to estimate parameters for just about any existing model whether or not you have the model's source code. PEST is able to "take control" of a model, running it as many times as it needs while adjusting its parameters until the discrepancies between selected model outputs and a complementary set of field or laboratory measurements is reduced to a minimum in the weighted least-squares sense.

## 26. PESTAN

(Pesticide Transport Model)

PESTAN, Pesticide Transport, is a U.S. EPA program for evaluating the transport of organic solutes through the vadose zone to groundwater. PESTAN uses an analytical solution to calculate organic movement based on a linear isotherm, first-order degradation and hydrodynamic dispersion. Input data includes water solubility, infiltration rate, bulk density, sorption constant, degradation rates, saturated water content, characteristic curve coefficient, saturated hydraulic conductivity and dispersion coefficient.

## 27. Processing Modflow (PMWIN)

(Graphical Interface for MODFLOW, MODPATH, PMPATH, MT3D, PEST, and UCODE)

Processing MODFLOW for Windows (PMWIN) is a complete simulation system. It comes complete with a professional graphical preprocessor and postprocessor, the 3-D finite-difference ground-water models MODFLOW-88, MODFLOW-96, and MODFLOW 2000; the solute transport models MT3D, MT3DMS, RT3D and MOC3D; the particle tracking model PMPATH 99; and the inverse models UCODE and PEST-ASP for automatic calibration. A 3D visualization and animation package, 3D Groundwater Explorer, is also included.

## 28.   POLLUTE

(Finite-Layer Contaminant Migration Model - Landfill Design)

POLLUTE can be used for fast, accurate, and comprehensive contaminant migration analysis. It implements a "1½-dimensional" solution to the advection-dispersion equation. Unlike finite-element and finite-difference formulations, POLLUTE does the numerical stability problems of alternate approaches. Landfill designs that can be considered range from simple systems on a natural clayey aquitard to composite liners with multiple barriers and multiple aquifers. In addition to advective-dispersive transport, POLLUTE can consider adsorption, radioactive and biological decay, phase changes, and transport through fractures. The Graphical User Interface makes the editing, execution, and printing of data easy and flexible. This interface also has options to quickly design landfills with primary composite barriers or primary and secondary composite barriers.

## 29.   PRINCE

(7 Mass Transport and 3 Flow Models)

PRINCE is a well-known software package of ten analytical groundwater models originally developed as part of an EPA 208 study. There are seven one-, two- and three-dimensional mass transport models and three two-dimensional flow models in PRINCE. These groundwater models have been rewritten from the original mainframe FORTRAN codes in graphics-rich and PC-friendly C. Two popular analytical models have been added to the original collection, and the ability to import digitized or AutoCAD-produced DXF site map files has been added. The result is a widely acclaimed, user-friendly, menu-driven package with built-in high resolution graphics for X-Y, 2-D contour and 3-D surface plots.

## 30.   PRZM3

(Pesticide Transport Model - Exposure Assessments)

PRZM3 links two models, PRZM and VADOFT to predict pesticide transport and transformation down through the crop root and vadose (unsaturated) zone to the water table. PRZM3 incorporates soil temperature simulation, volatilization and vapor phase transport in soils, irrigation simulation, and microbial transformation. PRZM is a one-dimensional finite-difference model which uses a method of characteristics (MOC) algorithm to eliminate numerical dispersion. VADOFT is a one-dimensional finite-element code that solves Richards' equation for flow in the unsaturated zone. The user may make use of constituting relationships between pressure, water content, and hydraulic conductivity to solve the flow equation. PRZM3 is capable of simulating multiple pesticides or parent-daughter relationships. PRZM3 is also capable of estimating probabilities of concentrations or fluxes in or from various media for the purpose of performing exposure assessments. PRZM and VADOFT are linked together with the aid of a flexible execution supervisor that allows the user to build models that are tailored to site-specific situations. Monte Carlo pre and postprocessors are provided in order to perform probability-based exposure assessments.

## 31.    RBCA Tier 2 Analyzer

(2D Groundwater Flow and Biodegradation Model)

The RBCA Tier 2 Analyzer is a two-dimensional groundwater model with a comprehensive selection of contaminant transport simulation capabilities including single or multiple species sequential decay reactions such as reductive dechlorination of PCE instantaneous or kinetic-limited BTEX biodegradation with single or multiple electron acceptors and equilibrium or non-equilibrium sorption.

## 32.    SEAWAT

(Three-Dimensional Variable-Density Ground-Water Flow)

The SEAWAT program was developed to simulate three-dimensional, variable- density, transient ground-water flow in porous media. The  source code for SEAWAT was developed by combining MODFLOW and MT3DMS into a single program that solves the coupled flow and solute-transport equations. The SEAWAT code follows a modular structure, and thus, new capabilities can be added with only minor modifications to the main program. SEAWAT reads and writes standard MODFLOW and MT3DMS data sets, although some extra input may be required for some SEAWAT simulations. This means that many of the existing pre- and post-processors can be used to create input data sets and analyze simulation results. Users familiar with MODFLOW and MT3DMS should have little difficulty applying SEAWAT to problems of variable-density ground-water flow.

## 33.    SESOIL

(Long-Term Pollutant Fate and Migration in the Unsaturated Zone)

SESOIL is a seasonal compartment model which simulates long-term pollutant fate and migration in the unsaturated soil zone. SESOIL describes the following components of a user-specified soil column which extends from the ground surface to the ground-water table.

- Hydrologic cycle of the unsaturated soil zone.
- Pollutant concentrations and masses in water, soil, and air phases.
- Pollutant migration to groundwater.
- Pollutant volatilization at the ground surface.
- Pollutant transport in washload due to surface runoff and erosion at the ground surface.

SESOIL estimates all of the above components on a monthly basis for up to 999 years of simulation time. It can be used to estimate the average concentrations in groundwater. The soil column may be composed of up to four layers, each layer having different soil properties which affect the pollutant fate. In addition, each soil layer may be subdivided into a maximum of 10 sublayers in order to provide enhanced resolution of pollutant fate and migration in the soil column. The following pollutant fate processes are accounted for: Volatilization, Adsorption, Cation Exchange, Biodegradation, Hydrolysis and Complexation.

## 34.    SLAEM / MLAEM

(Analytic Element Models - Model Regional Groundwater Flow in Systems of Confined Aquifers, Unconfined Aquifers and Leaky Aquifers)

The AEM family of computer programs, SLAEM, MLAEM/2, and MLAEM, are based on the Analytic Element Method developed by Dr. O.D.L. Strack. The computer programs are intended for modeling regional groundwater flow in systems of confined, unconfined, and leaky aquifers. SLAEM (Single Layer Analytic Element Model) is the single-layer version of the program. MLAEM/2 (Multi Layer Analytic Element Model) can access two layers while the number of layers supported by MLAEM is limited only by hardware. All programs run under Microsoft Windows 95, 98 and NT. The programs are native windows applications and are accessed via a modern and flexible Graphical User Interface (GUI), as well as via a command-line interface. The latter capability makes it easy to drive the program from other programs such as Arc-View, ARC/INFO, and PEST. The programs create files from data entered graphically via the GUI; these files can be read in later. The programs read DXF-files and produce BNA files that may be read by other programs such as SURFER.

## 35.    SOLUTRANS

(3-D Analytic Solute Transport Model)

SOLUTRANS is a 32-bit Windows program for modeling three-dimensional solute transport based on the solutions presented by Leij et al. for both equilibrium and non-equilbrium transport. The interface and input requirements are so simple that it only takes a few minutes to develop models and build insight about complex solute transport problems. With SOLUTRANS you can, in a matter of minutes, model solute transport from a variety of source configurations and build important insights about key processes. SOLUTRANS offers a quick and simple alternative to complex, time-consuming 3-D numerical flow and transport models.

## 36.    SUTRA

(2D/3D Saturated/Unsaturated Transport Model)

SUTRA is a 2D/3D groundwater saturated-unsaturated transport model, a complete saltwater intrusion and energy transport model. SUTRA simulates fluid movement and transport of either energy or dissolved substances in a subsurface environment. SUTRA employs a two-dimensional hybrid finite-element and integrated finite-difference method to approximate the governing equations that describe the two interdependent processes that are simulated: (1) fluid density-dependent saturated or unsaturated groundwater flow and either (2a) transport of a solute in the groundwater, in which the solute may be subject to equilibrium adsorption on the porous matrix and both first-order and zero-order production or decay, or (2b) transport of thermal energy in the groundwater and solid matrix of the aquifer.

## 37.    SVFlux 2D

(Saturated/Unsaturated Automated 2D/3D Seepage Modeling Software)

SVFlux 2D represents the next level in seepage analysis software. Designed to be simple and effective, the software offers features designed to allow the user to focus on seepage solutions, not convergence problems or difficult mesh creation. Great care has been taken to model geometry CAD-style input after the popular AutoCAD(TM) software. Freeform boundary equations and an optional soil database of over 6,000 soils to choose from further simplify model design. The finite element solution makes use of fully automatic mesh generation and mesh refinement to solve the problem quickly as well as indicating zones of critical gradient.

## 38.    SWIFT

(3-D Model to Simulate Groundwater Flow, Heat, Brine and Radionuclide Transport)

SWIFT is a fully-transient, three-dimensional model to simulate groundwater flow, heat (energy), brine and radionuclide transport in porous and fractured geologic media. The primary equations for fluid (flow), heat and brine are coupled by fluid density, viscosity and porosity. In addition to transient analysis, SWIFT offers a steady-state option for coupled flow and brine. The equations are solved using central or backward spatial and time weighting approximations by the finite-difference method. In addition to Cartesian, cylindrical grids may be used. Contaminant transport includes advection, dispersion, sorption and decay, including chains of constituents. Both dual-porosity and discrete-fracture representations along with rock matrix interactions may be simulated. The nonlinearities resulting from water table and variable density are solved iteratively.

## 39.    SWIMv1/SWIMv2

(Soil water infiltration and movement model - simulate soil water balances)

SWIMv1 (Soil Water Infiltration and Movement model version 1) is a software package for simulating water infiltration and movement in soils. SWIMv1 consists of a menu-driven suite of three programs that allow the user to simulate soil water balances using numerical solutions of the basic soil water flow equations. As in the real world, SWIMv1 allows addition of water to the system as precipitation and removal by runoff, drainage, evaporation from the soil surface and transpiration by vegetation. SWIMv1 helps researchers and consultants understand the soil water balance so they can assess possible effects of such practices as tree clearing, strip mining and irrigation management. SWIMv1 is valuable for scientists and consultants involved in land planning and land management. For example, if a development is being considered which involves tree clearing, SWIMv1 can be used to indicate salinity or surface runoff problems that could result from a change in the soil water balance associated with the removal of the trees.

SWIMv2 (Soil Water Infiltration and Movement model version 2) is a mechanistically-based model designed to address soil water and solute balance issues associated with both production and the environmental consequences of production. SWIMv2 employs fast, numerically-efficient techniques for solving Richards' equation for water flow and the convection-dispersion equation for solute transport and is suitable for personal computer applications. The model deals with a one-dimensional vertical soil profile which may be vertically inhomogeneous but is assumed to be horizontally uniform. It can be used to simulate runoff, infiltration, redistribution, solute transport and redistribution of solutes, plant uptake and transpiration, evaporation, deep drainage and leaching.

## 40.    TWODAN

(2-D Analytic Ground-Water Flow Model)

TWODAN is a popular and versatile analytic ground-water flow model for Windows. TWODAN has a suite of advanced analytic modeling features that allow you to model everything from a single well in a uniform flow field to complex remediation schemes with numerous wells, barriers, surface waters, and heterogeneities. TWODAN has many capabilities: heterogeneities, impermeable barriers, resistant barriers, and transient solutions, to name a few. The analytic method of TWODAN demands minimal input, and the new seamless Windows interface can be quickly mastered. TWODAN can be used for modeling remedial design alternatives, wellhead capture zones, and regional aquifer flow. TWODAN combines advanced analytic elements with a user interface. Compare It is a good tool for most remediation design, capture zone analysis, and regional modeling problems.

## 41.    VAM2D

(2-D Variably-Saturated Groundwater Analysis Model)

VAM2D (Variably Saturated Analysis Model in Two (2) Dimensions) is a two-dimensional finite-element groundwater model that simulates transient or steady-state groundwater flow and contaminant transport in porous media. VAM2D analyzes unconfined flow problems using a rigorous saturated-unsaturated modeling approach using efficient numerical techniques. Accurate mass balance is maintained in VAM2D even when simulating highly nonlinear soil moisture relations. Hysteresis effects in the water retention curve can also be simulated. A wide range of boundary conditions can be treated in VAM2D including seepage faces, water-table conditions, recharge, infiltration, evapotranspiration, and pumping and injection wells. The contaminant transport option can account for advection, hydrodynamic dispersion, equilibrium sorption, and first-order degradation. Transport of a single species or multiple parent-daughter components of a decay chain can be simulated. The VAM2D code can perform simulations using an areal plane, cross section, or axisymmetric configuration.

## 42.    Visual Groundwater

(3-D Visualization and Animation of Site Data and Modeling Results)

Visual Groundwater is a 3-D visualization software package which can be used to deliver high-quality, three-dimensional presentations of subsurface characterization data and groundwater modeling results. Visual Groundwater combines state-of-the-art graphical tools for 3-D visualization and animation with a data management system specifically designed for borehole investigation data. Visual Groundwater also comes with a data conversion utility to create 3-D data files using random X, Y, Z data, and gridded data sets. Three-dimensional images of complex site characterization data and modeling results can be easily created.

## 43.    VISUAL HELP

(Modeling Environment for the U.S. EPA HELP Model for Evaluating and Optimizing Landfill Designs)

Visual HELP for Windows 95/98/2000/NT is an advanced hydrological modeling environment available for designing landfills, predicting leachate mounding and evaluating potential leachate contamination. Visual HELP combines the latest version of the HELP model with an easy-to user interface and powerful graphical features for designing the model and evaluating the modeling results. Visual HELP's user-friendly interface and flexible data handling procedures provides convenient access to both the basic and advanced features of the HELP model. This completely-integrated modeling HELP environment allows the user to graphically create several profiles representing different parts of a landfill; automatically generate statistically-reliable weather data (or create your own); run complex model simulations; visualize full-color, high-resolution results; and prepare graphical and document materials for your report.

## 44.    Visual MODFLOW

(Integrated Modeling Environment for MODFLOW, MODPATH, MT3D)

Visual MODFLOW provides professional 3D groundwater flow and contaminant transport modeling using MODFLOW-2000, MODPATH, MT3DMS and RT3D. Visual MODFLOW Pro seamlessly combines the standard Visual MODFLOW package with WinPEST and the Visual MODFLOW 3D-Explorer to give the most complete and powerful graphical modeling environment available. This fully-integrated groundwater modeling environment allows to:

- Graphically design the model grid, properties and boundary conditions,
- Visualize the model input parameters in two or three dimensions,
- Run the groundwater flow, pathline and contaminant transport simulations,
- Automatically calibrate the model using WinPEST or manual methods, and
- Display and interpret the modeling results in three-dimensional space using the Visual MODFLOW 3D-Explorer

## 45.    VISUAL PEST

(Graphical Model-Independent Parameter Estimation and Optimization)

Visual PEST combines the powerful parameter estimation capabilities of PEST2000 with the graphical processing and display features of WinPEST. PEST2000 is the latest version of PEST, the pioneer in model-independent parameter estimation. Since it was first released over six years ago, PEST has gained extensive use all over the world in many different fields. During this time it has undergone continued development with the addition of many new features that have improved its performance and utility to a level that makes it uniquely applicable in just about any modeling environment. PEST is now used extensively for automated model calibration and data interpretation in groundwater and surface-water hydrology, geophysics, geotechnical, mechanical and mining engineering as well as many other fields.

## 46.   VLEACH

(One-Dimensional Finite-Difference Vadose Zone Leaching Model)

VLEACH, a One-Dimensional Finite-Difference Vadose Zone Leaching Model, is a U.S. EPA program which describes the movement of an organic contaminant within and between three phases: (1) as a solute dissolved in water, (2) as a gas in the vapor phase, and (3) as an adsorbed compound in the solid phase. The leaching is simulated in a number of distinct, user-defined polygons vertically divided into a series of user-defined cells. At the end of the simulation, the results from each polygon are used to determine an area-weighted ground-water impact for the modeled area. VLEACH is a computer program for estimating the impact due to the mobilization and migration of a sorbed organic contaminant located in the vadose zone on the underlying groundwater resource. A graphical user interface for VLEACH is now available in the WHI UnSat Suite.

## 47.   VS2DT

(Flow and Solute Transport in Variably-Saturated, Single-Phase Flow in Porous Media)

VS2DT is a USGS program for flow and solute transport in variably-saturated, single-phase flow in porous media. A finite-difference approximation is used in VS2DT to solve the advection-dispersion equation. Simulated regions include one-dimensional columns, two-dimensional vertical cross sections, and axially-symmetric, three-dimensional cylinders. The VS2DT program options include backward or centered approximations for both space and time derivatives, first-order decay, equilibrium adsorption (Freundlich or Langmuir) isotherms, and ion exchange. Nonlinear storage terms are linearized by an implicit Newton-Raphson method. Relative hydraulic conductivity in VS2DT is evaluated at cell boundaries using full upstream weighting, arithmetic mean or geometric mean. Saturated hydraulic conductivities in VS2DT are evaluated at cell boundaries using distance-weighted harmonic means. A graphical user interface for VS2DT is now available in the WHI UnSat Suite.

## 48.   WHI UnSat Suite

(A Compilation of the Most Popular Unsaturated Zone Models)

The WHI UnSat Suite Plus combines SESOIL, VLEACH, PESTAN, VS2DT and HELP in a revolutionary graphical environment specifically designed for simulating one-dimensional groundwater flow and contaminant transport through the unsaturated zone. All five of these models are compiled and optimized to run as native Windows applications and are seamlessly integrated within the WHI UnSat Suite modeling environment. Professional Applications of the WHI UnSat Suite Plus

- Simulate long-term pollutant fate and transport (VOCs, PAHs, pesticides and heavy metals) in the unsaturated zone under seasonally variable conditions using SESOIL.
- Predict vertical migration of volatile hydrocarbons through the vadose zone using VLEACH.
- Estimate agricultural pesticide migration through the unsaturated zone using PESTAN.
- Simulate groundwater flow and transport processes through heterogeneous, unsaturated soil using VS2DT.
- Predict seasonal recharge rates through heterogeneous soil conditions under variable weather conditions using the HELP model.
- Generate up to 100 years of statistically reliable climatalogical data for virtually any location in the world using the WHI Weather Generator.

## 49.   WinFlow

(Analytical Steady-State and Transient Groundwater Flow Model)

WinFlow is a powerful yet easy-to-use groundwater flow model. WinFlow is similar to Geraghty & Miller's popular QuickFlow model which was developed by one of the authors of QuickFlow. WinFlow is a true Windows program incorporating a multiple document interface (MDI). WinFlow is an interactive analytical model that simulates two-dimensional steady-state and transient groundwater flow. The steady-state module in WinFlow simulates groundwater flow in a horizontal plane using analytical functions developed by Strack. The transient module uses equations developed by Theis and by Hantush and Jacob for confined and leaky aquifers, respectively. Each module uses the principle of superposition to evaluate the effects from multiple analytical functions (wells, etc.) in a uniform regional flow field.

## 50.   WinTran

(Groundwater Flow and Finite-Element Contaminant Transport Model)

WinTran couples the steady-state groundwater flow model from WinFlow with a contaminant transport model. The transport model has the feel of an analytic model but is actually an embedded finite-element simulator. The finite-element transport model is constructed automatically by WinTran but displays numerical criteria (Peclet and Courant numbers) to allow the user to avoid numerical or mass balance problems. Contaminant mass may be injected or extracted using any of the analytic elements including wells, ponds, and linesinks. In addition, constant concentration elements have been added. WinTran displays both head and concentration contours, and concentration may be plotted versus time at selected monitoring locations. The transport model includes the effects of dispersion, linear sorption (retardation), and first-order decay. WinTran aids in risk assessment calculations by displaying the concentration over time at receptor or observation locations. WinTran will display the breakthrough curves after the simulation is finished.

Considering the large variability and the quick development of groundwater models, a new, more sophisticated model can often replace a previously applied model. Additionally, the reconsideration of the conceptual model and the regeneration of the mesh may need a new allocation of the parameters. Therefore, it is important that model data (information) are stored independently from a given model, with a preference for GIS-based databases. Considerable development in the field of user-friendly GIS and data base servers makes the set-up and the modification of models easier and more time-effective. One such model is FEFLOW which incorporates mathematical modelling with GIS-based data exchange interfaces.

The input data for a groundwater model include natural and artificial stress, and parameters, dimensions, and physico-chemical properties of all aquifers considered in the model. A finer level of detail of the numerical approximation (solution) greatly increases the data requirements. Input data for aquifers are common values such as transmissivities, aquitard resistances, abstraction rates, groundwater recharges, surface water levels etc. The most common output data are groundwater levels, fluxes, velocities and changes in these parameters due to stress put into the model.

**Concluding Remarks**

Mathematical models are tools, which are frequently used in studying groundwater systems. In general, mathematical models are used to simulate (or to predict) the groundwater flow and in some cases the solute and/or heat transport. Predictive simulations must be viewed as estimates, dependent upon the quality and uncertainty of the input data. Models may be used as predictive tools, however field monitoring must be incorporated to verify model predictions. The best method of eliminating or reducing modelling errors is to apply good hydrogeological judgement and to question the model simulation results. If the results do not make physical sense, find out why.

**References**

1.    Anderson, M.P. and W.W. Woessner (1992). Applied  Groundwater Modeling. Academic Press, Inc., San Diego, CA., 381 p.

2.    American Society for Testing and Materials (1993). Standard Guide for Application of a Ground-Water Flow Model to a Site-Specific Problem. ASTM Standard D 5447-93, West Conshohocken, PA, 6 p.

3.    American Society for Testing and Materials (1995). Standard Guide  for Subsurface Flow and Transport Modeling. ASTM  Standard D 5880-95, West Conshohocken, PA, 6 p.

4.    Bear, J., and A. Verruijt (1987). Modeling Groundwater Flow  and Pollution. D. Reidel Publishing Company, 414 p.

5.    Franke, O.L., Bennett, G.D., Reilly, T.E., Laney, R.L., Buxton, H.T., and Sun, R.J. (1991). Concepts and Modeling in Ground-Water Hydrology -- A Self-Paced Training Course. U.S. Geological Survey Open-File Report 90-707.

6.      Kashyap, Deepak (1989). Mathematical Modelling for Groundwater Management –
        Status in India. Indo-French Seminar on Management of Water Resources, 22-24
        September, 1989, Festival of France-1989, Jaipur, pp. IV-59 to IV-75.

7.      Kinzelbach, W. (1986). Groundwater Modeling: An Introduction with Sample
        Programs in BASIC. Elsevier, New York, 333 p.

8.      Kumar, C. P. (1992). Groundwater Modelling – In. Hydrological Developments in
        India Since Independence. A Contribution to Hydrological Sciences, National
        Institute of Hydrology, Roorkee, pp. 235-261.

9.      Kumar, C. P. (2001). Common Ground Water Modelling Errors and Remediation.
        Journal of Indian Water Resources Society, Volume 21, Number 4, October 2001, pp.
        149-156.

10.     McDonald, M.G. and A.W. Harbaugh (1988). A Modular Three-Dimensional Finite-
        Difference Ground-Water Flow Model. USGS TWRI Chapter 6-A1, 586 p.

11.     Pinder, G.F., and J.D. Bredehoeft (1968). Application of the Digital Computer for
        Aquifer Evaluation. Water Resources Research, Vol. 4, pp. 1069-1093.

12.     Wang, H.F. and M.P. Anderson (1982). Introduction to Groundwater Modeling.
        W.H. Freeman and Company, San Francisco, CA, 237 p.

19.     Website - Kumar Links to Hydrology Resources:
        http://www.angelfire.com/nh/cpkumar/hydrology.html

20.     Website - Scientific Software Group:
        http://www.scisoftware.com/

21.     Website - USGS Ground-Water Software:
        http://water.usgs.gov/nrp/gwsoftware/

# 14-A Modular Three-Dimensional Groundwater Model (MODFLOW)

## Introduction

Groundwater systems are affected by natural processes and human activity, and require targeted and ongoing management to maintain the condition of groundwater resources within acceptable limits, while providing desired economic and social benefits. Groundwater management and policy decisions must be based on knowledge of the past and present behaviour of the groundwater system, the likely response to future changes and the understanding of the uncertainty in those responses.

The location, timing and magnitude of hydrologic responses to natural or human-induced events depend on a wide range of factors - for example, the nature and duration of the event that is impacting groundwater, the subsurface properties and the connection with surface water features such as rivers and oceans. Through observation of these characteristics, a conceptual understanding of the system can be developed, but often observational data is scarce (both in space and time), so our understanding of the system remains limited and uncertain.

It is not possible to see into the sub-surface, and observe the geological structure and the groundwater flow processes. The best we can do is to construct bores, use them for pumping and monitoring, and measure the effects on water levels and other physical aspects of the system. It is for this reason that groundwater flow models have been, and will continue to be, used to investigate the important features of groundwater systems, and to predict their behaviour under particular conditions.

Groundwater models provide additional insight into the complex system behaviour and (when appropriately designed) can assist in developing conceptual understanding. Furthermore, once they have been demonstrated to reasonably reproduce past behaviour, they can forecast the outcome of future groundwater behaviour, support decision-making and allow the exploration of alternative management approaches. However, there should be no expectation of a single 'true' model, and model outputs will always be uncertain. As such, all model outputs presented to decision-makers benefit from the inclusion of some estimate of how good or uncertain the modeller considers the results.

Models also form an integral part of decision support systems in the process of managing water resources, salinity and drainage, and should not be regarded as just an end point in themselves. The development and evaluation of resource management strategies for sustainable water allocation, and for control of land and water resource degradation, are heavily dependent on groundwater model predictions. Regional scale groundwater flow modelling studies are commonly used for water resource evaluation and to help quantify sustainable yields and allocations to end-users.

Typical model purposes include:

> Improving hydrogeological understanding (synthesis of data);
> Aquifer simulation (evaluation of aquifer behaviour);
> Designing practical solutions to meet specified goals (engineering design);
> Optimising designs for economic efficiency and account for environmental effects (optimisation);
> Evaluating recharge, discharge and aquifer storage processes (water resources assessment);
> Predicting impacts of alternative hydrological or development scenarios (to assist decision-making);
> Quantifying the sustainable yield (economically and environmentally sound allocation policies);
> Resource management (assessment of alternative policies);
> Sensitivity and uncertainty analysis (to guide data collection and risk-based decision-making);
> Visualisation (to communicate aquifer behaviour).

An important step in the modelling process is a formal software selection process in which all possible options are considered. This step has often been short-circuited in the past. In many cases, modellers have immediately adopted MODFLOW, developed by the US Geological Survey (USGS) (Harbaugh et al., 2000).

## MODFLOW - Modular Three-Dimensional Finite-Difference Groundwater Model

MODFLOW is the name that has been given to the USGS Modular Three-Dimensional Groundwater Flow Model. Because of its ability to simulate a wide variety of systems, its extensive publicly available documentation, and its rigorous USGS peer review, MODFLOW has become the worldwide standard groundwater flow model. MODFLOW is used to simulate systems for water supply, containment remediation and mine dewatering. When properly applied, MODFLOW is the recognized standard model.

MODFLOW is a three-dimensional finite-difference groundwater model that was first published in 1984. It has a modular structure that allows it to be easily modified to adapt the code for a particular application. Many new capabilities have been added to the original model. Harbaugh (2005) documents a general update to MODFLOW, which is called MODFLOW-2005 in order to distinguish it from earlier versions. MODFLOW-2005 is written primarily in Fortran 90. Only the GMG solver package is written in C. The code has been used on UNIX-based computers and personal computers running various forms of the Microsoft Windows operating system.

The main objectives in designing MODFLOW were to produce a program that can be readily modified, is simple to use and maintain, can be executed on a variety of computers with minimal changes, and has the ability to manage the large data sets required when running large problems. The MODFLOW report includes detailed explanations of physical and mathematical concepts on which the model is based and an explanation of how those concepts were incorporated in the modular structure of the computer program.

MODFLOW is most appropriate in those situations where a relatively precise understanding of the flow system is needed to make a decision. MODFLOW was developed using the finite-difference method. The finite-difference method permits a physical explanation of the concepts used in construction of the model. Therefore, MODFLOW is easily learned and modified to represent more complex features of the flow system.

*MODFLOW Input Requirements*

A large amount of information and a complete description of the flow system are required to make the most efficient use of MODFLOW. In situations where only rough estimates of the flow system are needed, the input requirements of MODFLOW may not justify its use. To use MODFLOW, the region to be simulated must be divided into cells with a rectilinear grid resulting in layers, rows and columns. Files must then be prepared that contain hydraulic parameters (hydraulic conductivity, transmissivity, specific yield, etc.), boundary conditions (location of impermeable boundaries and constant heads), and stresses (pumping wells, recharge from precipitation, rivers, drains, etc.).

*MODFLOW Simulation*

MODFLOW-2005 simulates steady and nonsteady flow in an irregularly shaped flow system in which aquifer layers can be confined, unconfined, or a combination of confined and unconfined. Flow from external stresses, such as flow to wells, areal recharge, evapotranspiration, flow to drains, and flow through river beds, can be simulated. Hydraulic conductivities or transmissivities for any layer may differ spatially and be anisotropic (restricted to having the principal directions aligned with the grid axes), and the storage coefficient may be heterogeneous. Specified head and specified flux boundaries can be simulated as a head dependent flux across the model's outer boundary that allows water to be supplied to a boundary block in the modelled area at a rate proportional to the current head difference between a "source" of water outside the modelled area and the boundary block.

The governing three-dimensional flow equation used by MODFLOW (McDonald and Harbaugh, 1988 and Harbaugh, et al., 2000) combines Darcy's Law and the principle of conservation of mass via Equation (1).

$$\frac{\partial}{\partial x}(K_{xx}\frac{\partial h}{\partial x}) + \frac{\partial}{\partial y}(K_{yy}\frac{\partial h}{\partial y}) + \frac{\partial}{\partial z}(K_{zz}\frac{\partial h}{\partial z}) + q_s = S_s\frac{\partial h}{\partial t} \qquad ...(1)$$

where $K_{xx}$, $K_{yy}$ and $K_{zz}$ are the values of hydraulic conductivity along the $x$, $y$ and $z$ coordinate axes oriented parallel to the major axes of hydraulic conductivity [L/T], $h$ is the hydraulic head [L], $q_s$ is the volumetric flux of groundwater sources and sinks per unit volume [1/T] with positive values indicating flow into the groundwater system, $S_s$ is specific storage [1/L], and $t$ [T] is time. MODFLOW solves a volume-averaged form of Equation (1).

The groundwater flow equation is solved using the finite-difference approximation. Figure 2 presents a sample of MODFLOW three-dimensional grid. The flow region is subdivided into blocks in which the medium properties are assumed to be uniform. In plan view, the blocks are made from a grid of mutually perpendicular lines that may be variably spaced. Model layers can have varying thickness. A flow equation is written for each block, called a cell.

Several solvers are provided for solving the resulting matrix problem; the user can choose the best solver for the particular problem. Flow-rate and cumulative-volume balances from each type of inflow and outflow are computed for each time step.

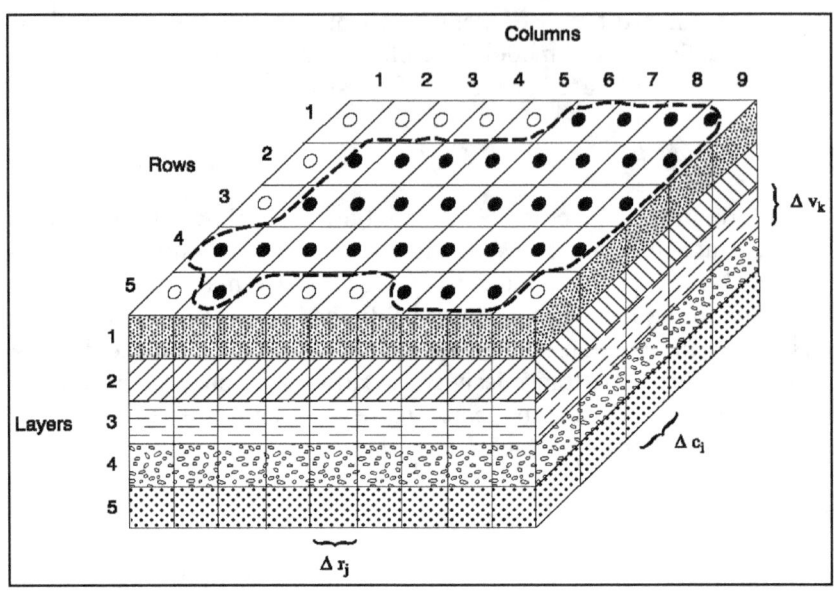

Figure 1: MODFLOW 3-D Grid

*MODFLOW Packages*

The modular structure of MODFLOW consists of a Main Program and a series of highly-independent subroutines called modules. The modules are grouped in packages. Each package deals with a specific feature of the hydrologic system which is to be simulated such as flow from rivers or flow into drains or with a specific method of solving linear equations which describe the flow system such as the Strongly Implicit Procedure or Preconditioned Conjugate Gradient. The division of MODFLOW into modules permits the user to examine specific hydrologic features of the model independently. This also facilitates development of additional capabilities because new modules or packages can be added to the program without modifying the existing ones. The input/output system of MODFLOW was designed for optimal flexibility. MODFLOW-2005 version includes the following functionality that is documented in Harbaugh (2005).

| | | |
|---|---|---|
| BAS | : | Basic Package |
| BCF | : | Block-Centered Flow Package |
| LPF | : | Layer-Property Flow Package |
| HFB | : | Horizontal Flow Barrier Package |
| CHD | : | Time-Variant Specified-Head Option |
| RIV | : | River Package |
| DRN | : | Drain Package |
| WEL | : | Well Package |
| GHB | : | General Head Boundary Package |
| RCH | : | Recharge Package |
| EVT | : | Evapotranspiration Package |

| SIP | : | Strongly Implicit Procedure Package |
|-----|---|-------------------------------------|
| PCG | : | Preconditioned Conjugate Gradient Package |
| DE4 | : | Direct solver |

The following functionality is also included. This functionality is documented in separate reports for use in earlier versions of MODFLOW. Conversion of this functionality to work with MODFLOW-2005 is documented in separate files that are provided with the MODFLOW-2005 distribution.

| STR | : | Streamflow-Routing Package |
|-----|---|----------------------------|
| FHB | : | Flow and Head Boundary Package |
| IBS | : | Interbed Storage Package |
| GMG | : | Geometric MultiGrid Solver Package |
| HUF | : | Hydrogeologic-Unit Flow Package |
| MNW1 | : | Version 1 of Multi-Node Well Package |
| MNW2 | : | Version 2 of the Multi-Node Well Package |
| ETS | : | Evapotranspiration with a Segmented Function Package |
| DRT | : | Drains with Return Flow Package |
| RES | : | Reservoir Package |
| SUB | : | Subsidence Package |
| OBS | : | Observation Process |
| SFR | : | Streamflow-Routing Package |
| LAK | : | Lake Package |
| UZF | : | Unsaturated Zone Package |
| GAG | : | Gage Package |
| SWT | : | Subsidence and Aquifer-System Compaction Package |
| LMT | : | Link to the MT3DMS contaminant-transport model |
| HYDMOD: | | Hydrograph capability |
| PCGN | : | Preconditioned Conjugate Gradient solver with improved nonlinear control |

The following packages are also included in most versions of MODFLOW.

**TLK1 (Transient Leakage)** - The TLK1 package is a new method of simulating transient leakage in the MODFLOW model. It solves the equations that describe the flow components across the upper and lower boundaries of confining units. The exact equations are approximated to allow efficient solution for the flow components. The flow components are incorporated into the finite-difference equations for model cells that are adjacent to confining units. Confining-unit properties can differ from cell to cell and a confining unit need not be present at all locations; however, a confining unit must be bounded above and below by model layers in which head is calculated or specified.

**IBS1 (Compaction Package)** - This addition to MODFLOW permits calculation of both elastic and inelastic release of water from fine-grained beds. This is especially useful in areas where land surface is subject to subsidence.

**CHD1 (Time-Variant Specified-Head Package)** - This package for MODFLOW permits specification of fixed head for boundary cells that vary from time step to time step during a stress period.

**STR1 (Streamflow Routing Package)** - The Stream package permits representation of intermittent streams in MODFLOW. It is especially useful in systems in the headwaters of small streams. The program limits the amount of groundwater recharge to the available streamflow. It permits two or more streams to merge into one with flow in the merged stream equal to the sum of the tributary flows. The program also permits diversions from streams.

**PCG2 (Preconditioned Conjugate Gradient Solver)** - PCG2 uses the preconditioned conjugate gradient method to solve the equations produced by MODFLOW for hydraulic head. Linear or nonlinear flow conditions may be simulated. PCG2 includes two preconditioning options: modified incomplete Cholesky preconditioning which is efficient on scalar computers; and polynomial preconditioning which requires less computer storage and, with modifications that depend on the computer used, is most efficient on vector computers. Convergence of the solver is determined using both head-change and residual criteria. Nonlinear problems are solved using Picard iterations.

**ZONEBUDGET** - The MODFLOW Zonebudget package calculates sub-regional water budgets using results from the USGS MODFLOW model. It uses cell-by-cell flow data saved by the model in order to calculate the budgets. Sub-regions of the modelled region are designated by zone numbers. The user assigns a zone number for each cell in the model. Composite zones can also be defined as combinations of the numeric zones.

**BCF3** - As originally published, MODFLOW could simulate the desaturation of variable-head model cells which resulted in their conversion to no-flow cells but could not simulate the resaturation of cells. That is, a no-flow cell could not be converted to variable head. However, such conversion is desirable in many situations. For example, one might wish to simulate pumping that desaturates some cells followed by the recovery of water levels after pumping is stopped. This program allows cells to convert from no-flow to variable-head. A cell is converted to variable head based on the head at neighbouring cells.

**GFD1 (Generalized Finite-Difference Package)** - This package for the advanced user of MODFLOW permits specification of interblock conductance. It is essential for use with RAD-MOD.

**RAD-MOD** - A preprocessor for assembling files needed to use MODFLOW to simulate radial flow towards a well. Although MODFLOW permits simulation of flow toward a well, it does so with a rectilinear grid. RAD-MOD permits simulation using a two-dimensional cross section.

**Horizontal Flow Barrier Package** - This package for MODFLOW simulates thin, vertical low-permeability geologic features that impede the horizontal flow of groundwater. These geologic features are approximated as a series of horizontal-flow barriers conceptually situated on the boundaries between pairs of adjacent cells in the finite-difference grid. The key assumption underlying this package is that the width of the barrier is negligibly small in comparison with the horizontal dimensions of the cells in the grid. Barrier width is not explicitly considered in the package but is included implicitly in a hydraulic characteristic defined as either (1) barrier transmissivity divided by barrier width if the barrier is in a constant-transmissivity layer or (2) barrier hydraulic conductivity divided by barrier width if the barrier is in a variable-transmissivity layer. Furthermore, the barrier is assumed to have zero storage capacity. Its sole function is to lower the horizontal branch conductance between the two cells that is separates.

## MODFLOW Graphical User Interfaces

For MODFLOW graphical user interfaces and complete modelling environments, the following MODFLOW products are available:

> GMS (Groundwater Modelling System)
> Visual MODFLOW
> Processing MODFLOW for Windows (PMWIN)
> Groundwater Vistas
> Argus ONE

*GMS (Groundwater Modelling System)* is a sophisticated groundwater modelling environment. The Groundwater Modelling System is a comprehensive package which provides tools for every phase of a groundwater simulation including site characterization, model development, post-processing, calibration, and visualization. GMS supports TINs, solids, borehole data, 2D and 3D geostatistics, and both finite element and finite difference models in 2D and 3D. Currently supported models include MODFLOW, MODPATH, MT3D, RT3D, FEMWATER, SEEP2D, SEAM3D, PEST, UCODE and UTCHEM. Due to the modular nature of GMS, a custom version of GMS with desired modules and interfaces can be configured.

*Visual MODFLOW* interface has been specifically designed to increase modelling productivity and decrease the complexities typically associated with building three-dimensional groundwater flow and contaminant transport models. The interface is divided into three separate modules: the Input Module, the Run Module, and the Output Module. When a file is opened or created, we will be able to seamlessly switch between these modules to build or modify the model input parameters, run the simulations, and display the results (in plan view or full-screen cross section).

*Processing MODFLOW for Windows* (PMWIN) is another complete simulation system. It comes with a professional graphical pre-processor and post-processor, the 3-D finite-difference groundwater models MODFLOW-88, MODFLOW-96, and MODFLOW 2000; the solute transport models MT3D, MT3DMS, RT3D and MOC3D; the particle tracking model PMPATH 99; and the inverse models UCODE and PEST-ASP for automatic calibration. A 3D visualization and animation package, 3D Groundwater Explorer, is also included.

*Groundwater Vistas* is a Windows modelling environment for the MODFLOW family of models that allows for the quantification of uncertainty. The approach used by Stochastic MODFLOW is the Monte Carlo technique, a common method employed by groundwater professionals for assessing risk. In the past, however, these risk assessments have relied primarily on simple analytical solutions and calculations. Here is a practical tool for assessing risk using more complex and real-world groundwater models.

- Monte Carlo versions of MODFLOW, MODPATH & MT3D
- Groundwater Vistas performs all pre-processing and post-processing
- Monte Carlo models may be launched directly from Groundwater Vistas
- Virtually any aquifer property or boundary condition can be sampled using normal, lognormal, uniform, log-uniform, or triangular distributions
- Hydraulic conductivity and leakance can use geostatistical simulation results
- Monte Carlo simulations may be conditioned to remove unrealistic results
- Compute mean and standard deviation for head, drawdown
- Support for SWIFT-3D Flow and Transport Model

*MODFLOW Graphical User Interface* (MODFLOW-GUI) for Argus ONE adds support for the U.S. Geological Survey's MODFLOW-2000 and the Reservoir, Transient Leakage, Interbed Storage, Lake, and Gage packages. It can also import MODFLOW-88 and MODFLOW-96 models. A utility program, GW Chart, was developed in conjunction with the MODFLOW GUI and is used for post-processing of the output of MODFLOW. Also, in conjunction with the development of the MODFLOW GUI, three utility Plug-In Extensions (PIE's) were developed. One PIE facilitates importing gridded data into Argus ONE. This is helpful when attempting to reproduce an existing model in Argus ONE. Another PIE allows editing of data points on data layers. A third utility PIE allows expressions to be evaluated at specific X, Y coordinates. A series of example models were created and can be used to learn about the features of MODFLOW, Argus ONE, and the MODFLOW GUI.

The MODFLOW GUI is from the USGS, the actual developers of MODFLOW. As a result, the MODFLOW GUI is always the first to support new technologies introduced into MODFLOW. MODFLOW-GUI supports the following packages: BAS5, BAS6, BCF5, BCF6, LPF, WEL5, RIV5, DRN5, GHB5, RCH5, EVT5, SOR5, SIP5, PCG2, DE4, STR1, HFB1, HFB6, FHB1, IBS1, RES1, TLK1, ETS1, DRT1, LMG1, OBS, HOB, GBOB, DROB, RVOB, ADOB, SEN, PES, MOC3D, MODPATH, and ZONEBDGT.

## Concluding Remarks

A groundwater model is a simplified representation of a groundwater system. While groundwater models are, by definition, a simplification of a more complex reality, they have proven to be useful tools over several decades for addressing a range of groundwater problems and supporting the decision-making process.

Limitations and uncertainties exist in any modelling study in regard to our hydrogeological understanding, the conceptual model design, and model calibration and prediction simulations, as well as recharge and evapotranspiration estimation and simulation. There are also limitations associated with the capabilities of the existing groundwater modelling software packages to adequately represent the complexities of any given hydrogeological system, and particularly in regard to surface-groundwater interaction. These limitations are best addressed by careful scoping of proposed modelling approaches at the outset and review at various stages throughout the project. It is also important that modellers properly document model limitations at the proposal stage and in technical reports, as well as outlining possible methods of resolving them by subsequent work programmes of data acquisition and analysis and/or modelling. In some cases, the limitations may be so severe that there may be little value in putting the effort into a modelling study until more data and hydrogeological understanding is obtained, or until new technical methods are developed.

**References**

1.  Bear, J., Beljin, M. S., Rose, R. (1992). Fundamentals of Groundwater Modelling. EPA Ground Water Issue.

2.  California Environmental Protection Agency Website, Groundwater Modelling, http://www.waterboards.ca.gov/water_issues/programs/land_disposal/gw_modelling.shtml

3.  EUGRIS Website (Portal for Soil and Water Management in Europe), Predictive Modelling, http://www.eugris.info/FurtherDescription.asp?Ca=2&Cy=0&T=Predictive%20modelling&e=19

4.  Harbaugh, A. W., E. R. Banta, M. C. Hill, and M. G. McDonald (2000). MODFLOW-2000, The U.S. Geological Survey Modular Ground-Water Model - User Guide to Modularization Concepts and the Ground-Water Flow Process. USGS Open-File Report 00-92. Reston, Virginia: U.S. Geological Survey, 2000.

5.  Harbaugh, A. W. (2005). MODFLOW-2005. The U.S. Geological Survey Modular Groundwater Model - The Groundwater Flow Process: U.S. Geological Survey Techniques and Methods 6-A16.

6.  McDonald, M. G., and Harbaugh (1988). A. W. A Modular Three-Dimensional Finite-Difference Ground-Water Flow Model: Techniques of Water-Resources Investigations of the United States Geological Survey, Book 6, Chapter A1, 586 p.

7.  Murray-Darling Basin Commission (2000). Groundwater Flow Modelling Guideline. November 2000, 133 p.

8.  Sinclair Knight Merz and National Centre for Groundwater Research and Training (2012). Australian Groundwater Modelling Guidelines. Waterlines Report Series No. 82, 2012, 191 p.

9.  The Scientific Software Group Website, MODFLOW Description, **Error! Hyperlink reference not valid.**

10. USGS Website, MODFLOW and Related Programs, http://water.usgs.gov/nrp/gwsoftware/modflow.html/

11. Wikipedia Website, Groundwater Model, http://en.wikipedia.org/wiki/Groundwater_model/

# 15-Pitfalls and Sensitivities in Groundwater Modelling

## Introduction

A groundwater model is a computer-based representation of the essential features of a natural hydrogeological system that uses the laws of science and mathematics. Its two key components are a conceptual model and a mathematical model. The conceptual model is an idealised representation (usually graphical) of our hydrogeological understanding of the essential flow processes of the system. A mathematical model is a set of equations, which, subject to certain assumptions, quantifies the physical processes active in the aquifer system(s) being modelled.

While the model itself obviously lacks the detailed reality of the groundwater system, the behaviour of a valid model approximates that of the aquifer(s). A groundwater model provides a scientific means to synthesise the available data into a numerical characterisation of a groundwater system. The model represents the groundwater system to an adequate level of detail, and provides a predictive tool to quantify the effects on the system of specified hydrological, pumping or irrigation stresses.

Typical model purposes include improving hydrogeological understanding (synthesis of data); aquifer simulation (evaluation of aquifer behaviour); designing practical solutions to meet specified goals (engineering design); optimising designs for economic efficiency and account for environmental effects (optimisation); evaluating recharge, discharge and aquifer storage processes (water resources assessment); predicting impacts of alternative hydrological or development scenarios (to assist decision-making); quantifying the sustainable yield (economically and environmentally sound allocation policies); resource management (assessment of alternative policies); sensitivity and uncertainty analysis (to guide data collection and risk-based decision-making); and visualisation (to communicate aquifer behaviour).

There are very few published and accepted guidelines on groundwater flow modelling. The notable example is the suite of Standard Guides (1993, 1994, 1995, 1996) from the American Society for Testing and Materials (ASTM), which are reasonably well-accepted standard practice guidelines. The groundwater modelling guideline documents are quite consistent in regard to the accepted general approach to groundwater modelling (with greater or lesser emphasis on certain aspects, depending on the application of the guideline) which may be summarised as:

* Define purpose of study and objectives
* Develop conceptual model
* Select model approach/code (analytical or numerical and software package)
* Develop and calibrate model
* Assess parameter sensitivity and calibration uncertainty
* Complete prediction scenarios and sensitivity/uncertainty analysis
* Report
* Post-audit (at some time in future)

## Pitfalls and Sensitivities

Pitfalls and sensitivities are often specific to a model (STOWA, 1999). However, an attempt has been made to sum up the frequently occurring problems. There are many model programs available for a numerical approach to the groundwater flow. These programs are based on the elementary conservation of equations (Darcy's law and the continuity equation). The differences are mainly found in the method of discretization (finite differentiations, finite elements and analytical elements) and the way in which the user can define the boundary conditions.

Two classes of model programs can be distinguished in the groundwater quality models. The first class comprises the 'lumped' model programs in which chemical aspects are defined by strongly simplified parameters (dispersion, sorption, retardation). Well-known model programs in this class are: MT3D, HST3D, RT3D, MODWALK. Just like the groundwater flow models, the differences between the various programs lie mainly in the method of discretization and the way in which the user can define the boundary conditions. The second class of model programs explicitly describes the chemical reactions. Representatives of this class are FREEQM and CHARON. This type of model program can not be used without considerable chemical expertise.

*(a) Model Setup*

Conceptual Model

A conceptual model is constructed prior to numerical modelling. This conceptual model defines, among other things, the global structures of the subsoil and the substances found in it. The hydraulic properties play a particularly important role for the groundwater flow. The choices made in the conceptual model with regard to the limits of the model field, both horizontally and vertically, are generally not changed in the further course of the modelling process. The choice of the location and type of the boundary can greatly influence the model results and therefore requires effective underpinning.

The layers are often classified into aquifers and separating layers. The method of binding of these layers greatly influences the simulated flow field, and thereby the model results. Another important aspect is the existence of hydraulic short circuits or blockages. These phenomena occur in separating layers, fracture systems (open or closed), sand and gravel banks and dams. Information on these phenomena is often not explicitly included or is insufficiently known. Non-recognition of these structures can lead to incorrect interpretation of the results during the further course of the modelling process. Moreover, the negligence of density variation can lead to completely incorrect flow directions and calibrated constants. In the coastal areas in particular, the effects of density variation due to deviations in the salt content will have to be taken into account. This also applies at waste dump sites and other locations with strongly polluted groundwater.

For the total sediment discharge, it is particularly the estimation of local heterogeneity which is important. These help to determine the travelling times and breakthrough curves. Also vital is a good hypothesis of the geo-hydrochemical processes. Examples of important aspects are the sorption processes (balance or imbalance, linear or non-linear), the presence of decomposition, the geochemical conditions (for example rich or poor in oxygen), the presence of organic matter in the sediment, etc. The limited observation material is often not adequate to be able to distinguish between various processes in the calibration phase, so that the modelling is strongly dependent on the expertise in the conceptual phase.

## Choice of Model Program

The choice of model program is not particularly critical for the final results of a model for groundwater flow. However, the modeller must be aware of the underlying assumptions and the usage limitations of the various model programs. The top system (small surface waters, drainage and unsaturated zone) may be defined in more detail in one model than in another, for example.

In the modelling of total sediment discharge, the differences between the various model programs may be very relevant, however. There is numerical dispersion in most of the model programs based on finite elements and finite differentials. The mass balance is often not guaranteed in model programs based on the finite elements approach. This may result in major errors, particularly when there are strong gradients in the concentration, due to the model incorrectly distributing the flux and the concentration over a large surface area.

Besides physical and chemical considerations, practical considerations also play a role in the choice of model program The more organised the input of parameters and options, the less the chance of practical usage errors. Moreover, the calculating time and the memory required may also play a role in the more complex (unsteady) problems.

*(b) Analyse the Model*

## Sensitive Parameters

In groundwater flow models, the sensitive factors often depend on the objective of the model. For many models, the degrees of resistance of separating layers are sensitive parameters as they are more difficult to estimate (variation of 10 or 100) than, for example, the permeability of aquifers (variation of 2). Local holes in separating layers have much larger regional effects than local areas with high permeability. Conversely, areas with a much higher local resistance in the separating layer, hardly have any influence at all, while local areas with high resistance in aquifers are generally relevant.

The lack of flow over a separating layer means that the value of the resistance is not important, but does make it almost impossible to determine the resistance. Once flow has been applied over this layer in a scenario (though extraction, for example), the resistance may well be a dominating factor (if it is relatively great, for instance).

The most sensitive parameters for total sediment discharge models depend on the type of substance being described and on the local situation. For those substances which absorb to organic matter, the retardation factor and the organic matter content can often be noted as being sensitive parameters.

## Discretization

The spatial and temporal discretization must be small enough to minimise numerical errors. Generally speaking, this means that smaller steps must be taken when the gradients become steeper. The steepness of the gradients is partly determined by the hydraulic properties of the system and can often be characterised by the composite leakage factor. Elements must generally be smaller than this leakage factor. This therefore also applies to surface waters where there are hardly any resistance layers at all. The errors made in such cases can lead to both incorrect flows in the model and to incorrect hydraulic parameters in the calibration.

Many total sediment transport models are very sensitive to the grid size in terms of the numerical dispersion. The discretization suitable for a flow model is not automatically also suitable for a total sediment discharge model based on the speed field of the flow model in question. The time step and grid distances can be chosen independently of one another. Some model programs automatically determine the (maximum) time step. If this is not the case, major numerical errors may occur.

Reduction of the grid distances and the time steps will not automatically produce better model results. A very fine grid may give the impression of a very detailed and therefore accurate model. Unless information is added on the right scale however, the only added value of a finer grid is its ability to prevent numerical errors. In combination with overly detailed parameterization, a finer grid may even provide less information. Conversely, too coarse an elements network can lead to 'stable' calibration results, which are incorrect due to insufficient possibilities for simulation of variations in hydraulic head and flux.

One of the main pitfalls in the modelling of boundaries is the definition of a closed boundary or fixed hydraulic head at the location of a catchment boundary under an infiltration area. That boundary and therefore the flow and the hydraulic head change as the circumstances change. The same applies in modelling a freshwater-salt water boundary area, this can not be seen as a closed boundary either if the circumstances in its vicinity change.

Special attention is required for the discretization in the vertical. A quasi 3D approach is an effective one for modelling for regional flow. Quasi 3D means that the vertical differences in the hydraulic head within an aquifer are neglected in the calculations. This does not mean that there can be no vertical flow component within the aquifer. In such cases, great care must be exercised with the schematization in aquifers and separating layers, as incorrect connections can lead to major errors in the model. Moreover, in local problems and in flows in heterogenic packages, this approach may lead to relevant errors, particularly in the calculated flow distribution and the total sediment discharge.

## Parameterization

A numerical groundwater flow model or a groundwater transport model comprises many spatially distinguishable units (blocks, elements). In principle, each unit is attributed a value for the parameters (permeability, storage coefficient, dispersion coefficient, sorption etc). This results in many (often tens of thousands of) degrees of freedom. Given the limited availability of information, both in terms of the geological definition as the observations of the dynamics (hydraulic head and concentration measurements), it is essential that the number of degrees of freedom be reduced. This takes place through a certain form of processing of the parameter values. Common methods include zoning, whereby a certain zone is attributed the same value and a geostatistic interpolation.

A pitfall of parameterization is that the structures which are modelled are too detailed and can not be adjusted in the course of the modelling, due to a lack of field observations, so that they in fact begin to lead a life of their own. Another important point is that the scale on which information on the parameter values is available is not always in keeping with the model discretization. This applies to geological information from drilling, for example, and to geohydrological information from pumping tests. The parameters in the grid blocks must become 'block effective' values. The point observations therefore need to be scaled up. The method of scaling may be very sensitive, particularly when there is great heterogeneity. If a 'simple' mathematical average or a linear interpolation of the point values is used in such cases, major errors may be the result, particularly with regard to the breakthrough curves and residence times. It is difficult to determine beforehand which method should be used, but there must in any case be careful verification in relation to observations from the field.

## Boundary Conditions

An important component of the modelling process is formulation of the boundary conditions. These are not only the conditions concerning the physical external boundaries of the model but also the conditions concerning the so-called interior boundaries (extraction points, drains etc). It is essential to include as many observations of fluxes (discharges) alongside the potentials (measured or estimated hydraulic heads and groundwater levels). Hydraulic parameters in a model without given fluxes can not, in principle, be defined as they may otherwise be attributed any possible value (and therefore also useless ones) in a calibration procedure.

## Calibration and Scaling up

Measurements with which a model is calibrated (hydraulic head, fluxes and concentrations) are often point observations in relation to the 'block effective' values which the model calculates. When, there is great small-scale variability in particular, it is very questionable to what extent a calculated head or concentration in a grid block or element must meet the measured point value. The trend must generally be in keeping, though even this need not always be the case. Prior effective (geostatistic) analysis of the representativeness of the measuring points is therefore always recommendable.

## Calibration and Minimization Criteria

Calibration (both manually and using calibration programs) is carried out by changing parameter values in order to gear the model results to observations. A squares sum or a variance is often used as a measure for calibration. A large number of calibration procedures is based on unweighted criteria. In other words, all measurements are attributed equal importance. This can lead to imbalanced calibration, for example in areas where there are clusters of observations with a great deal of superfluous information. The cluster then weighs disproportionately heavy in relation to a single measuring point at a different location. If the scope of the variables' values is very diverse (in concentration measurements, for example), there is a risk that a peak in the observations will be disproportionately weighted.

Steady models are commonly used in the geohydrological practice. These models neglect the effect of storage and can only describe an 'average' flow, and are therefore usually only applied for modelling of the quantity. The principle behind steady models is that a balanced situation is described, i.e. a situation in which the effects of changes in time can be neglected with regard to the effects to be calculated. There are a number of pitfalls here, particularly for the description of the groundwater flow (flow direction and residence times). There is often less sensitivity for the effects in terms of groundwater levels or hydraulic head.

The observations used in the calibration come from a dynamic system. Observations from a so-called 'average hydrological year' are often applied, and this can lead to serious errors (also in terms of hydraulic head) in systems with a long-term 'memory' (great inertia). For calibration of the hydraulic parameters of the quantity, it is often more advisable to look for an almost steady state (i.e. a state in which the storage variation is negligible) such as that found at the end of the rainy season, and also to take the average of this over a number of years.

In order to calibrate a steady model effectively (i.e. usable for calculation of an average flow situation), the steady values must be calculated for the observations from the dynamic system (hydraulic head, concentrations, precipitation, surface water levels etc). A rule of thumb in this is to take the average over approximately four times the correlation length (the period within which the variables to be modelled still show cohesion).

If there is little dynamics in the calibration phase, it is difficult to determine the parameters which influence time-dependent behaviour (storage). If possible (in groundwater decontamination for example), great dynamic variation should be applied to the starting state, for the benefit of calibration of the model.

*(c) Use the Model*

In the user phase (forecasting), the model results are often sensitive to different parameters when compared with the calibration phase (present state). A well-known example is that a model calibrated for an average state can not effectively be applied to a dry or wet state. Another example is that the groundwater flow is often strongly dependent on the feed from the top system. Due to the complexity of the top system, many model studies invest a great deal of energy in its calibration. However, if the model is intended for analysis of the effects of a change in deep groundwater reclamation, the separating layers between the aquifers may be equally important.

One of the best-known pitfalls in separating layers is that the calibration phase is generally dominated by a small hydraulic gradient, to the extent that its resistance can not be calibrated, while that resistance may be a deciding factor when reclamation is increased in the new situation. Measures in the top system often lead to a change in the representative resistance of that top system. It also occurs that a change in the flow (direction and volume) alters the hydraulic properties of layers.

For total sediment discharge models too, the conditions (in terms of flow and discharge) may be quite different in the user phase than in the calibration phase. Diffusion and decomposition may have an important effect on the total sediment discharge in the calibration phase for example, while convective transport is much more important in the user phase. What is needed in that case is modelling outside of the scope of the calibration. Generally speaking, the dynamic variation is different and more limited in the starting state than in the user phase. A change in the direction of flow (horizontal and/or vertical) can have major consequences for the parameter values determined by the calibration (for example those for retardation and sorption).

## Concluding Remarks

Groundwater models provide a scientific and predictive tool for determining appropriate solutions to water allocation, surface water - groundwater interaction, landscape management or impact of new development scenarios. However, if the modelling studies are not well designed from the outset, or the model doesn't adequately represent the natural system being modelled, the modelling effort may be largely wasted, or decisions may be based on flawed model results, and long term adverse consequences may result. Considering the common pitfalls and sensitivities, normally encountered during groundwater modelling studies, may help in improving the model conceptualisation and understanding the uncertainty in model results.

## References

1.  American Society for Testing and Materials (1993). Standard Guide for Application of a Ground-Water Flow Model to a Site-Specific Problem. ASTM Standard D 5447-93, West Conshohocken, PA.

2.  American Society for Testing and Materials (1993). Standard Guide for Comparing Ground-Water Flow Model Simulations to Site-Specific Information. ASTM Standard D 5490-93, West Conshohocken, PA.

3.  American Society for Testing and Materials (1994). Standard Guide for Defining Boundary Conditions in Ground-Water Flow Modeling. ASTM Standard D 5609-94, West Conshohocken, PA.

4.  American Society for Testing and Materials (1995). Standard Guide for Subsurface Flow and Transport Modeling. ASTM Standard D 5880-95, West Conshohocken, PA.

5.  American Society for Testing and Materials (1996). Standard Guide for Calibrating a Ground-Water Flow Model Application, ASTM Standard D5981-96, West Conshohocken, PA.

6.  Anderson, M.P. and Woessner, W.W. (1992). Applied Groundwater Modeling. Academic Press, Inc., San Diego, CA., 381 p.

7.  McDonald, M.G. and Harbaugh, A.W. (1988). A Modular Three-Dimensional Finite-Difference Ground-Water Flow Model. USGS TWRI Chapter 6-A1, 586 p.

8.   STOWA Report 99-05 (1999). Good Modelling Practice Handbook. Dutch Department of Public Works, Institute for Inland Water Management and Waste Water Treatment, Report 99.036.

# 16-Common Groundwater Modelling Errors

## Introduction

Groundwater models are used to predict the future changes in hydraulic heads or the migration pathway and concentrations of contaminants in groundwater. The accuracy of model predictions depends upon the degree of successful calibration and verification of the model in determining transport flow directions, and the applicability of the groundwater flow and solute transport equations to the problem being simulated. Errors in the predictive model, even though small, can result in gross errors in solutions projected forward in time. This chapter presents an overview of the common errors in any groundwater modelling study and ways to remove them.

## Model Conceptualization Errors

### 1. Inappropriate model selection

Model selection depends upon the modelling goals and the hydrogeological conditions of the site. It is inappropriate to attempt to define the optimum placement and pumping rate for an extraction well and monitoring system with a simple model incapable of representing known hydrogeological features of a complex site, such as rivers, lakes, variable aquifer thickness or hydraulic conductivity, 3-D groundwater flow, or multiple aquifers. Conversely, wellhead protection areas are frequently delineated by reverse particle tracking on a confirmed potentiometric surface. A buffer area to account for dispersion processes and other uncertainties is often added to the final delineation. In these cases, use of a MODFLOW-based model may significantly increase data requirements, modelling effort, and cost without a proportional improvement in delineation accuracy.

Within these guidelines, model selection is upto consultant, although the use of Windows-based software packages is encouraged. The model must also be documented, tested and accepted by the hydrogeological community.

### 2. Selection of inappropriate boundary conditions

Specified head boundaries should be assigned to represent site features and as appropriate to replicate a potentiometric surface. Ringing the model with constant head boundaries in an attempt to replicate a measured potentiometric surface generally indicates a lack of understanding of the hydrogeological conditions of the site.

### 3. Excessive discretization (excessive number of hydraulic conductivity zones with no physical basis)

Hydraulic conductivity zones should be based on aquifer tests and geological characterizations. Assigning hydraulic conductivity zones solely to improve calibration, with no physical rationale is not justified. A large number of hydraulic conductivity zones generally indicates overcalibration, a lack of site-specific data, or a lack of understanding of groundwater modelling data requirements.

## 4. Lack of far-field data

The model domain must be large enough so that the boundary conditions do not distort the aquifer's response to stresses in the area of interest. It is recognized that most data will generally be from the area of interest. However, there should be some basis for specifying boundary conditions and parameters in modelled areas outlying the area of interest. If there are no far-field data and assumptions are made, the effect of assumed conditions on model response must be evaluated.

## 5. Oversimplification of problem (2-D model when obviously 3-D flow)

Two-dimensional models should be used for problems where the direction of groundwater flow is obviously in two dimensions (e.g. radial flow to a well or a single aquifer with relatively small vertical hydraulic head or contaminant concentration gradients). Three-dimensional models should generally be used where multiple aquifers are present or the vertical movement of groundwater or contaminants is important.

## 6. Placing model boundaries too close to area of interest, which may include pumping centre

Model boundaries must not inhibit the aquifer responses to expected stresses in the area of interest. It is not possible to quantify what "too close" is, but checking the reasonableness of the flow through any suspicious boundary zone is recommended.

## 7. Lack of understanding of site hydrogeological processes (excessive recharge, aquifer thickness/permeability not known, fully penetrating streams, isotropic/homogenous conditions)

To accurately model a site, the site's hydrogeological characteristics must be properly defined by field work. Model parameters, and spatial variations of them, should not be assigned to improve calibration statistics without some physical justification. Text book values for some parameters, such as recharge and effective porosity may be used, although they should be used conservatively.

## Data Input Errors

### 1. Inconsistent parameter units

Parameter units must be self-consistent if the groundwater modelling program does not allow specifying the units for each parameter. With these programs, do not mix days and seconds, cubic meter per second and cubic meter per day, or cm per year and meter per day. If, for example, day is selected for time unit and meter for length unit, then recharge and hydraulic conductivity must be in meter per day, pumping rates must be in cubic meter per day, constant head boundaries and grid dimensions must be in meter, etc.

## 2. Incorrect sign for pumping or recharge

Most, if not all, groundwater modelling programs expect pumping to be specified as a negative value, injection and recharge as positive values.

## 3. Well not specified correctly (well screen assigned incorrectly, pumping schedule not accurate, pumping rate not appropriate for problem)

Well screen: Although not required in MODFLOW, some preprocessing software requires a well screen to be specified. The well screen should accurately represent the actual well screen elevations. If no well screen is defined, no water can be pumped or injected.

Pumping: The modelled pumping rate for an existing well must match the actual pumping rate when simulating current conditions. This does not mean, for example, that a well which pumps 1000 gpm for six hours per day must have a transient pumping schedule to duplicate that time table. A steady-state pumping rate of 250 gpm is equivalent for most groundwater studies. Monthly or yearly averaging may or may not be appropriate, depending on the purpose or scope of the model.

## 4. Aquifer stresses (pumping, recharge, evapotranspiration, etc.) not specified over entire transient simulation period

The added complexity of transient models provides increased opportunities for mistakes and can also mask data input errors, such as failing to define pumping rates, recharge rates, river stages, etc. for the full time period. The key is to understand what the results mean. If, for example, the results indicate that water mining began after six years of a 10-year simulation period, it may be checked whether the recharge rates were defined for years 6 through 10.

## 5. Using interpolated input data

If the interpolated, potentiometric surface indicates a point source or sink without a physical basis, fix it or apply engineering judgement to the results.

## 6. Forcing questionable data to fit

Always apply good engineering judgement to the data. If, for example, a hydraulic head contour is forced to make a sharp turn to fit an observation at a single well, that observation may not be accurate. Similarly, if early-time aquifer test data do not fit a curve, chances are the pump had not settled down, there was some well bore storage, or something similar that caused the discrepancy (Figure 1).

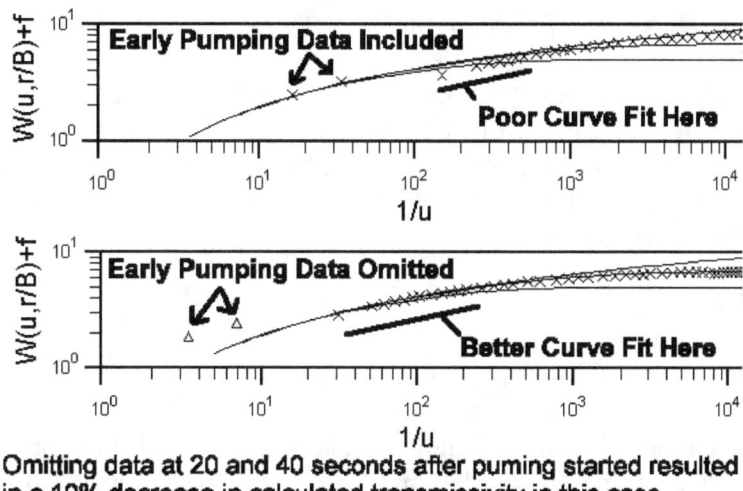

Omitting data at 20 and 40 seconds after puming started resulted in a 10% decrease in calculated transmissivity in this case.

Figure 1:     Possible Error in Analysing Pumping Test Data

## Calibration Errors

*1. Forcing a fit either by using unrealistic data values or overdiscretizing a aquifer or aquitard layer*

If the model can only achieve reasonable calibration statistics using unreasonable data values or by artificially assigning numerous zones of hydraulic conductivity, recharge, etc., the hydrogeology of the site has not been fully understood. Proper characterization of the site in and near the area of interest is necessary.

*2. Target wells clustered in a small portion of the model - i.e. lack of far field calibration data*

Target wells must be distributed over the model domain in order for the calibration statistics to be meaningful to the entire model.

*3. Target wells too close to, or within, specified head boundaries*

If a target well is too close to, or within, specified head boundaries, the aquifer's response at the well to stresses will be artificially limited by the boundary. Target wells should be distributed over the model domain and generally away from the constraints of specified head boundaries.

*4. Using interpolated data distribution rather than point data*

Interpolated data carries an uncertainty with it from the interpolation process. It is unacceptable to calibrate a model to interpolated data for hydraulic head, solute concentration, etc. Calibrating to model output from another model of the same site is also unacceptable. The only meaningful comparison is to actual, measured data points.

## 5. Misinterpreting mass balance information

The mass balance report includes statistics on the percent discrepancy between water added to the model and water removed from the model. The solver used within the model will always attempt to minimize this discrepancy. Thus, this is an indicator of solver accuracy for the specified model inputs, but can not indicate how well the model replicates the hydrogeological characteristics of the site.

## Simulation Results Errors

### 1. Omitting results inconsistent with preconceptions

An example of this type of mistake might be claiming that 100% capture of a plume is achieved in the model, but model simulations show that some particles from the source area are not captured. Other examples include ignoring the distribution of contaminants and trying to demonstrate a different flow direction inconsistent with measured hydraulic heads or groundwater chemistry.

### 2. Not incorporating data variability or uncertainty into the analysis

In spite of best efforts, model parameters can never be precisely known. Variations in measured data are either an indication of measurement uncertainty or the result of real physical differences. Use the range of data values that reflect data variability to bracket the model results. A conservative approach, which recognizes the uncertainty inherent in modelling, displays a firm understanding of the goals of groundwater modelling.

### 3. Blind acceptance of model output

All modellers should always question the model predictions. The accuracy of computer modelling is no better than the accuracy of the data used in the analysis. Make sure the model results agree with your understanding of the site hydrogeology and sound hydrogeological principles.

## Concluding Remarks

Predictive simulations must be viewed as estimates, dependent upon the quality and uncertainty of the input data. Models may be used as predictive tools, however field monitoring must be incorporated to verify model predictions. The best method of eliminating or reducing modelling errors is to apply good hydrogeological judgement and to question the model simulation results. If the results do not make physical sense, find out why.

## References

1.     Anderson, M.P. and W.W. Woessner (1992). Applied Groundwater Modeling. Academic Press, Inc., San Diego, CA., 381 p.

2.   American Society for Testing and Materials (1993). Standard Guide for Application of a Ground-Water Flow Model to a Site-Specific Problem. ASTM Standard  D 5447-93, West Conshohocken, PA, 6 p.

3.   American Society for Testing and Materials (1995). Standard Guide for Subsurface Flow and Transport Modeling. ASTM Standard  D 5880-95, West Conshohocken, PA, 6 p.

4.   Bear, J., and A. Verruijt (1987). Modeling Groundwater Flow and Pollution. D. Reidel Publishing Company, 414 p.

5.   Franke, O. L., Bennett, G. D., Reilly, T. E., Laney, R. L., Buxton, H. T. and Sun, R. J. (1991). Concepts and Modeling in Ground-Water Hydrology -- A Self-Paced Training Course. U.S. Geological Survey Open-File Report 90-707.

6.   Groundwater Modeling Program, Land and Water Management, Hydrologic Studies Unit, Department of Environmental Quality, State of Michigan, 2000.

7.   Kashyap, Deepak (1989). Mathematical Modelling for  Ground Water Management – Status in India. Indo-French Seminar on Management of Water Resources, 22-24 September, 1989,  Festival of France-1989, Jaipur, pp. IV-59 to IV-75.

8.   Kinzelbach, W. (1986). Groundwater Modeling: An Introduction with Sample Programs in BASIC. Elsevier, New York, 333 p.

9.   Kumar, C. P. (1992). Ground Water Modelling – In. Hydrological Developments in India Since Independence. A Contribution to Hydrological Sciences, National Institute of Hydrology, Roorkee, pp. 235-261.

10.  McDonald, M.G. and A.W. Harbaugh (1988). A Modular  Three-Dimensional Finite-Difference Ground-Water Flow Model, USGS TWRI Chapter 6-A1, 586 p.

# 17-Modelling of Unsaturated Flow

## Introduction

Subsurface formations containing water may be divided vertically into several horizontal zones according to how large a portion of the pore space is occupied by water. Essentially, we have a zone of saturation in which all the pores are completely filled with water, and an overlaying zone of aeration in which the pores contain both gases (mainly air and water vapour) and water. The latter zone is called the unsaturated zone. Sometimes the term soil water is used for the water in this zone.

The water movements in the unsaturated zone, together with the water holding capacity of this zone, are very important for the water demand of the vegetation, as well as for the recharge of the ground water storage. A fair description of the flow in the unsaturated zone is crucial for predictions of the movement of pollutants into ground water aquifers.

For analytical studies on soil moisture regime, critical review and accurate assessment of the different controlling factors is necessary. The controlling factors of soil moisture may be classified under two main groups viz. climatic factors and soil factors. Climatic factors include precipitation data containing rainfall intensity, storm duration, interstorm period, temperature of soil surface, relative humidity, radiation, evaporation, and evapotranspiration. The soil factors include soil matric potential and water content relationship, hydraulic conductivity and water content relationship of the soil, saturated hydraulic conductivity, and effective medium porosity. Besides these factors, the information about depth to water table is also required.

Most of the processes involving soil-water interactions in the field, and particularly the flow of water in the rooting zone of most crop plants, occur while the soil is in an unsaturated condition. Unsaturated flow processes are in general complicated and difficult to describe quantitatively, since they often entail changes in the state and content of soil water during flow. Such changes involve complex relations among the variable soil wetness, suction, and conductivity, whose inter-relations may be further complicated by hysteresis.

The formulation and solution of unsaturated flow problems very often require the use of indirect methods of analysis, based on approximations or numerical techniques. For this reason, the development of rigorous theoretical and experimental methods for treating these problems was rather late in coming. In recent decades, however, unsaturated flow has become one of the most important and active topics of research and this research has resulted in significant theoretical and practical advances.

## Soil Water Flow

The one-dimensional partial differential equation which describes the movement of moisture through unsaturated porous media subject to appropriate boundary and initial conditions has many field applications in the water environment. In hydrology, it describes the infiltration process that links the surface and sub-surface waters on land. In soil physics, it describes the capillary rise as well as drainage and evaporation of moisture in soils. In environmental pollution, it describes the longitudinal dispersion of pollutants in water courses. Therefore, the problem of seeking solutions to this equation has become a subject of concern for investigators from many different disciplines.

Downward infiltration into an initially unsaturated soil generally occurs under the combined influence of suction and gravity gradients. As the water penetrates deeper and the wetted part of the profile lengthens, the average suction gradient decreases, since the overall difference in pressure head (between the saturated soil surface and the unwetted soil inside the profile) divides itself along an ever-increasing distance. This trend continues until eventually the suction gradient in the upper part of the profile becomes negligible, leaving the constant gravitational gradient in effect as the only remaining force moving water downward. Since the gravitational head gradient has the value of unity (the gravitational head decreasing at the rate of 1 cm with each centimeter of vertical depth below the surface), it follows that the flux tends to approach the hydraulic conductivity as a limiting value. In a uniform soil (without crust) under prolonged ponding, the water content of the wetted zone approaches saturation. However, in practice, because of air entrapment, the soil-water content may not attain total saturation but some maximal value lower than saturation which has been called 'satiation'. Total saturation is assured only when a soil sample is wetted under vacuum.

The dynamics of soil water is cast in the form of mathematical expressions that describe the hydrological relations within the system. The governing equations define a mathematical model. The entire model has usually the form of a set of partial differential equations, together with auxiliary conditions. The auxiliary conditions must describe the system's geometry, the system parameters, the boundary conditions and, in case of transient flow, also the initial conditions. Operations with such a mathematical model are called simulation.

If the governing equations and auxiliary conditions are simple, an exact analytical solution may be found. Otherwise, a numerical approximation is applicable. The numerical simulation models are by far the most applied ones.

*Constitutive Equations*

The relationships that govern the flow of water in unsaturated soil are quasi-linear equations of the parabolic type. Since the coefficients in these equations are functions of the dependent variables, exact analytical solutions for specific boundary conditions are extremely difficult to obtain.

Darcy's equation for vertical flow is

$$q = -K\frac{\partial H}{\partial z} = -K\frac{\partial}{\partial z}(h-z)$$

... (1)

where q is the flux, H the total hydraulic head, h the soil water pressure head, z the vertical distance from the soil surface downward (i.e., the depth), and K the hydraulic conductivity. At the soil surface, $q = i$, the infiltration rate. In an unsaturated soil, h is negative. Combining this formulation of Darcy's equation (1) with the continuity equation $\partial\theta/\partial t = -\partial q/\partial z$ gives the general flow equation

$$\frac{\partial \theta}{\partial t} = \frac{\partial}{\partial z}\left(K\frac{\partial H}{\partial z}\right) = \frac{\partial}{\partial z}\left(K\frac{\partial h}{\partial z}\right) - \frac{\partial K}{\partial z}$$

$$\ldots (2)$$

If soil moisture content $\theta$ and pressure head h are uniquely related, then the left-hand side of equation (2) can be written

$$\frac{\partial \theta}{\partial t} = \frac{d\theta}{dh}\cdot\frac{\partial h}{\partial t}$$

which transforms equation (2) into

$$C\frac{\partial h}{\partial t} = \frac{\partial}{\partial z}\left(K\frac{\partial h}{\partial z}\right) - \frac{\partial K}{\partial z}$$

$$\ldots (3)$$

where C (= $d\theta/dh$) is defined as the specific (or differential) water capacity (i.e., the change in water content in a unit volume of soil per unit change in matric potential).

Alternatively, we can transform the right-hand side of equation (2) once again using the chain rule to render

$$\frac{\partial h}{\partial z} = \frac{dh}{d\theta}\cdot\frac{\partial \theta}{\partial z} = \frac{1}{C}\cdot\frac{\partial \theta}{\partial z}$$

We thus obtain

$$\frac{\partial \theta}{\partial t} = \frac{\partial}{\partial z}\left(\frac{K}{C}\cdot\frac{\partial \theta}{\partial z}\right) - \frac{\partial K}{\partial z}$$

or
$$\frac{\partial \theta}{\partial t} = \frac{\partial}{\partial z}\left(D.\frac{\partial \theta}{\partial z}\right) - \frac{\partial K}{\partial z}$$

$$\ldots (4)$$

Where, D is the soil water diffusivity. Equations (2), (3) and (4) can all be considered as forms of the Richards equation.

Note that the above three equations contain two terms on their right-hand sides, the first term expressing the contribution of the suction (or wetness) gradient and the second term expressing the contribution of gravity. Whether the one or the other term predominates depends on the initial and boundary conditions and on the stage of the process considered. For instance, when infiltration takes place into an initially dry soil, the suction gradients at first can be much greater than the gravitational gradient and the initial infiltration rate into a horizontal column tends to approximate the infiltration rate into a vertical. On the other hand, when infiltration takes place into an initially wet soil, the suction gradients are small from the start and become negligible much sooner. The effects of ponding depth and initial wetness can be significant during early stages of infiltration, but decrease in time and eventually tend to vanish in a very deeply wetted profile.

The following simplifications can be introduced to find analytical solutions: K is an analytical function of θ or h; hysteresis is neglected; the medium is homogeneous and isotropic; the flow is considered to be stationary or a succession of steady-state situations (quasistationary approach); the gravity force is neglected. The first two assumptions linked with the third one have resulted in a great number of analytical solutions. The gravity force is often neglected in describing the infiltration process in originally dry soil, resulting in analytical solutions as derived by e.g. Philip (1957, 1958).

The classical Richards-flow theory (Richards, 1931), upon which most simulation models are based, holds for stable flow conditions only. Yet instability of flow has been observed under a wide variety of circumstances such as abrupt and gradual increases of hydraulic conductivity with depth, compression of air ahead of the wetting front and water repellency of the solid phase. Another example of non-Richards type of flow is the preferential flow through non-capillary macropores. With classical flow theories one may then underestimate the velocity and depth of water/solute transport.

## Numerical Approach

With the advance of digital computers, emphasis has shifted drastically from the classical approach of analytical solutions to the rapidly developing field of numerical analysis. At present, numerical approximations are possible for complex, compressible, non-homogeneous and anisotropic flow regions having various boundary configurations.

Numerical methods are based on subdividing the flow region into finite segments bounded and represented by a series of nodal points at which a solution is obtained. This solution depends on the solutions of the surrounding segments and on an appropriate set of auxiliary conditions. In recent years, a number of numerical methods have been introduced. The methods most appropriate to the problem of soil water dynamics are finite difference method, finite element method and boundary element method. The finite difference method has been discussed below.

### Finite Difference Methods

Finite difference methods (Remson et al., 1971), either explicit or implicit, belong to the most frequently used techniques in modelling unsaturated flow conditions. The most simple type of finite differencing, the explicit one, orders the differencing operators in such a manner that the resulting finite difference equation contains only one unknown, and consequently, may be solved simply and directly. The explicit method is computationally simple but it has one serious drawback. In order to attain reasonable accuracy, the length of the interval in space must be kept small. To get a stable solution, the time step has to be small compared with the space interval. Thus, it is necessary to have a large number of time steps when using the simple explicit method.

Implicit solution methods generally use much larger time steps than explicit ones, but their stability depends upon the degree of nonlinearity of the differential equation. There are a great number of methods to solve an implicit set of algebraic equations, such as linearization, predictor-corrector or iteration methods.

In dealing with unsaturated flow problems that involve more than one space dimension and a grid with many nodal points, it is often necessary to use a mixed scheme that relies on simultaneous displacements along one space dimension and on successive displacements along the remaining space dimensions. This leads to the method of successive over relaxation (SOR). In the case of isotropic conditions, faster convergence may be sometimes achieved by using the iterative alternating direction implicit procedure (ADIPIT).

The advantage of the finite difference method is its simplicity and efficiency in treating the time derivatives. On the other hand, the method is rather incapable to deal with complex geometries of flow regions. A slow convergence, a restriction to bilinear grids and difficulties in treating moving boundary conditions are other serious drawbacks of the method.

*Discretization Schemes*

Different discretization schemes can be used using explicit or implicit methods. In the explicit method, a series of linearized independent equations is solved directly, while in the implicit method, a system of simultaneous linear algebraic equations (involving tridiagonal coefficient matrix with zero elements outside the diagonals) has to be solved. For a given grid point at a given time, the values of the coefficients K(h) or K($\theta$) and C(h) or D($\theta$) can be expressed either from their values at the preceding time step (explicit linearization) or from a prediction at time (t+1/2$\Delta$t) using a method described by Douglas and Jones, 1963 (implicit linearization).

The following discretization schemes (Equation 3) can be used for the various models.

Model 1 : Explicit Scheme Solved Directly

$$h_i^{j+1} = h_i^j + \frac{\Delta t}{C_i^j \Delta z}\left[ K_{i+1/2}^j \left( \frac{h_{i+1}^j - h_i^j}{\Delta z} - 1 \right) - K_{i-1/2}^j \left( \frac{h_i^j - h_{i-1}^j}{\Delta z} - 1 \right) \right]$$

... (5)

where, j refers to time, and i refers to depth and

$$K_{i+1/2}^j = \frac{K_{i+1}^j + K_i^j}{2};$$

$$K_{i-1/2}^j = \frac{K_i^j + K_{i-1}^j}{2}.$$

Defining D$_{max}$ as the maximum value of the soil water diffusivity in the soil profile at time t, the scheme is stable when (Haverkamp et al., 1977)

$$\Delta t < \frac{r(\Delta z)^2}{D_{max}}$$

... (6)

where, $\Delta Z$ is the layer thickness and r an arbitrary chosen coefficient.

The method is limited by extensive use of computer time when the water content approaches saturation and $\Delta t$ becomes very small ($D_{max}$ becomes large).

## Model 2 : Implicit Scheme with Explicit Linearization

$$C_i^j \frac{h_i^{j+1} - h_i^j}{\Delta t} = \frac{1}{\Delta z}\left[ K_{i+1/2}^j \left( \frac{h_{i+1}^{j+1} - h_i^{j+1}}{\Delta z} - 1 \right) - K_{i-1/2}^j \left( \frac{h_i^{j+1} - h_{i-1}^{j+1}}{\Delta z} - 1 \right) \right]$$

*or*  $$C_i^j \frac{h_i^{j+1} - h_i^j}{\Delta t} = \frac{1}{\Delta z}\left[ \left( \frac{K_{i+1}^j + K_i^j}{2} \right)\left( \frac{h_{i+1}^{j+1} - h_i^{j+1}}{\Delta z} - 1 \right) - \left( \frac{K_i^j + K_{i-1}^j}{2} \right)\left( \frac{h_i^{j+1} - h_{i-1}^{j+1}}{\Delta z} - 1 \right) \right]$$

Rearranging the terms, we get

$$\left[ -F_2 \frac{\Delta t}{(\Delta z)^2} \right]h_{i-1}^{j+1} + \left[ C_i^j + (F_1 + F_2)\frac{\Delta t}{(\Delta z)^2} \right]h_i^{j+1} - \left[ F_1 \frac{\Delta t}{(\Delta z)^2} \right]h_{i+1}^{j+1} = C_i^j h_i^j + (F_2 - F_1)\frac{\Delta t}{\Delta z}$$

... (7)

where,

$$F_1 = K_{i+1/2}^j = \frac{K_{i+1}^j + K_i^j}{2};$$

$$F_2 = K_{i-1/2}^j = \frac{K_i^j + K_{i-1}^j}{2}.$$

## Model 3 : Implicit Scheme with Implicit Linearization (Prediction - Correction)

From equation (3), we have

$$C \frac{\partial h}{\partial t} = \frac{\partial}{\partial z}\left[ K\left( \frac{\partial h}{\partial z} - 1 \right) \right]$$

$$or \quad C\frac{\partial h}{\partial t} = \frac{\partial K}{\partial z}\left(\frac{\partial h}{\partial z}-1\right)+K\frac{\partial^2 h}{\partial z^2}$$

$$or \quad \frac{C}{K}\frac{\partial h}{\partial t} = \frac{\partial^2 h}{\partial z^2}+\frac{1}{K}\frac{\partial K}{\partial z}\left(\frac{\partial h}{\partial z}-1\right)$$

$$\ldots (8)$$

<u>Prediction</u> (Estimation of $C_i^{\,j}$ and $K_i^{\,j}$):

From equation (8), by taking time step as $\Delta t/2$, we have

$$\frac{2C_i^{\,j}}{K_i^{\,j}}\cdot\frac{h_i^{j+1/2}-h_i^{j}}{\Delta t} = \frac{h_{i+1}^{j+1/2}-2h_i^{j+1/2}+h_{i-1}^{j+1/2}}{(\Delta z)^2}+\frac{1}{K_i^{\,j}}\cdot\frac{K_{i+1}^{\,j}-K_{i-1}^{\,j}}{2\Delta z}\left[\frac{h_{i+1}^{\,j}-h_{i-1}^{\,j}}{2\Delta z}-1\right]$$

Rearranging the terms, we get

$$-\frac{\Delta t}{(\Delta z)^2}h_{i-1}^{j+1/2}+\left[\frac{2C_i^{\,j}}{K_i^{\,j}}+\frac{2\Delta t}{(\Delta z)^2}\right]h_i^{j+1/2}-\frac{\Delta t}{(\Delta z)^2}h_{i+1}^{j+1/2} = \frac{2C_i^{\,j}}{K_i^{\,j}}h_i^{\,j}+\frac{1}{2}\frac{K_{i+1}^{\,j}-K_{i-1}^{\,j}}{K_i^{\,j}}\frac{\Delta t}{\Delta z}\left[\frac{h_{i+1}^{\,j}-h_{i-1}^{\,j}}{2\Delta z}-1\right]$$

$$\ldots (9)$$

<u>Correction</u> (Estimation of $h_i^{\,j}$ ):

From equation (8), by taking time step as $\Delta t$, we have

$$\frac{C_i^{j+1/2}}{K_i^{j+1/2}}\cdot\frac{h_i^{j+1}-h_i^{j}}{\Delta t} = \frac{1}{2}\left[\frac{h_{i+1}^{j+1}-2h_i^{j+1}+h_{i-1}^{j+1}}{(\Delta z)^2}+\frac{h_{i+1}^{\,j}-2h_i^{\,j}+h_{i-1}^{\,j}}{(\Delta z)^2}\right]+\frac{1}{K_i^{j+1/2}}\cdot\frac{K_{i+1}^{j+1/2}-K_{i-1}^{j+1/2}}{2\Delta z}\left[\frac{h_{i+1}^{j+1/2}-h_{i-1}^{j+1/2}}{2\Delta z}-1\right]$$

Rearranging the terms, we get

$$-\frac{1}{2}\frac{\Delta t}{(\Delta z)^2}h_{i-1}^{j+1}+\left[\frac{C_i^{j+1/2}}{K_i^{j+1/2}}+\frac{\Delta t}{(\Delta z)^2}\right]h_i^{j+1}-\frac{1}{2}\frac{\Delta t}{(\Delta z)^2}h_{i+1}^{j+1}$$

$$=\frac{C_i^{j+1/2}}{K_i^{j+1/2}}h_i^{\,j}+\frac{1}{2}\frac{\Delta t}{(\Delta z)^2}\left[h_{i+1}^{\,j}-2h_i^{\,j}+h_{i-1}^{\,j}\right]+\frac{1}{2}\frac{K_{i+1}^{j+1/2}-K_{i-1}^{j+1/2}}{K_i^{j+1/2}}\frac{\Delta t}{\Delta z}\left[\frac{h_{i+1}^{j+1/2}-h_{i-1}^{j+1/2}}{2\Delta z}-1\right]$$

$$\ldots (10)$$

## Model 4 : Crank-Nicolson Scheme

$$C_i^{j+1/2} \frac{h_i^{j+1} - h_i^j}{\Delta t} = \frac{1}{\Delta z} \left[ K_{i+1/2}^{j+1/2} \left( \frac{h_{i+1}^{j+1/2} - h_i^{j+1/2}}{\Delta z} - 1 \right) - K_{i-1/2}^{j+1/2} \left( \frac{h_i^{j+1/2} - h_{i-1}^{j+1/2}}{\Delta z} - 1 \right) \right]$$

where,

$$h_i^{j+1/2} = \frac{h_i^j + h_i^{j+1}}{2};$$

$$K_{i-1/2}^{j+1/2} = F_1 = \left( \frac{K_i^j K_{i-1}^j}{K_i^j + K_{i-1}^j} \right) + \left( \frac{K_i^{j+1} K_{i-1}^{j+1}}{K_i^{j+1} + K_{i-1}^{j+1}} \right);$$

$$K_{i+1/2}^{j+1/2} = F_2 = \left( \frac{K_i^j K_{i+1}^j}{K_i^j + K_{i+1}^j} \right) + \left( \frac{K_i^{j+1} K_{i+1}^{j+1}}{K_i^{j+1} + K_{i+1}^{j+1}} \right);$$

$$C_i^{j+1/2} = F_3 = \frac{C_i^j + C_i^{j+1}}{2}.$$

Rearranging the terms, we get

$$-\frac{1}{2} F_1 \frac{\Delta t}{(\Delta z)^2} h_{i-1}^{j+1} + \left[ F_3 + \frac{1}{2}(F_1 + F_2) \frac{\Delta t}{(\Delta z)^2} \right] h_i^{j+1} - \frac{1}{2} F_2 \frac{\Delta t}{(\Delta z)^2} h_{i+1}^{j+1}$$

$$= \frac{1}{2} F_1 \frac{\Delta t}{(\Delta z)^2} h_{i-1}^j + \left[ F_3 - \frac{1}{2}(F_1 + F_2) \frac{\Delta t}{(\Delta z)^2} \right] h_i^j + \frac{1}{2} F_2 \frac{\Delta t}{(\Delta z)^2} h_{i+1}^j + (F_1 - F_2) \frac{\Delta t}{\Delta z}$$

$$\ldots (11)$$

When equation (7), (9), (10) or (11) is applied at all nodes, the result is a system of simultaneous linear algebraic equations with a tridiagonal coefficient matrix with zero elements outside the diagonals and unknown values of h. In solving this system of equations, a so-called direct method can be used by applying a tridiagonal algorithm of the kind discussed by Remson et al. (1971).

### Initial and Boundary Conditions

Initial conditions must be defined when transient soil water flow is modelled. Usually values of matric head or soil water content at each nodal point within the soil profile are required. However, when these data are not available, water contents at field capacity or those in equilibrium with the ground water table might be considered as the initial ones.

## Upper boundary conditions

While the potential evapotranspiration rate from a soil depends only on crop and atmospheric conditions, the actual flux through the soil surface and the plants is limited by the ability of the soil matrix to transport water. Similarly, if the potential rate of infiltration exceeds the infiltration capacity of the soil, part of the water runs off, since the actual flux through the top layer is limited by moisture conditions in the soil. Consequently, the exact boundary conditions at the soil surface can not be estimated a priori and solutions must be found by maximizing the absolute flux. The minimum allowed pressure head at the soil surface, $h^{lim}$ (time dependent) can be determined from equilibrium conditions between soil water and atmospheric vapour.

The possible effect of ponding has been neglected so far. In case of ponding, usually the height of the ponded water as a function of time is given. However, when the soil surface is at saturation then the problem is to define the depth in the soil profile where the transition from saturation to partial saturation occurs.

In most of the dynamic transient models, the surface nodal point is treated during the first iteration as a prescribed flux boundary and matric head h is computed. If $h^{lim} \leq h \leq 0$, the upper boundary condition remains a flux boundary during the whole iteration. If not, the surface nodal point is treated as a prescribed pressure head in the following iteration. Then in case of infiltration, h = 0 and in case of evaporation h $= h^{lim}$. The actual flux is then calculated explicitly and is subject to the condition that actual upward flux through the soil-air interface is less than or equal to potential evapotranspiration (time dependent).

If the relative humidity (f) and the temperature of the air (T) as a function of time are known, and if it may be assumed that the pressure head at the soil surface is at equilibrium with the atmosphere, then h(0,t) can be derived from the thermodynamic relation (Edlefson and Anderson, 1943):

$$h(0,t) = \frac{RT(t)}{Mg} \ln[f(t)]$$

... (12)

where R is the universal gas constant (8.314 x 10 erg/mole/K), T is the absolute temperature (K), g is acceleration due to gravity (980.665 cm/s ), M is the molecular weight of water (18 gm/mole), f is the relative humidity of the air (fraction) and h is in bars. Knowing h(0,t), θ(0,t) can be derived from the soil water retention curve.

## Lower boundary conditions

At the lower boundary, one can define three different types of conditions: (a) Dirichlet condition, the pressure head is specified; (b) Neumann condition, the flux is specified; and (c) Cauchy condition, the flux is a function of a dependent variable.

The phreatic surface (place, where matric head is atmospheric) is usually taken as lower boundary of the unsaturated zone in the case where recorded water table fluctuations are known a priori. Then the flux through the bottom of the system can be calculated. In regions with a very deep ground water table, a Neumann type of boundary condition is used.

## Dirichlet condition

Easy recording of changes in phreatic surface in case of present ground water table is the main advantage of specifying a matric head zero as the bottom boundary. A drawback is that with shallow ground water tables (less than 2 m below soil surface) the simulated effects of changes in phreatic surface are extremely sensitive to variations in the soil hydraulic conductivity.

The nodal points in a soil profile usually have fixed positions and probably none of them will coincide with the water table level. The nodal point, where the matric head is prescribed, is often the one immediately beneath the phreatic level. When large fluxes across the lower boundary occur, an error is introduced by this approximation.

## Neumann condition

A flux as lower boundary condition is usually applied in cases where one can identify a no-flow boundary (e.g. an impermeable layer) or a free drainage case. In the latter case the flux is always directed downward and the gradient $dh/dz = 1$, so the Darcian flux is equal to the hydraulic conductivity at the lower boundary.

## Cauchy condition

This type of boundary condition is used when unsaturated flow models are combined with models for regional ground water flow or when the effects of surface water management are to be simulated. Writing the lower boundary flux as function of the phreatic surface, which is in this case the dependent variable, one can incorporate relationships between the flux to/from the drainage system and the height of the phreatic surface. This flux-head relation can be obtained from drainage formulae or from regional ground water flow models.

With the lower boundary conditions, the connection with the saturated zone can be established. In this way, effects of activities influencing the regional ground water system upon, for instance, crop evapotranspiration can be simulated. The coupling between the two systems is possible by considering the phreatic surface as an internal moving boundary with one-way or two-way relationships. When the Cauchy condition is linked with a one-dimensional vertical flow model, one can consider such a solution as quasi-two-dimensional, since both vertical and horizontal flow are calculated.

## Required Input Data

Simulation of water dynamics in the unsaturated zones requires input data concerning the model parameters, the geometry of the system, the boundary conditions and, when simulating transient flow, initial conditions. With geometry parameters, the dimensions of the problem domain are defined. With the physical parameters, the physical properties of the system under consideration are described. With respect to the unsaturated zone, it concerns the soil water characteristic, $h(\theta)$ and the hydraulic conductivity, $K(\theta)$.

For a proper description of the unsaturated flow, a correct description of the two hydraulic functions, $K(\theta)$ and $h(\theta)$, is important. The hydraulic conductivity, $K(\theta)$, decreases strongly as the moisture content $\theta$, decreases from saturation. The experimental procedure for measuring $K(\theta)$ at different moisture contents is rather difficult and not very reliable. Alternative procedures have been suggested to derive the $K(\theta)$ function from more easily measurable characterizing properties of the soil. In many studies, the hydraulic conductivity of the unsaturated soil is defined as product of a non-linear function of the effective saturation, and hydraulic conductivity at saturation. The relation is given by

$$K(\theta) = K_s \left( \frac{\theta - \theta_r}{\theta_s - \theta_r} \right)^n$$

... (13)

where,

$K_s$ = hydraulic conductivity at saturation;
$\theta_s$ = saturated water content; and
$\theta_r$ = residual water content.

The value of n is found to be 3.5 for coarse textured soils. n will vary with soil type. In literature, established empirical correlation between n and soil characteristic is available.

The relationship between the soil water pressure head $h(\theta)$ and moisture content $\theta$, usually termed as the water retention curve or the soil moisture characteristic, is basically determined by the textural and the structural composition of the soil. Also the organic matter content may have an influence on the relationship. A characteristic feature of the water retention curve is that suction head (-h) decreases fairly rapidly with increasing moisture content. Hysteresis effects may appear, and, instead of being a single valued relationship, the h-$\theta$ relation consists of a family of curves. The actual curve will have to be determined from the history of wetting and drying.

If root water uptake is also modelled, the parameters defining the relation between root water uptake and soil water status should be given, together with crop specifications. In case a functional flux-head relationship is used as lower boundary condition, the parameters describing the interaction between surface water and ground water and, if necessary, the vertical resistance of poorly permeable layers have to be supplied.

## References

1.    Dahlblom, Peter (1987). Mathematical Modelling of Soil Water Movement at Varpinge. Report No. 3114, Department of Water Resources Engineering, Institute of Science and Technology, Lund University, Lund.

2.    Douglas, J. J., and B. F. Jones (1963). On Predictor-Corrector Method for Non Linear Parabolic Differential Equations. J. Siam, Volume 11, pp. 195-204.

3.    Feddes, R. A., Kabat, P., Van Bakel, P. J. T., Bronswijk, J. J. B. and Halbertsma, J. (1988). Modelling Soil Water Dynamics in the Unsaturated Zone - State of the Art. Journal of Hydrology, Vol. 100, pp. 69-111.

4.    Haverkamp, R., M. Vauclin, J. Touma, P. J. Wierenga, and G. Vachaud (1977). A Comparison of Numerical Simulation Models for One-Dimensional Infiltration. Soil Sci. Soc. Am. J., Volume 41, pp. 285-294.

5.    Hillel, Daniel (1980). Applications of Soil Physics. Academic Press, New York, 385 p.

6.    Kumar, C. P. (1994). Review of Analytical Procedures of Unsaturated Flow. Hydrology Journal, Indian Association of Hydrologists, Roorkee, Vol. XVII, No. 3 & 4, July-December 1994, pp. 59-71.

7.    Kumar, C. P. (1996). Development of a Soil Moisture Prediction Model. Journal of Applied Hydrology, Association of Hydrologists of India, Vol. IX, No. 3 & 4, July & Oct. 1996, pp. 1-15.

8.    Kumar, C. P. (1996). Prediction of Evaporation Losses from Shallow Water Table using a Numerical Model. Journal of The Institution of Engineers (India), Volume 77, August 1996, pp. 103-107.

9.    Kumar, C. P. (1998). A Numerical Simulation Model for One-Dimensional Infiltration. ISH Journal of Hydraulic Engineering, Volume 4, No. 1, March 1998, pp. 5-15.

10.   Kumar, C. P. (1998). Estimation of Ground Water Recharge due to Rainfall by Modelling of Soil Moisture Movement. ISH Journal of Hydraulic Engineering, Volume 4, No. 2, September 1998, pp. 55-62.

11.   Philip, J. R. (1957). The Theory of Infiltration. Soil Sci., Volume 83, pp. 345-357 and 435-448.

12.   Philip, J. R. (1958). The Theory of Infiltration 6. Soil Sci., Volume 85, pp. 278-286.

13.   Philip, J. R. (1969). Theory of Infiltration. Advances in Hydroscience, Volume 5, pp. 215-305, Academic Press, NY.

14.   Remson, I., Hornberger, G. M. and Molz, F. J. (1971). Numerical Methods in Subsurface Hydrology. Wiley-Interscience, New York, 389 pp.

15.   Richards, L. A. (1931). Capillary Conduction of Liquids through Porous Mediums. Physics, Vol. 1, pp. 318-333.

# 18-Hydrological Problems in Coastal Areas

## Introduction

Man has always preferred deltaic and coastal areas for settlement. In ancient times, mainly because of the abundance of food from fisheries and the presence of fertile soils for agriculture. Even now-a-days, a great deal of the population of the earth is still concentrated in these areas. Favourable conditions for a good infrastructure have promoted transport, trade and the development of industry. However, these areas face many hydrological problems like steady subsidence, poor drainage and extensive flood danger, scarcity of fresh drinking water, tendency of the main flow to shift away to entirely different areas creating a constant problem for water traffic, and shifting and extending of shipping channels. Man's activities interfere with natural processes and many physical and biological changes are produced or expected, such as salt water intrusion.

All coastal areas will in some way be affected by various environmental impacts. The most significant impact will be changes along the shoreline and changes in rivers and estuaries. In the long term, changes in the geohydrological system in coastal areas will also become obvious. These will finally induce changes in water management as well.

## Deltaic Environments

A delta, in the geographic sense, is a deltoid-shaped area determined by the major bifurcations of a river and resulting from relatively rapid deposition of river-borne sediment into a more or less still standing body of water. A delta, in the geologic sense, is defined as a deposit of sediments partly subaerial and made by a stream at the place of entrance into a permanent body of water. The major portion of the deltaic sediments is deposited subaqueously in the permanent body of water where waves and currents aid in the transportation and deposition.

The environment of a large delta building into the sea is transitional between the normal marine and continental, and the deposits of each may be expected to interlens with the other. In exploring deltaic deposits for hydrocarbons, it is necessary to consider the possible types of deltas which may have existed. Delta types appear to be determined to a large extent by the nature and size of the drainage area controlling the sediment volume, and the hydrography and oceanography of the basin of deposition.

No satisfactory classification of deltas has been established. Tidal-estuarine, arcuate, cuspate, lobate and birdfoot deltas have been mentioned in the literature. Some deltas are composites of several of these types such as the Mississippi Delta which is a composite of birdfoot, lobate and cuspate types. The Rhone is a composite arcuate and cuspate delta. The Amu Dar'ya is lobate and cuspate.

Beaches and barrier islands fringe deltaic plains where littoral currents are strong, and waves and tidal ranges are large. The Niger, an arcuate type delta, is an example. Beaches and barriers appear to develop on the flanks of distributaries of many deltas which are building in water bodies with small tidal ranges, relatively large waves and relatively strong littoral currents. The cuspate Nile Delta is an example. Beaches and barrier islands are relatively uncommon on the margins of deltas which are building in water bodies with small tidal ranges and waves and weak shore currents. The lobate Lena and the Mississippi birdfoot deltas are examples. Tidal-estuarine deltas are probably deltas in youth and are associated with streams depositing small quantities of sediment in a body of water. The Amazon appears to be an example. The Irrawaddy and the Ganges deltas represent an undescribed type and appear to have resulted from the very rapid deposition of very large amount of sediments. The water of these two rivers head in areas of considerable relief which are near the delta.

The Amazon, Me'Kong and Irrawaddy may represent relative stages of delta maturity. Rivers having large changes in gradient and carrying large sediment loads form "alluvial fan type deltas" in interior drainage basins. The upstream parts of the Amu Dar'ya, the Nile and also the Colorado of the southwestern United States are examples.

In India, the east flowing rivers which drain into the Bay of Bengal, produce vast deltas but not all these rivers form deltas. The main deltas formed by these rivers are the Mahanadi Delta, the Godavari-Krishna Delta, the Cauvery Delta and the Bengal Delta.

## Hydrological Problems in Coastal Areas

Man has always preferred deltaic and coastal areas for settlement. In ancient times, mainly because of the abundance of food from fisheries and presence of fertile soils for agriculture. Even nowadays, a great deal of the population of the earth is still concentrated in these areas. Favourable conditions for a good infrastructure have promoted transport, trade and the development of industry. However, living in these areas has drawbacks also, some of them being danger of flooding; waterlogging and drainage congestion; salt water intrusion; and effects of cyclones, aquaculture along the coasts, dams, embankments etc.

Coastal areas are among the most intensively used areas on all continents. Of today's world population of approximately six thousand million, about 60% live in low-lying coastal areas. Over the next 100 years, the world population is likely to double. Of the 11 thousand million people living in the year 2100, it is assumed that 75 % will inhabit low-lying coastal areas. This means, there will be an increase from 3.6 thousand million to 8.25 thousand million inhabitants. Average population densities, and hence human impacts, are consequently around ten times higher in coastal areas than in continental interiors.

Moreover, the coastal areas are not only among the most densely populated regions of the earth, they are also subject to extraordinarily intensive uses such as industrial and commercial sites, agriculture, recreation and tourism. The increase in population will be accompanied by intensification of these land uses. This intensive exploitation will result in considerable impacts on the hydrological conditions in the coastal areas. The physical, biological and chemical conditions of estuaries are significantly influenced by river inputs and their load of dissolved and particulate matter. The structural sensitivity of rivers and estuaries gives particular cause for concern. Many shoreline features, deltas and estuaries are inherently unstable with their status of dynamic equilibrium maintained by both biological and physical processes. Changes in coastal vegetation and changes in land use in the whole catchment area can affect the characteristics of the coastline and estuaries. Here, deforestation in the headwater areas of the river basins should primarily be mentioned since deforestation may modify the runoff behaviour and the transport of sediments downstream. River training, in addition, nearly always accelerates the river flow and thus increases the erosion.

### Effects of Storm Surges and River Flow

Storm surges affect low-lying coastal areas along shallow seas and if they penetrate into rivers, estuaries and deltas, they interact with river floods and thus cause extra high water levels in the lower river reaches. Storm surges are generated in low pressure areas in the atmosphere which, in a first instance, lift the water level locally causing strong winds depending on the air pressure gradients. By transferring the wind energy to the water, the surges are amplified, an effect which is most pronounced in shallow coastal water.

There are two zones in the world where these phenomena are of importance: (a) the tropical cyclone zone consisting of two belts north and south of the equator and (b) the northern and southern temperate zones of the extra-tropical cyclones. Tropical cyclones originate from local heating of the sea surface, leading to a relatively small but intensive low air pressure area which starts moving over the ocean until it hits a coastal area somewhere. The lifetime of the accompanying storm surge can be expressed in hours. Extra-tropical cyclones are also related to atmospheric depressions but the phenomena act on a longer time scale (i. e. with a lower velocity). The duration of storm surges, thus generated, may be a number of tidal cycles (as a rule half day). Although the probability of a storm surge coinciding with a river flood is relatively small, nevertheless if it occurs, the effects can be disastrous.

In the upper reaches of estuaries and the lower portions of rivers, floods occur from a combination of fresh water and marine causes, in particular when a storm surge in the sea coincides with a fresh water flood downstream. Flooding may also occur in creeks and tide-locked water courses when fresh water is unable to discharge due to sustained high marine water levels. It is well known that both causative phenomena (i.e. surges and river floods) are initiated by weather conditions i.e. wind and low pressure in the marine case and rainfall in the fresh water case. Both are modified by topography of the estuary and of the river basin. In global terms, hurricanes are encountered in the tropical belt where storm surges and flooding conditions are most severe. However, other combinations of conditions have resulted in disastrous floods in temperate regions. Flooding is also aggravated by other factors such as tides, seiches and wind waves.

If the same low pressure system causes heavy rainfall in the estuary catchment, the flooding can be much aggravated by fluvial flooding. The fluvial input is a function of catchment factors, in particular the lag time. Heavy rainfall inland sometimes coincides with a storm surge propagating upstream into a river, the runoff from which causes a flood wave which propagates down the river. Hence storm surge forecasting can involve an upstream directed storm surge wave and a downstream directed flood wave.

Data have to be measured, gauges have to be installed and there must be data analysis before it is possible to make a forecast. For real time forecasting, there must be some means of transmitting the concerned data to the forecasting centre. Efforts should also be made to develop models. Different methods are applied with regard to water level forecasting. There are empirical statistical methods, hydrodynamical methods and hybrid modelling techniques with different levels of quality. The techniques used, depend on the demands of the users, their ability to handle these techniques and the data available to feed these models. Data acquisition and transfer is one of the most important pre-suppositions not only for routine forecasting but also to prove and calibrate the relevant methods before they are established. As most of the events to be forecasted are generated by storms, sudden air pressure changes or heavy rainfall, a well operating weather forecast service may also be a pre-requisite.

As to the methods already established, especially in the developed countries, a trend towards using numerical models is to be seen. There is an increasing demand for more precise weather forecasts. Special attention is drawn to possible climatic changes such as sea level rise which might affect coastal areas seriously. Effectiveness of the used methods may differ from place to place. Upto now, no method has been found to be perfect and the available methods still have to be improved. Even if the method works sufficiently, the technical equipment may be a weak link. A breakdown in the telecommunication system or deficiency in the data measuring system, means loss of necessary information. Data sampling has to be improved which means that hydrological and meteorological networks have to be extended.

## Consequences of Sea Level Changes

It is evident that a rise in sea level will increase the risk of flooding. Many measures have been taken to minimise this risk such as the use of natural levees; construction of dams, dikes and coastal embankments; and prevention of scouring of coastal and river embanking to avoid a sudden collapse. Moreover, a sea level rise will also affect the other aspects of the environment, though on a different time scale. Although a relative sea level rise may show very different local effects because of e.g. the geological and hydro(geo)logical situation, the phenomenon of sea level rise on its own will obviously be felt all over the world.

All coastal areas will in some way be affected by various environmental impacts brought about by a possible sea level rise. The most significant impact will be changes along the shoreline and changes in rivers and estuaries. In the long term, changes in the geohydrological system in coastal areas will also become obvious. These will finally induce changes in water management as well.

Investigations on the phenomena affecting sea level change, which are essential for impact and response strategies, should be promoted more intensively in the future. Much effort is being channeled into the study of slow and gradual sea level variation whereas the much more important and relevant causes of sea level change, such as cyclones, are either ignored or paid little attention. One storm surge event can cause more erosion in a few hours than a gradual sea level change could do on a much longer time scale. The term sea level rise should be replaced by sea level variation since sea level is decreasing at some locations around the globe. All the tide gauge networks around the globe should be properly related to each other through the identification of common datums. There is a need for development of detailed numerical models and observational programmes to assess sea level changes and impacts in the local areas.

## Salt Water Intrusion

Saline water is the most common pollutant in fresh ground water. Intrusion of saline water occurs where saline water displaces or mixes with fresh water in an aquifer. In a coastal aquifer, there is direct contact between continental fresh water and marine salt water. Besides difference in viscosity between the two fluids, there exist a density change that depends mainly on salinity differences (Custodio and Bruggeman, 1987). In a stable system, the fresh water floats on the salt water and a landward sloping interface exists between them. The salt ground water body adopts the form of wedge resting on the aquifer floor. The fresh water thickness decreases from the wedge toe towards the sea. Since the fresh water flow thickness decreases seaward, the slope of the piezometric head or the water table must increase towards the coast. Thus, the interface is concave upwards.

Salt water intrusion into fresh ground water formations generally results from activities of man. In coastal aquifers, salt water intrusion occurs due to invasion of sea water. This situation commonly occurs in coastal aquifers in hydraulic continuity with sea when pumping of wells disturbs the natural hydrodynamic balance. Ground water abstraction reduces the coastal fresh water discharge and alters the dynamic equilibrium. The final result is a deeper and more penetrating sea water wedge. When abstraction is greater than actual recharge, no final equilibrium position can be attained, but the sea water intrudes very deeply. Fresh water abstractions create local water head drawdowns and when the well is located over the salt water wedge, a salt water upconing develops.

A rapid recharge of fresh water increases the fresh water head. Such situation occurs in water table aquifers where direct recharge by rainfall infiltration is possible and recharge also occurs by irrigation return flows, floods, river channels etc. Both fresh water and salt water flow towards the sea and the mixing zone moves slowly seaward and downwards increasing its slope. The reverse is true in dry period. Sea tides produce the same effect. Many other activities such as urbanisation, industrialisation, suppression of irrigated areas, river regulations with upstream dams and deforestation can lead to an increase in the potential for sea water encroachment.

**Concluding Remarks**

The present problems of the sustainable use and development of coastal areas are already critical in many countries. The problems result from the conflict between different human uses of coastal land and near-shore waters, over exploitation of renewable coastal resources, discharge of wastes and effluents to coastal waters, increasing hydrographic stresses such as storm and sea level rise and the rapid growth of coastal population. There is a certain awareness of the negative impact of many local activities but there are not always alternatives to give people access to a minimum income, especially in developing countries.

Water is such a primary good that often secondary effects (subsidence, sediment starvation, salt water intrusion) and sometimes even primary effects (major diversion from basin to basin) are ignored initially. A reactive response is a common policy. A change to an active planning approach is required. There is an urgent need to increase public and political awareness of the impacts on coastal zones.

**References**

1.    Custodio, E. and Bruggeman, G. A. (1987). Groundwater Problems in Coastal Areas. A contribution to the International Hydrological Programme, Studies and Reports in Hydrology, No. 45, UNESCO, Paris.

2.    IHP-IV Project H-2.2 (1990-95). Hydrology, Water Management and Hazard Reduction in low-lying Coastal Regions and Deltaic Areas in Particular with regard to Sea Level Changes. Report by Karl Hofius, Secretary, Federal Republic of Germany, National Committee for IHP/OHP.

3.      Kumar, C. P. (1997). The Problems associated with Deltas with emphasis on Salt Water Intrusion. Status Report, UNDP Project (IND/90/003), National Institute of Hydrology, Roorkee.

4.      Kumar, C. P. (1998). The Modelling of Salt Water Intrusion. Training Report in the field of Deltaic Hydrology, UNDP Project (IND/90/003), National Institute of Hydrology, Roorkee.

# 19-Modelling of Sea Water Intrusion with SUTRA

## Introduction

In many coastal areas, the development and management of fresh groundwater resources are seriously constrained by the presence of sea water intrusion. Sea water intrusion is a natural phenomenon that occurs as a consequence of the density contrast between fresh and saline groundwater. Normally, the denser saline water forms a deep wedge that is separated from the fresh water body by a transition zone of variable density. In some cases, this wedge can extend for many kilometers inland. Providing conditions to remain unperturbed, the saline water body will remain stationary, its position being largely defined by the fresh water potential and hydraulic gradient. However, when the aquifer is disturbed by pumping of the fresh water, sea level change or by changing recharge conditions, the saline water body will gradually move until a new equilibrium condition is achieved. Problems arise when saline water from the deep saline wedge enters pumping wells and affects the water quality. Most commonly, this occurs at individual wells where heavy pumping lowers the fresh water potential in the immediate vicinity of the well and causes saline water to be drawn upwards to the well, a phenomenon known as upconing. This type of problem is often very localized and can be rectified by distributing production amongst a group of smaller shallower wells. A similar but potentially more serious problem occurs when a coastal aquifer is overdeveloped on a regional scale. This results in lowering of the fresh water potential throughout the area and progressive and extensive invasion of the aquifer by sea water. In some heavily exploited aquifers, the inflow of sea water may represent a significant component of the aquifer's flow budget.

## Ghyben - Herzberg Principle

Until fairly recently, most research on the relationship between fresh and saline groundwater in coastal aquifers has been based on the analytical solutions. Ghyben (1888) and Herzberg (1901) independently developed similar formulations on sea water intrusion, widely known as the Ghyben-Herzberg principle. Their formulation was based on the hydrostatic equilibrium between fresh water and saline water (assumed immiscible) in a U-shaped tube. It states that the fresh water-saline water interface in an aquifer occurs at a depth of Z below mean sea level, as represented by

$$Z = \frac{\rho_f}{\rho_s - \rho_f} h_f \qquad \qquad \dots (1)$$

where, $\rho_f$ is the density of fresh water (M/L$^3$), $\rho_s$ is the density of saline water (M/L$^3$), and $h_f$ is the elevation of the fresh water level above mean sea level (L).

Substitution of $\rho_f$ (1000 kg/m$^3$) and $\rho_s$ (1025 kg/m$^3$) in equation (1) shows that $Z = 40 \, h_f$. In other words, the depth to the saline water interface below mean sea level is 40 times the elevation of water table above sea level ($h_f$). It also follows from equation (1) that if water table in an unconfined coastal aquifer is lowered by 1 m, the fresh water-saline water interface will rise by 40 m. This simple formulation has been used widely by hydrogeologists, but it is an inadequate representation, because it describes a steady state equilibrium and does not take into account important mechanisms such as advection and dispersion.

## Numerical Modelling

In practice, the spatial relationship between fresh and saline groundwater in coastal and island aquifers is complex and management of the fresh water resource may be a difficult and sensitive issue. The aquifer system is rarely near equilibrium and the fresh and saline water bodies are normally separated by a transition zone created due to chemical diffusion and mechanical mixing. Under these conditions, the response of the saline water body to pumping is difficult to predict and depends on various factors including aquifer geometry and properties (intrinsic permeability, anisotropy, porosity, dispersivity); abstraction rates and depths; recharge rate; and distance of pumping wells from the coastline. Sophisticated tools are required to quantify the aquifer response to these factors. With the advent of large scale and widely available computing resources, the numerical approach to sea water intrusion analysis has moved to the forefront.

Currently, several solute transport models, suitable for the simulation of sea water intrusion and upconing of saline water beneath pumping sites, are commercially available. These include SUTRA (Voss, 1984), SEAWAT (Langevin et al., 2007), HST3D (Kipp, 1987) and SALTFLOW (Molson and Frind, 1994). These models provide solutions of two simultaneous, non-linear, partial differential equations that describe the conservation of mass of fluid and conservation of mass of salt in porous media. SUTRA (Saturated-Unsaturated Transport) is a computer program that simulates fluid movement and the transport of either energy or dissolved substances in a subsurface environment. The SEAWAT program is a coupled version of MODFLOW and MT3DMS designed to simulate three dimensional, variable-density, saturated groundwater flow. HST3D (Heat and Solute Transport in 3 Dimensions) employs three-dimensional finite-difference approximations of the governing equations. This model is capable of simulating an aquifer with irregular geometry. SALTFLOW is also three-dimensional but utilize a finite-element approximation of the governing equations for an aquifer that is subject to the intrusion of sea water. Further details about SUTRA are presented below.

## Saturated-Unsaturated TRAnsport Model (SUTRA)

SUTRA is a finite-element simulation model for saturated-unsaturated, fluid-density-dependent groundwater flow with energy transport or chemically-reactive single-species solute transport. SUTRA may be employed for areal and cross-sectional modelling of saturated groundwater flow systems and for cross-sectional modelling of unsaturated zone flow. Solute transport simulation using SUTRA may be employed to model natural or man-induced chemical species transport including processes of solute sorption, production and decay and may be applied to analyse groundwater contaminant transport problems and aquifer restoration designs. In addition, solute transport simulation with SUTRA may be used for modelling of variable density leachate movement and for cross-sectional modelling of salt water intrusion in aquifers in near-well or regional scales with either dispersed or relatively sharp transition zones between fresh water and salt water. SUTRA energy transport simulation may be employed to model thermal regimes in aquifers, subsurface heat conduction, aquifer thermal energy storage systems, geothermal reservoirs, thermal pollution of aquifers and natural hydrogeologic convection systems.

SUTRA is written in Fortran 77 including optional Fortran 90 statements allowing dynamic array allocation. SUTRA requires that files needed for the simulation be defined prior to execution. A Name File (SUTRA.FIL) is used for this purpose. For each file (upto six), the unit number is specified on one record followed by a record specifying the file name. Generally, the program is easily installed on most computer systems. The code has been used on a wide variety of computers, ranging from UNIX-based computers to DOS-based 386 computers with as little as 640K of RAM. The latest version 2.2 of SUTRA allows specification of time-dependent boundary conditions without programming FORTRAN code, and offers optional output files that summarize specifications and computed results at boundary condition nodes.

SUTRA is included with a number of utility codes for both pre- and post-processing in a package called SutraSuite. SutraGUI is a pre-processor that is applicable to both 2D and 3D problems. SutraPrep is a text-based preprocessor that creates 3D input datasets for SUTRA. There are three post-processors for 2D problems: SutraPlot, ModelViewer and SutraGUI. There are two post-processors for 3D problems: SutraPlot creates rotatable 3D plots of the mesh, 2D contours of results, and plots of velocity vectors; ModelViewer creates rotatable 3D color visualizations of results and plots of velocity vectors.

Two additional post-processing tools for both 2D and 3D simulations are available. GW_Chart can graphically display SUTRA fluid, solute and energy budgets, as well as hydrographs for SUTRA observation node output, showing pressure, concentration, and temperature as a function of time. The utility CheckMatchBC aids in setting boundary condition conductances for a SUTRA simulation; it checks the match of specified and simulated pressures, concentrations and temperatures and reports the number of matching digits. GW_Chart and CheckMatchBC are distributed with the installation files for SutraGUI.

*Purpose and Scope*

SUTRA (Saturated-Unsaturated Transport) is a computer program that simulates fluid movement and the transport of either energy or dissolved substances in a subsurface environment. The original version of SUTRA was released in 1984 (Voss, 1984). The latest version 2.2 adds the ability to specify time-dependent sources and boundary conditions (without programming) and to output information pertaining to source and boundary condition nodes in a convenient format. The code employs a two- or three dimensional finite-element and finite-difference method to approximate the governing equations that describe the two interdependent processes that are simulated:

- Fluid density-dependent saturated or unsaturated groundwater flow; and
- Either (a) transport of a solute in the ground water, in which the solute may be subject to: equilibrium adsorption on the porous matrix, and both first-order and zero-order production or decay; or (b) transport of thermal energy in the ground water and solid matrix of the aquifer.

SUTRA may also be used to simulate simpler subsets of the above processes. As the primary calculated result, SUTRA provides fluid pressures and either solute concentrations or temperatures, as they vary with time, everywhere in the simulated subsurface system.

SUTRA flow simulation may be employed for two-dimensional (2D) areal, cross sectional and three-dimensional (3D) modeling of saturated groundwater flow systems, and for cross sectional and 3D modeling of unsaturated zone flow. Solute-transport simulation using SUTRA may be employed to model natural or man-induced chemical-species transport including processes of solute sorption, production, and decay. For example, it may be applied to analyze groundwater contaminant transport problems and aquifer restoration designs. In addition, solute-transport simulation with SUTRA may be used for modeling of variable-density leachate movement, and for cross sectional modeling of saltwater intrusion in aquifers at near-well or regional scales, with either dispersed or relatively sharp transition zones between freshwater and saltwater. SUTRA energy-transport simulation may be employed to model thermal regimes in aquifers, subsurface heat conduction, aquifer thermal-energy storage systems, geothermal reservoirs, thermal pollution of aquifers, and natural hydrogeologic convection systems.

Mesh construction, which is quite flexible for arbitrary geometries, employs quadrilateral finite elements in 2D Cartesian or radial-cylindrical coordinate systems, and hexahedral finite elements in 3D systems. Permeabilities may be anisotropic and may vary in both direction and magnitude throughout the system, as may most other aquifer and fluid properties. Boundary conditions, sources and sinks may be time dependent. A number of input data checks are made to verify the input data set. An option is available for storing intermediate results and restarting a simulation at the intermediate time. Output options include fluid velocities, fluid mass and solute mass or energy budgets, and time-varying observations at points in the system. Both the mathematical basis for SUTRA and the program structure are highly general, and are modularized to allow for straightforward addition of new methods or processes to the simulation. The FORTRAN-90 coding stresses clarity and modularity rather than efficiency, providing easy access for later modifications.

*Governing Equations*

The simulation of sea water intrusion requires the solution of partial differential equations that describe conservation of mass of fluid and conservation of mass of solute. These are summarized below (Voss, 1984).

<u>Conservation of mass of fluid</u>

The fluid mass balance in a saturated porous medium can be expressed as :

$$\frac{\partial(\varepsilon\rho)}{\partial t} = -\Delta.(\varepsilon\rho V) + Q_p \qquad\qquad \dots (2)$$

where $\varepsilon$ (x,y,z,t) is porosity (dimensionless); $\rho$ (x,y,z,t) is fluid density (M/L$^3$); $Q_p$ (x,y,z,t) is fluid mass source [M/(L$^3$T)]; V (x, y, z, t) is fluid velocity (L/T); x, y and z are cartesian coordinate variables (L); t is time (T); and $\Delta$ is [($\partial/\partial$x)i + ($\partial/\partial$y)j + ($\partial/\partial$z)k]. The term on the left hand side of equation (2) expresses the change in fluid mass contained in the void space of the local volume with time. The first term on the right hand side of equation (2) represents the contribution to local fluid mass change due to excess of fluid inflows over outflows. The second term ($Q_p$) accounts for external additions of fluid.

The fluid mass balance (equation 2) can also be represented by:

$$( \rho \, S_{op}) \frac{\partial p}{\partial t} + \left[ \varepsilon \frac{\partial \rho}{\partial C} \right] \frac{\partial C}{\partial t} - \Delta. \left[ \left( \frac{\varepsilon \rho k}{\mu} \right).( \Delta p - \rho g) \right] = Q_p$$

... (3)

where $S_{op} = [(1-\varepsilon) \, \alpha + \varepsilon \, \beta]$ is specific pressure storativity (LT$^2$/M); $\alpha$ is porous matrix compressibility (LT$^2$/M); $\beta$ is fluid compressibility (LT$^2$/M); C is solute mass fraction or mass of solute (M$_s$) per mass of fluid (M$_s$ /M); k (x,y,z) is solid matrix permeability tensor (L$^2$); $\mu$ (x,y,z,t) is fluid viscosity (M/(LT)), p (x,y,z,t) is fluid pressure [M/(LT$^2$)], and g is the gravity vector (L/T$^2$).

Conservation of mass of solute

The solute mass balance for a single species stored in solution is expressed as :

$$\partial \frac{( \varepsilon \rho C)}{\partial t} = - \Delta.( \varepsilon \rho V \, C) + \Delta. \left[ \varepsilon \rho ( D_m I + D). \Delta C \right] + Q_p C^*$$

... (4)

where $D_m$ is apparent molecular diffusivity of solutes in solution in a porous medium (L$^2$/T); I is the identity tensor (dimensionless); D is the dispersion tensor (L$^2$/T); and C$^*$ is the solute mass fraction of fluid sources (M$_s$ /M). The term on the left hand side of equation (4) expresses the change in solute mass with time in a volume due to mechanisms represented by terms on the right hand side. The first term on the right hand side of equation (4), involving fluid velocity (V), represents advection of solute mass into or out of the local volume. The second term, involving molecular diffusivity of solute ($D_m$) and dispersivity (D), expresses the contribution of solute diffusion and dispersion to the local changes in solute mass. The diffusion contribution is based on a physical process driven by concentration gradients, and is often negligible at the field scale. The last term accounts for dissolved-species mass added by a fluid source with concentration C$^*$.

The mechanical dispersion tensor D is related to the velocity of groundwater flow. For an isotropic porous medium in two spatial dimensions, it is expressed as follows:

$$D = \begin{bmatrix} D_{xx} & D_{xy} \\ D_{yx} & D_{yy} \end{bmatrix}$$

The tensor D is symmetric and its elements are:

$$D_{xx} = \frac{1}{V^2}(d_L V_x^2 + d_T V_y^2)$$

$$D_{yy} = \frac{1}{V^2}(d_T V_x^2 + d_L V_y^2)$$

$$D_{xy} = D_{yx} = \frac{1}{V^2}(d_L - d_T)(V_x V_y)$$

... (6)

where $V(x,y,t) = (V_x^2 + V_y^2)^{1/2}$ is the magnitude of fluid velocity (L/T); $V_x(x,y,t)$ and $V_y(x,y,t)$ are the components of velocity in the x and y directions (L/T); and $d_L(x,y,t)$ and $d_T(x,y,t)$ are, respectively, the longitudinal and transverse dispersion coefficients (L²/T). The longitudinal and transverse dispersion coefficients describe dispersive fluxes along the direction of fluid flow and perpendicular to it. The coefficients $d_L$ and $d_T$ are velocity dependent, namely

$$d_L = \alpha_L V$$
$$d_T = \alpha_T V$$

...(7)

and $\alpha_L(x,y)$ and $\alpha_T(x,y)$ are the longitudinal and transverse dispersivity, respectively, of the solid matrix (L). Studies have shown that field observed dispersion coefficients are orders of magnitude larger than those observed in small scale laboratory tests. Similarly, field observations show that the dispersion coefficient increases with displacement distance at a given site.

*Numerical Methods*

SUTRA may be employed in one-dimensional or two-dimensional analyses. Flow and transport simulation may be either steady state which requires only a single solution step, or transient which requires a series of time steps in the numerical solution. Single-step steady state solutions are usually not appropriate for non-linear problems with variable density, saturation, viscosity and non-linear sorption.

SUTRA simulation is based on a hybridization of finite-element and integrated finite-difference methods employed in the framework of a method of weighted residuals. The method is robust and accurate when employed with proper spatial and temporal discretization. Standard finite-element approximations are employed only for terms in the balance equations which describe fluxes of fluid mass, solute mass and energy. All other non-flux terms are approximated with a finite-element mesh version of the integrated finite-difference methods. The hybrid method is the simplest and most economical approach which preserves the mathematical elegance and geometric flexibility of finite-element simulation while taking advantage of finite-difference efficiency.

The complex density-dependent groundwater flow and mass transport models provide stable and accurate results when employed with proper spatial and temporal discretization. In general, spatial discretization requires a fine mesh in areas where either accurate results are required or where parameters vary greatly over short distances; however, the spatial discretization should also be consistent with the dispersivity parameters. Voss (1984) recommends a grid spacing of less than four times the dispersivity in each direction, whereas Daus et al. (1985) and Molson and Frind (1994) recommend more stringent criteria, in which the grid Peclet Number (ratio of the spatial discretization and the dispersion length) and the Courant Number (ratio of the advective distance during one time step to the spatial discretization) should match the following constraints:

$$P_x = \frac{\Delta x}{\alpha_L} \leq 2, \quad P_y = \frac{\Delta y}{\alpha_T} \leq 2, \quad P_z = \frac{\Delta z}{\alpha_T} \leq 2 \qquad \ldots (8)$$

$$C_x = \frac{V_x \Delta t}{\Delta x} \leq 1, \quad C_y = \frac{V_y \Delta t}{\Delta y} \leq 1, \quad C_z = \frac{V_z \Delta t}{\Delta z} \leq 1 \qquad \ldots (9)$$

where $P_x$, $P_y$ and $P_z$ are the Peclet Numbers; $C_x$, $C_y$ and $C_z$ are the Courant Numbers; $\Delta x$, $\Delta y$ and $\Delta z$ are the grid spacings; $\alpha_L$ and $\alpha_T$ are the longitudinal and transverse dispersivity, respectively; and $\Delta t$ is the time step.

SUTRA employs a new method for calculation of fluid velocities. Fluid velocities, when calculated with standard finite-element methods for systems with variable fluid density, may display spurious numerically generated components within each element. These errors are due to fundamental numerical inconsistencies in spatial and temporal approximations for the pressure gradient and density, gravity terms which are involved in velocity calculation. Spurious velocities can significantly add to the dispersion of solute or energy. This false dispersion makes accurate simulation of all but systems with very low vertical concentration or temperature gradients impossible, even with fine vertical spatial discretization. Velocities as calculated in SUTRA, however, are based on a new, consistent, spatial and temporal discretization. The consistently evaluated velocities allow stable and accurate transport simulation (even at steady state) for systems with large vertical gradients of concentration or temperature. An example of such a system, that SUTRA successfully simulates, is a cross-sectional regional model of a coastal aquifer wherein the transition zone between horizontally flowing fresh water and deep stagnant salt water is relatively narrow.

The time discretization used in SUTRA is based on a backward finite-difference approximation for the time derivatives in the balance equations. Some non-linear coefficients are evaluated at the new time level of solution by projection while others are evaluated at the previous time level for non-iterative solutions. All coefficients are evaluated at the new time level for iterative solutions. The finite-element method allows the simulation of irregular regions with irregular internal discretization. This is made possible through use of quadrilateral elements with four corner nodes. Coefficients and properties of the system may vary in value throughout the mesh.

SUTRA includes an optional numerical method based on asymmetric finite-element weighting functions which results in "upstream weighting" of advective transport and unsaturated fluid flux terms. Although upstream weighting has typically been employed to achieve stable, non-oscillatory solutions to transport problems and unsaturated flow problems, the method is not recommended for general use, as it merely changes the physical system being simulated by increasing the magnitude of the dispersion process. A practical use of the method is, however, to provide a simulation of the sharpest concentration of temperature variations possible with a given mesh. This is obtained by specifying a simulation with absolutely no physical diffusion or dispersion and with 50% upstream weighting. The result may be interpreted as the solution with the minimum amount of dispersion possible for a stable result in the particular mesh in use.

In general simulation analyses of transport, upstream weighting is discouraged. The non-upstream methods are also provided by SUTRA and are based on symmetric weighting functions. These methods are robust and accurate when the finite-element mesh is properly designed for a particular simulation and should be used for most transport simulations.

*Data Requirement*

The most essential types of data required are salinity records with depth in a number of observation wells, hydro-dispersive parameters (or atleast detailed description of lithology, by which estimates of hydraulic conductivity can be made from similar type of areas), and tidal lags and heights at various points in the region of interest. It is also necessary to have recharge information. This involves not only the knowledge of rainfall but also how much of it enters the groundwater system and how much is drawn off by vegetation. Other useful information would include an accurate topographic map, a land use map, water supply data including extraction data, local knowledge and experience, and estimates of expected changes in regional rainfall patterns.

Two SUTRA data files are required: (i) SUTRA input data and (ii) initial conditions of pressure and concentration or temperature for the simulation. Re-programming of subroutines BCTIME and UNSAT is required to implement time-dependent boundary conditions and unsaturated flow functions, respectively. A graphical postprocessor, SUTRA-PLOT, developed by Souza (1987) is available for use with SUTRA. This postprocessor facilitates interpretation of the simulation results.

**Concluding Remarks**

The coastal areas often contain some of the most densely populated areas in the world. The availability of flat land, communication arteries, easy sea transportation, good soils and high productivity of organic matter explains this fact. Intrusion of salt water into heavily exploited aquifers is a serious problem being faced in coastal areas. The development of groundwater resources, therefore, requires careful management in coastal areas.

The importance of field data can not be over-emphasized. No model can replace a comprehensive field program which provides the required data. If reasonably good data are available, numerical models such as SUTRA, can be employed to provide an important means for guiding management decisions.

## References

1.    Daus, T., Frind, E. O. and Sudicky, E. A. (1985). A Comparative error analysis in finite element formulation of the advection-dispersion equation. Advances in Water Resources, Vol. 8, No. 2, pp. 86-95.

2.    Ghyben, W. B. (1888). Nota in verband met de voorgenomen putboring nabij Amsterdam: The Hague, Netherlands, Tijdschrift van Let Koninklijk Instituut van Ingenieurs, pp. 8-22.

3.    Herzberg, A. (1901). Die Wasserversorgung einiger Nordseebader: Wasserversorgung. Vol. 44, pp. 815-819 and 842-844.

4.    Kipp, K. L. (1987). HST3D - A computer code for simulation of heat and solute transport in three dimensional groundwater flow systems. U.S. Geological Survey, Water Resources Investigations Report 86-4095, 517p.

5.    Langevin, C.D., Thorne, D.T., Jr., Dausman, A.M., Sukop, M.C., and Guo, W. (2007). SEAWAT Version 4: A Computer Program for Simulation of Multi-Species Solute and Heat Transport: U.S. Geological Survey Techniques and Methods. Book 6, Chapter A22, 39 p.

6.    Kumar, C. P. (1998). The Modelling of Salt Water Intrusion. Training Report in the field of Deltaic Hydrology, UNDP Project (IND/90/003), National Institute of Hydrology, Roorkee.

7.    Molson, J. W. and Frind, E. O. (1994). SALTFLOW - Density-dependent flow and mass transport model in three dimensions. User Guide, Waterloo Centre for Groundwater Research, University of Waterloo, Waterloo, Ontario, 68p.

8.    Souza, W. R. (1987). Documentation of a graphical display program for the saturated-unsaturated transport (SUTRA) finite-element simulation model. U.S. Geological Survey, Water Resources Investigations Report 87-4245, 122 p.

9.    Voss, Clifford I. (1984). A finite-element simulation model for saturated-unsaturated fluid-density-dependent groundwater flow with energy transport or chemically reactive single-species solute transport. U.S. Geological Survey, Water Resources Investigations Report 84-4369, 409p.

10.   Voss, Clifford I. and Provost, Alden M. (2010). SUTRA - A Model for Saturated-Unsaturated, Variable-Density Ground-Water Flow with Solute or Energy Transport. Water-Resources Investigations Report 02-4231, Reston, Virginia, 291 p.

# 20-Challenges in Numerical Modelling of Salt Water Intrusion

## Introduction

Salt water intrusion into coastal aquifers is often a result of over-exploitation of groundwater or urbanization that prevents infiltration of runoff water. Due to this, predictions of complicated hydrological and hydro-geological processes by numerical models have been an important subject amongst hydrologists and water resources engineers. In practice, however, the development of forecast models is not easy due to factors such as geological structure, hydro-geological parameters, recharge areas, rate of rainfall, pumping rates, irrigation depth on farm land, and so on.

Modelling of variable-density groundwater flow in coastal zones, where the density distribution is non-uniform, has obviously been improved during the last decades. The increase in computer capacity was one of the major reasons though research on the mathematical equations did speed up the progress. Models that simulated the interface between fresh and saline groundwater were very useful to roughly simulate the existence of variable-density groundwater. The movement of the interface through time due to external processes such as changes in extraction rates could be modelled. Solute transport models, however, replaced the interface models because these mixing models could more easily comprehend the concept of the transition zone in the subsoil. The interface models still have their educational means. Several benchmark problems were posed to validate the variable-density flow computer codes. Models that cope with 3D transient variable-density groundwater flow and coupled multi-component reactive solute transport will be the state-of-the-art in the coming years. Calibration techniques may be further improved, whereas an increase in data collection is still essential to get reliable predictive models.

Up to the 1980s, the behaviour of density dependent groundwater flow has been investigated by means of analogue models (e.g. the Hele Shaw model for multiple fluid flow to study waste water injection into a fresh-saline groundwater system) and analytical solutions. Since computers appeared on the scene, numerical models gained ground. At present, a large number of numerical models are available which are capable of handling fresh and saline groundwater flow in aquifer systems. Some good reviews of literature on fresh and saline groundwater and available computer codes and models are given in Reilly and Goodman (1985), Custodio and Bruggeman (1987) and (Bear *et al.*, 1999).

## Interface and Solute Transport Models

The Badon Ghijben (1889)-Herzberg (1901) principle describes the position of an interface between fresh and saline groundwater (Figure 1). Equation 1 represents the Badon Ghijben-Herzberg principle:

$$h = \alpha \, H \qquad \qquad \qquad \qquad ...(1)$$

where, H = depth of the fresh-salt interface below mean sea level (L), $\alpha = (\rho_s - \rho_f)/\rho_f$ = relative density difference and h = piezometric head of fresh water with respect to mean sea level (L). For ocean water $\rho_s = 1025$ kg/m$^3$ and fresh water $\rho_f = 1000$ kg/m$^3$, the relative density difference $\alpha = 1/40$. The equation is correct if there is only horizontal flow in the fresh water zone and the saline water is stagnant. Glover (1959) obtained an analytical solution for the exact position of the interface. Though the position of the interface is not correct at the outflow face, the use of the above equation still gives a rather good approximation of the real situation.

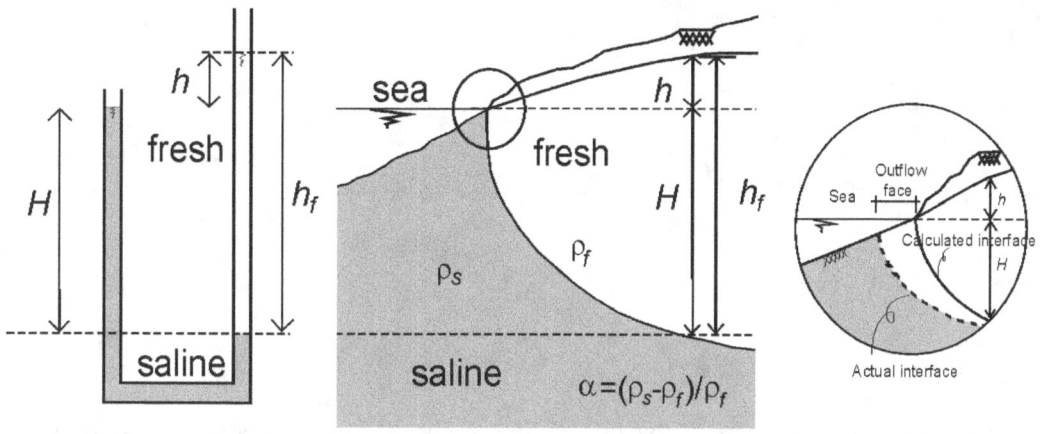

Figure 1.    The Badon Ghijben-Herzberg Principle: A Fresh-Salt Interface in Unconfined Coastal Aquifer

Interface models are based on the assumption that the interface between fresh and saline groundwater represents the actual situation: the well-known Badon Ghijben-Herzberg principle. The straightforward interface models can be applied as an educational means to gain a clear insight into the behaviour of fresh and saline groundwater in coastal aquifer systems. As such, these interface models are still widely applied. Nevertheless, some restrictions on the applicability of the principle should be considered:

• First, the principle only approximates the actual occurrence of fresh, brackish and saline groundwater in the subsoil. In fact, the brackish zone between fresh and saline groundwater should only be schematised by an interface when the maximum thickness of the brackish zone is only in the order of few meters. This condition applies only in rare situations where the freshwater lens is evolved by natural recharge, as occurs in some sand-dune areas or in (coral) islands.

• Second, the principle assumes a hydrostatic equilibrium, whereas in reality the aquifer system might considerably deviate from this equilibrium situation. In those cases, e.g. in freshwater bodies near the shoreline, the Badon Ghijben-Herzberg principle should not be applied, because the computed position of the interface deviates from the actual position as the coast is approached.

In many coastal aquifer systems, a relatively broad transition zone (or mixing zone, zone of dispersion or brackish zone between fresh, brackish and saline groundwater) is present because of various processes during geological history (e.g. regressions, transgressions), see Figure 2.

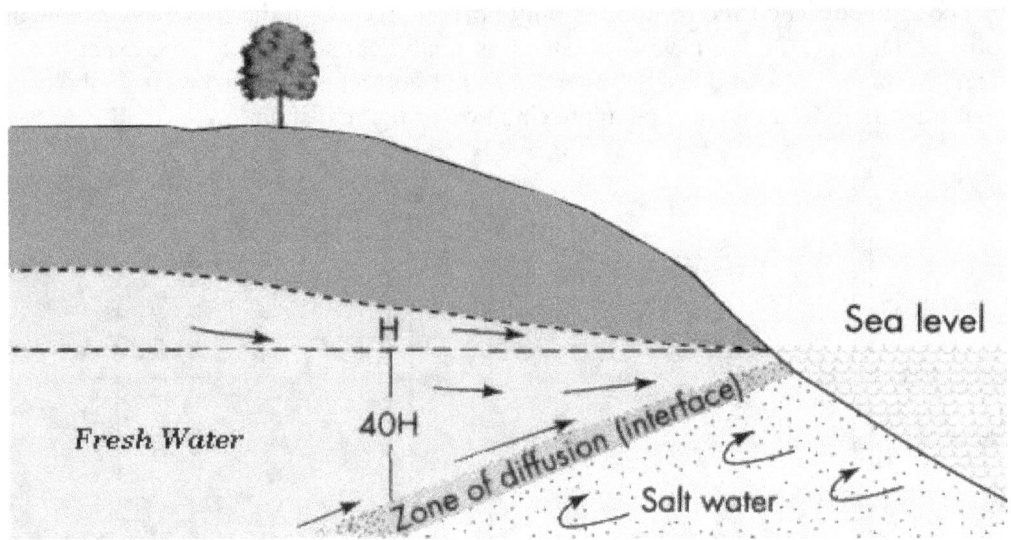

Figure 2.    Salt Water Intrusion in a Coastal Aquifer - The Mixing Concept: Circulation of Salt Water from the Sea to the Zone of Diffusion and Back to the Sea (Modified from Henry, 1964)

In addition, the transition zone also increases as a result of the circulation of brackish water due to inflow of saline groundwater (mixing with fresh groundwater due to hydrodynamic dispersion), the tidal regime and human activities, such as (artificial) recharge and groundwater extraction at high and variable rates (Cooper *et al.*, 1964). Under such conditions, more sophisticated models are required than just models with expressions for interfaces: models that take into account variable densities and the transport of solutes. These models are referred to as solute transport models or salt water intrusion models. They apply the advection-dispersion equation to convert solute concentration (or total dissolved solids) to density through the equation of state. They are able to simulate: salt water intrusion in coastal aquifers where mostly non-uniform density distributions occur; changes in solute concentration (e.g. near pumping wells due to upconing); and changes in storages of freshwater in sand-dune areas.

As solute transport models usually apply numerical schemes, they can also be utilised to simulate aquifer systems with complex hydrogeologic geometries and inversions of fresh and saline groundwater caused by (pre)historic events. Some examples of these early solute transport models are INTERCOMP (1976) and the models of Lebbe (1983) and Voss (SUTRA, 1984). The complete mathematical description of the solute transport models can be found in Galeati *et al.* (1992), Diersch (1996), Person *et al.* (1996), Holzbecher (1998), Ingebritsen and Sanford (1998), Bear *et al.* (1999) and Diersch and Kolditz (2002).

## Two and Three-Dimensional Models

The practical application of two-dimensional groundwater flow models is rather limited when real geometries are considered. In many real situations, groundwater flow perpendicular to the coastline is disturbed in such a way that the schematisation and modelling of the actual situation by a cross-section can not be allowed any more. Such situations occur, for instance, in the vicinity of singular wells where groundwater is extracted or infiltrated or in areas with complex hydrogeologic geometries.

Under these circumstances, 3D models should be applied. Obviously, 3D models naturally require even more effort to be understood, implemented and utilised effectively than 2D models. For instance, the problems that arise to visualise 3D groundwater flow (effective velocities) and solute transport on a (2D) monitor should not be underestimated, though nowadays many sophisticated and powerful GUI's (graphical user interfaces) are available.

## Constraints with Three-Dimensional Modelling of Salt Water Intrusion

Though 3D modelling of salt water intrusion in hydrogeologic systems is technically no problem anymore, some practical problems remain (Oude Essink and Boekelman, 1996, 1998): the data availability problem, the computer problem and the numerical dispersion problem.

### (a) Data Availability

A numerical model of salt water intrusion in a coastal hydrogeologic system must be calibrated and verified with available hydrogeological data in order to prove its predictive capability, accuracy and reliability. Regrettably, in many cases, reliable and sufficient data are very scarce. As such, 3D modelling is (and will probably always be) restricted seriously. Some common problems are:

- Upscaling of data from 1D (a point source such as a groundwater level from a observation well) and 2D (from a line source as borehole information) to 3D: in fact, the collection of data is one- or two-dimensional. This information must be extrapolated or interpolated to a 3D distribution of subsoil parameters. This upscaling obviously faces difficulties; especially inter- and extrapolation of concentration measurements to a 3D density field is difficult.

- Long time series of hydrochemical constituents: the calibration of a transient groundwater flow model with salinities changing over time is still rather laborious. As the flow of groundwater and subsequently the transport of hydrochemical constituents are slow processes, it takes many years before salinisation can actually be detected. As such, relative long time series of monitored salinities (of some tens of years or even more) are necessary to accurately calibrate transient 3D salt water intrusion in large-scale coastal hydrogeologic systems. Unfortunately, these time series are available only occasionally. As a consequence, the calibration of this transient process will be less reliable and less good.

It has, therefore, to be accepted that the availability of data will lag behind the developments in computer possibilities, and thus, will restrict practical applications to a certain extent.

*(b) Computing Resources*

Until some years ago, 3D modelling of salt water intrusion in large-scale coastal hydrogeologic systems was also difficult due to the memory problem (limited memory to store data of a 3D model) and the speed problem (limited computer speed to execute a (transient) 3D model). This situation has changed pretty fast. Nowadays, the memory in a computer can be increased with several hundreds Mb of EM RAM at low costs. In addition, powerful standard personal computers significantly reduce the execution time, though this obviously depends on the disk-speed of the computer, the size of the model, the efficiency of the compiler and the type of the output device.

*(c) Numerical Dispersion*

Numerical approximations of the derivatives of the non-linear solute transport equation may introduce truncation errors and oscillation errors (Figure 3). The truncation error has the appearance of an additional dispersion-like term, called numerical dispersion, which may dominate the numerical accuracy of the solution. Oscillations may occur in the solution of the solute transport equation as a result of over and undershooting of the solute concentration values. If the oscillation reaches unacceptable values, the solution may even become unstable.

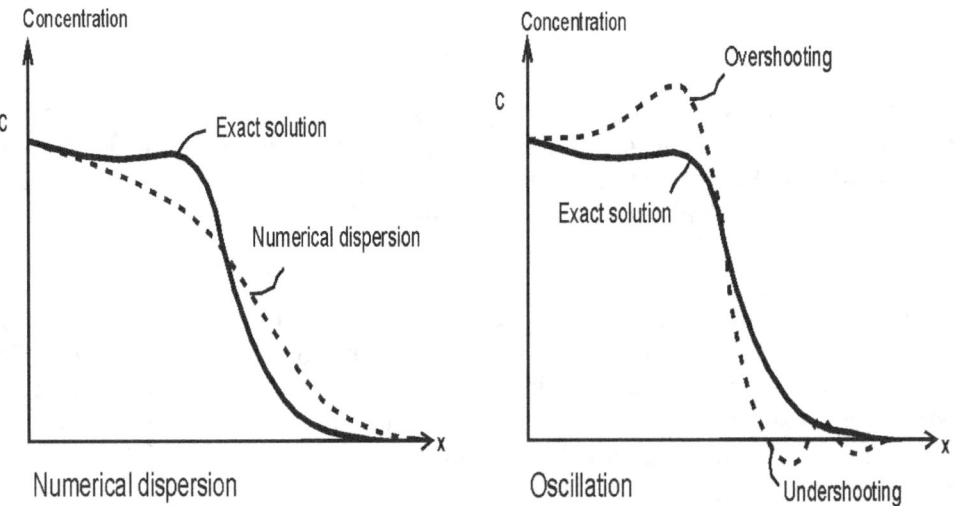

Figure 3.     Schematisation of Numerical Dispersion and Oscillation

There is a close relation between numerical accuracy (numerical dispersion) and stability (oscillation) (Peaceman, 1977; Pinder and Gray, 1977). In fact, numerical dispersion acts to stabilize the solution of the solute transport equation. Numerical dispersion spreads a sharp front, e.g. an interface between fresh and saline groundwater, by generating a solution that applies a greater dispersion than the true hydrodynamic dispersion. To suppress numerical dispersion, the numerical discretisation scheme (spatial as well as temporal) should be adapted. However, this scheme may lead to over and undershooting, and subsequently, oscillation may be amplified.

Therefore, the numerical discretisation scheme should be chosen carefully in order to control both numerical accuracy and stability.

Eigenvalue analyses of the advection-dispersion equation and Taylor series expansions are performed to demonstrate the numerical accuracy of a solution and the importance of the element dimension (Peaceman, 1977; Frind and Pinder, 1983; Bear and Verruijt, 1987). Approximations of the first-order derivatives generate errors in the order of magnitude of the second-order derivatives.

For the analysis of the truncation error, the so-called grid Peclet number $Pe_{grid}$ is defined:

$$Pe_{grid} = \left| \frac{V \Delta x}{D_h} \right| \qquad ...(2)$$

where, $Pe_{grid}$ = grid Peclet number, V = effective velocity of groundwater ($LT^{-1}$), $\Delta x$ = dimension of the element (L) and $D_h$ = hydrodynamic dispersion ($L^2 T^{-1}$). Grid Peclet number (and Courant number, physically interpreted as the ratio of the advective transport distance during one time step to the spatial discretisation) have been mentioned in various quantitative descriptions. Whether or not the numerical dispersion is suppressed, depends on the discretisation technique applied (Jensen and Finlayson, 1978; Voss and Souza, 1987). In order to obtain real and distinct eigenvalues, the spatial discretisation should meet the grid Peclet number condition (Daus *et al.*, 1985): $Pe_{grid} <= 2$ for finite difference algorithms; $Pe_{grid} <= 2$ for finite element algorithms, linear basic and $Pe_{grid} <= 4$ for finite element algorithms, quadratic basic.

This grid Peclet number condition imposes that the dimension of the element should be not greater than a few times the magnitude of the longitudinal dispersivity that represents hydrodynamic dispersion. Computer codes, which solve the advection-dispersion equation based on standard finite element or finite difference techniques, must satisfy this condition of spatial discretisation. Both widely used techniques have in common that they produce poor results at great grid Peclet numbers. As such, it is peculiar that this well-known fact does not have a broader attention in numerical modelling practices of groundwater contaminant transport (Uffink, 1990).

For example, if the (longitudinal) dispersivity is small (Gelhar *et al.*, 1992), e.g. in the order of decimeters for Holocene and Pleistocene deposits of marine and fluviatile origin, the dimensions of the elements must be in the order of meters. When the dimension of a 3D hydrogeologic system in question is in the order to tens of kilometers by tens of meters depth, many hundreds of millions of elements are needed to model large-scale coastal hydrogeologic systems properly. This amount is (very) difficult to achieve presently for effective transient modelling of variable-density groundwater flow.

## Salt Water Intrusion Models

There are quite a few computer codes available which can simulate density dependent groundwater flow and solute transport in order to model salt water intrusion. The interest is focused on the grid Peclet number condition.

$D^3F$ (Distributed Density Driven Flow) is one of the newest, sophisticated, computer codes to simulate density-driven groundwater flow (Fein *et al.*, 1998). It is based on the volume element method. Permeabilities, porosities, values for boundary and initial conditions can be defined as constants as well as simple spatial user-defined functions. In addition, a stochastic permeability distribution technique is available to consider small-scale heterogeneities. Fluid density and viscosity are functions of salt concentration and temperature.

**FAST-C** (2D/3D) (Holzbecher, 1998) can be applied to model density driven flow in porous media using the stream function formulation. Density variations are caused by temperature or salinity gradients. Steady state as well as transient modelling of confined aquifers are possible.

**FEFLOW** is a 3D computer code, which employs the finite element method (Diersch, 1996). The governing partial differential equations describe groundwater flow, where differences in density affect the fluid flow. Fluid density effects are caused by contaminant mass as well as temperature differences simultaneously, inducing thermohaline flow.

**HST3D** is a 3D finite difference code that can simulate heat and solute transport (Kipp, 1986). Modelling a large-scale coastal hydrogeologic system with a non-uniform density distribution with HST3D appears to be rather complicated when small longitudinal dispersivities should be applied. For heat transport applications, this code seems to be much more appropriate.

**MOCDENS3D** (Oude Essink, 1998, 2001) is MOC3D (Konikow *et al.*, 1996), but adapted for density differences. Buoyancy terms are introduced in the vertical effective velocities to take into account density effects. The adapted MODFLOW module solves the equation of density dependent groundwater flow. Solute transport is modelled by the method of characteristics (MOC) module in MOCDENS3D. The particle tracking method is used to solve advective solute transport and the finite difference method is used for hydrodynamic dispersive transport. As such, the movement of fresh, brackish and saline groundwater in porous media can be modelled in 3D. Numerical dispersion and oscillation is limited as the method of characteristics is applied.

**SEAWAT** (Guo and Bennett, 1998; Guo and Langevin, 2002) is a combination of MODFLOW and MT3DMS (Zheng and Wang, 1999). It is designed to simulate 3D variable-density ground water flow and solute transport. The program was developed by modifying MODFLOW subroutines to solve a variable-density form of the ground water flow equation and by combining MODFLOW and MT3DMS into a single program.

**SUTRA** is a well-documented two-dimensional finite element code (Voss, 1984). This code has become the widely accepted 2D variable-density groundwater flow model throughout the world (Voss and Souza, 1987; Souza and Voss, 1987). It can simulate density dependent groundwater flow with (thermal) energy transport or chemically reactive (single-species) solute transport. The

3D beta-version is available now. The grid Peclet number condition is applied here: $Pe_{grid} <= 4$. Narrow transition zones require a careful choice of spatial and temporal discretisation.

**SWICHA** is a 3D finite element code (Huyakorn *et al.*, 1987; Lester, 1991). It can simulate variable-density fluid flow and solute transport processes in saturated porous media. The applications range from simple 1D to complex 3D, coupled flow and solute transport. The solute transport equation may not converge to a solution if a so-called critical Peclet number is exceeded in an element. To solve this problem, SWICHA offers a trick at the user's option: artificial dispersion is added to the solute transport equation matrix. Spatial oscillations are then suppressed and the critical Peclet number is no longer exceeded in that element.

**SWIFT** (Sandia Waste-Isolation Flow and Transport model) is a 3D code to simulate groundwater flow, heat (energy), brine and radio nuclide transport in porous and fractured media (Ward, 1991; Ward and Benegar, 1998). It is stated that though grid Peclet numbers greater than 4 may cause problems (e.g. oscillations), a value of 10 or more is sometimes acceptable.

### Issues and Challenges

Some difficult issues related to 3D modelling of variable-density flow in a regional system and new challenges for the future are presented below:

- *Initial density matrix*: In a variable-density flow system, the initial density that should be implemented in the numerical model can highly determine the velocities and thus the transport of fresh, brackish and saline groundwater. An accurate 3D-density matrix can be critical for the accuracy of the model. Under certain hydrogeological conditions, the initial density distribution for a numerical model can be derived by just simulating the groundwater flow system under consideration for many (tens of) years until a steady-state situation from a density point of view is achieved. For instance, the initial (undisturbed) density distribution of a freshwater lens in a dune area can be derived by implementing natural fresh groundwater recharge in a completely saline aquifer system and simulate the system long enough. The obtained steady-state density distribution will probably be the original distribution without human activities as groundwater extractions. Unfortunately, many coastal aquifer systems are at present not in dynamic equilibrium, for instance due to long-term sea level fluctuations (Meisler *et al.*, 1984) or large-scale human activities with long-term effects on the groundwater system. For these specific cases, the best estimate of the 'initial' density distribution is the present measured one. In general, data of the density distribution is often scarce and 3D modelling of variable-density groundwater flow in coastal aquifers is often based on insufficient data of good quality.

- *Outflow face*: Velocities of fresh groundwater at the outflow face from the inland sand-dune area towards the sea can by enormous. Analytical formulae estimated the length of these faces to (tens of) meters (Glover, 1959; Veer, 1977). From a numerical point of view, modelling of these zones is difficult and restricts the application of regional models for long-term transient simulations of fresh, brackish and saline groundwater.

- *Concentration boundaries*: It is difficult to determine the concentration boundaries of regional transient variable-density groundwater flow models when long-term simulations are considered. Obviously, the verbs concentration boundaries imply that the concentration stay fixed, though incoming concentrations will probably vary. Therefore, these boundaries should be positioned far away from the area of interest, especially because these concentrations effect density and thus variable-density groundwater flow in the system.

- The 3D-density matrix highly determines the accuracy of the velocity field. Under normal circumstances, accurate 3D density distributions are not available. It is therefore essential for this type of groundwater flow modelling to increase (expensive) quality measurements of concentrations and/or electric conductivities.

- Processes such as sea level rise and increased human activities (e.g. increasing groundwater extractions, changes in land use) intensify the external stresses in many coastal aquifers. In addition, the interaction between the land surface and the sea through groundwater and solute fluxes jeopardise water resources in the coastal zone (ecological marine environment). Research on the combined effects of these processes is of interest to water (management) authorities.

- Optimalisation of a number of compensating techniques to prevent upconing and salt water intrusion in coastal aquifers could lead to a sustainable groundwater management. Examples are strategies of groundwater extraction schemes and technical measurements as creating physical barriers.

- The calibration of variable-density flow models is traditionally based on stagnant (freshwater) heads and chloride concentration records. Extending the calibration of these models could be possible when transient data of concentration and freshwater head, isotopes as well as geochemical information of pore waters are applied. In deep hydrogeologic systems, even thermal hydrogeological information should be taken into account. If necessary, thermohaline mathematical models should be used to cope with double-diffusive convection processes in hypersaline geothermal systems.

- Combining 3D transient variable-density groundwater flow models of coastal aquifers with multi-species reactive solute transport modules, which take into account kinetic geochemical interactions, would be the challenge for the coming years. Parallel computing will probably be necessary to account for all relevant hydrogeological and hydrogeochemical processes in the subsoil.

**Concluding Remarks**

Though 3D-modelling of salt water intrusion in large-scale coastal aquifers is technically possible, a number of practical problems arise.

- First of all, to suppress truncation and oscillation errors in the solution of the advection-dispersion equation, models based on the standard finite element method or finite difference method should satisfy the condition that the spatial discretization (that is the dimension of the grid block) should not be greater than a few times the magnitude of the (longitudinal) dispersivity that represents hydrodynamic dispersion. When the dimension of the aquifer system in question is large and the (longitudinal) dispersivity is small, the dimensions of the grid blocks must be in the order of (tens of) metres. This condition of spatial discretization considerably restricts the practical application of 3D computer codes with dispersive solute transport such as HST3D and SWICHA, to simulate 3D salt water intrusion in large-scale coastal (homogeneous) aquifers. However, an interconnected model of MODFLOW (adapted for simulating density dependent groundwater flow) and the transport model MT3D will probably avoid the numerical dispersion problem.

- Second, the required number of grid blocks to model large-scale coastal (homogeneous) aquifers is enormous, e.g. of several hundreds of thousands of grid blocks. Nowadays, the personal computer which is applied to execute a model with this number of grid blocks is still not fast enough, but it will be within a few years. The memory allocation is no problem any more, as computers with sufficient Mb of Extended Memory RAM are already available on a large scale.

- Third, the large number of groundwater data, required for calibration and verification, are not yet available in most cases. As such, data collection should be intensified. However, we have to accept that the collection of data will always lag behind the developments in computer possibilities.

## References

1. Badon Ghijben, W. and J. Drabbe (1889). Nota in verband met de voorgenomen putboring nabij Amsterdam. (in Dutch), Tijdschrift van het Koninklijk Instituut voor Ingenieurs 1888-1889: 8-22.

2. Bear, J., A.H.-D. Cheng, et al., Eds. (1999). Seawater Intrusion in Coastal Aquifers; Concept, Methods and Practices. Kluwer Academic Publishers: 625 p.

3. Bear, J. and A. Verruijt (1987). Modeling Groundwater Flow and Pollution. Dordrecht, The Netherlands, D. Reidel Publishing Company: 414 p.

4. Cooper, H.H., jr., F.A. Kohout, et al. (1964). Sea Water in Coastal Aquifers. 84 p.

5. Custodio, E. and G.A. Bruggeman, Eds. (1987). Groundwater Problems in Coastal Areas. Studies and Reports in Hydrology, UNESCO, International Hydrological Programme", Paris

6. Daus, A.D., E.O. Frind, et al. (1985). Comparative Error Analysis in Finite Element Formulations of the Advection-Dispersion Equation. Adv. Water Resour. 8: 86-95.

7.    Diersch, H.J.G. (1996). Interactive, Graphics-based Finite-Element Simulation System FEFLOW for Modeling Groundwater Flow, Contaminant Mass and Heat Transport Processes. FEFLOW User's Manual Version 4.5, Institute for Water Resources Planning and System Research, Ltd.

8.    Diersch, H.J.G. and O. Kolditz (2002). Variable-Density Flow and Transport in Porous Media: Approaches and Challenges. Adv.Water Resour. 25: 899-944.

9.    Fein, E., et al. (1998). D3F - A Simulator for Density driven Flow Modelling. User's Manual, GRS. Braunscheig, Germany.

10.   Frind, E.O. and G.F. Pinder (1983). The Principle Direction Technique for Solution of the Advection-Dispersion Equation. Proc. 10th IMACS World Congress on Systems Simulation and Scientific Computation, Concordia University, Montreal, Canada: 305-313.

11.   Galeati, G., G. Gambolati, et al. (1992). Coupled and Partially Coupled Eulerian-Lagrangian Model of Freshwater - Seawater Mixing. Water Resour. Res. 28(1): 149-165.

12.   Gelhar, L.W., C. Welty, et al. (1992). A Critical Review of Data on Field-Scale Dispersion in Aquifers. Water Resour. Res. 28(7): 1955-1974.

13.   Glover, R.E. (1959). The Pattern of Fresh-Water Flow in a Coastal Aquifer. J. Geophys. Res. 64: 457-459.

14.   Guo, W. and G.D. Bennett (1998). Simulation of Saline/Fresh Water Flows using MODFLOW. Proc. MODFLOW'98 Conference, Golden, Colorado, USA: 267-282.

15.   Guo, W. and C.D. Langevin (2002). SEAWAT - User's Guide to SEAWAT: A computer program for simulation of three-dimensional variable-density ground-water flow. U.S.G.S. Techniques of Water-Resources Investigations, Book 6, Chapter A7, 77 p. Tallahassee, Florida.

16.   Henry H. R. (1964). Effects of Dispersion of Salt Encroachment in Coastal Aquifers. Sea Water in Coastal Aquifers. U.S. Geological Survey Water Supply Paper 1613-C, C70–C84.

17.   Herzberg, A. (1901). Die Wasserversorgung einiger Nordseebaden, (in German). Zeitung für Gasbeleuchtung und Wasserversorgung 44: 815-819 and 842-844.

18.   Holzbecher, E. (1998). Modeling Density-driven Flow in Porous Media, Principles, Numerics, Software. Springer Verlag, Berlin Heidelberg: 286 p.

19.   Huyakorn, P.S., P.F. Andersen, et al. (1987). Saltwater Intrusion in Aquifers: Development and Testing of a Three- Dimensional Finite Element Model. Water Resour. Res. 23(2): 293-312.

20.    Ingebritsen, S.E. and W.E. Sanford (1998). Groundwater in Geological Processes. Cambridge University Press: 341 p.

21.    INTERCOMP (1976). A Model for calculating Effects of Liquid Waste Disposal in Deep Saline Aquifers. Resource Development and Engineering, Inc. U.S.G.S. Water-Resources Investigations Report 76-96: 263 p.

22.    Jensen, O.K. and B.A. Finlayson (1978). Solution of the Convection-Diffusion Equation using a Moving Coordinate System. Second Int. Conf. on Finite Elements in Water Resour., Imperial College, London: 4.21 - 32.

23.    Kipp, K.L., Jr. (1986). HST3D, A Computer Code for Simulation of Heat and Solute Transport in Three-dimensional Groundwater Flow Systems. IGWMC, International Ground Water Modeling Center. U.S.G.S. Water-Resources Investigations Report 86-4095

24.    Konikow, L.F., D.J. Goode, et al. (1996). A Three-Dimensional Method-of-Characteristics Solute-Transport Model (MOC3D). U.S.G.S. Water-Resources Investigations Report 96-4267: 87 p.

25.    Lebbe, L.C. (1983). Mathematical Model of the Evolution of the Fresh-Water Lens under the Dunes and Beach with Semi-diurnal Tides. Proc. 8th Salt Water Intrusion Meeting, Bari, Italy, Geologia Applicata e Idrogeologia, Vol. XVIII, Parte II: 211-226.

26.    Lester, B. (1991). SWICHA, A Three-Dimensional Finite-Element Code for Analyzing Seawater Intrusion in Coastal Aquifers. Version 5.05. GeoTrans, Inc., Sterling, Virginia, U.S.A., IGWMC, International Ground Water Modeling Center, Delft, The Netherlands: 178 p.

27.    Meisler, H., P.P. Leahy, et al. (1984). Effect of Eustatic Sea-Level Changes on Saltwater-Freshwater in the Northern Atlantic Coastal Plain. U.S. Geol. Surv. Water-Supply Paper 2255.

28.    Oude Essink, G.H.P. (1998). MOC3D adapted to Simulate 3D Density-Dependent Groundwater Flow. Proc. MODFLOW'98 Conference, Golden, Colorado, USA: 291-303.

29.    Oude Essink, G.H.P. (2001). Salt Water Intrusion in a Three-Dimensional Groundwater System in The Netherlands: A Numerical Study. Transport in Porous Media 43(1): 137-158.

30.    Oude Essink, G.H.P. (2003). Mathematical Models and Their Application to Salt Water Intrusion Problems. Proceedings, TIAC'03 - Coastal aquifers intrusion technology: Mediterranean Countries, Alicante, Spain, March 11-14, 2003: 57-77.

31. Oude Essink, G.H.P. and R.H. Boekelman (1996). Problems with Large-Scale Modelling of Salt Water Intrusion in 3D. Proceedings, 14th Salt Water Intrusion Meeting, Malmö, Sweden, June 17-21, 1996: p. 288-299.

32. Oude Essink, G.H.P. and R.H. Boekelman (1998). Problemas con el modelado a gran escala de la intrusion de agua  salada en 3D, (in Spanish). Boletin Geologicoy Minero: Hydrologia Subterranea 109(4): 403-420.

33. Peaceman, D.W. (1977). Fundaments of Numerical Reservoir Simulation. Developments in Petroleum Science 6, Elsevier Scientific Publishing Company. Amsterdam.

34. Person, M., J.P. Raffensperger, et al. (1996). Basin-Scale Hydrogeologic Modeling. Reviews of Geophysics 34(1): 61-87.

35. Pinder, G.F. and W.G. Gray (1977). Finite  Element Simulation in Surface and Subsurface Hydrology. Academic Press: 295 p.

36. Reilly, T.E. and A.S. Goodman (1985). Quantitative Analysis of Saltwater-Freshwater Relationships in Groundwater Systems  -  A Historical Perspective. Journal of Hydrology 80: 125-160.

37. Souza, W.R. and C.I. Voss (1987). Analysis of an Anisotropic Coastal Aquifer System using  Variable-Density Flow and Solute Transport Simulation. Journal of Hydrology 92: 17-41.

38. Uffink, G.J.M. (1990). Analysis of Dispersion by  the Random Walk Method. Delft University of Technology, 150 p.

39. Veer, P., van der (1977). Analytical Solution for  Steady Interface Flow in a Coastal Aquifer involving a Phreatic Surface with Precipitation. Journal of Hydrology 34: 1-11.

40. Voss, C.I. (1984). SUTRA -- A Finite Element  Simulation for Saturated-Unsaturated, Fluid-Density-Dependent Ground-Water Flow with Energy Transport or Chemically Reactive Single-Species Solute Transport. U.S.G.S. Water-Resources Investigations Report 84-4369: 409 p.

41. Voss, C.I. and W.R. Souza (1987). Variable Density Flow  and Solute Transport Simulation of Regional Aquifers containing a Narrow Freshwater-Saltwater Transition Zone. Water Resour. Res. 23(10): 1851-1866.

42. Ward, D.S. (1991). Data Input for SWIFT/386, version 2.50. Geotrans Technical Report, Sterling, Va.

43. Ward, D.S. and J. Benegar (1998). Data Input Guide for SWIFT-98, version 2.57. HSI Geotrans, Inc., Sterling, Va.

44. Zheng, C. and P.P. Wang (1999). MT3DMS, A Modular Three-Dimensional Multi-Species Transport Model for Simulation of Advection, Dispersion and Chemical Reactions of Contaminants in Groundwater Systems; Documentation and User's Guide. U.S. AERDCC Report SERDP-99-1, Vicksburg, MS.

# 21-Impact of Climate Change on Groundwater Resources

## Introduction

Water is indispensable for life, but its availability at a sustainable quality and quantity is threatened by many factors, of which climate plays a leading role. The Intergovernmental Panel on Climate Change (IPCC) defines climate as "the average weather in terms of the mean and its variability over a certain time-span and a certain area" and a statistically significant variation of the mean state of the climate or of its variability lasting for decades or longer, is referred to as climate change.

Evidence is mounting that we are in a period of climate change brought about by increasing atmospheric concentrations of greenhouse gases. Atmospheric carbon dioxide levels have continually increased since the 1950s. The continuation of this phenomenon may significantly alter global and local climate characteristics, including temperature and precipitation. Climate change can have profound effects on the hydrologic cycle through precipitation, evapotranspiration, and soil moisture with increasing temperatures. The hydrologic cycle will be intensified with more evaporation and more precipitation. However, the extra precipitation will be unequally distributed around the globe. Some parts of the world may see significant reductions in precipitation or major alterations in the timing of wet and dry seasons. Information on the local or regional impacts of climate change on hydrological processes and water resources is becoming more important. The effects of global warming and climatic change require multi-disciplinary research, especially when considering hydrology and global water resources.

The Intergovernmental Panel on Climate Change (IPCC) estimates that the global mean surface temperature has increased $0.6 \pm 0.2$ $^{\circ}$C since 1861, and predicts an increase of 2 to 4 $^{\circ}$C over the next 100 years. Global sea levels have risen between 10 and 25 cm since the late 19th century. As a direct consequence of warmer temperatures, the hydrologic cycle will undergo significant impact with accompanying changes in the rates of precipitation and evaporation. Predictions include higher incidences of severe weather events, a higher likelihood of flooding, and more droughts. The impact would be particularly severe in the tropical areas, which mainly consist of developing countries, including India.

Coupled atmosphere-ocean global climate models (GCMs) are used to estimate changes in climate. These physically-based numerical models simulate synoptic-scale climate and hydrological processes, and are forced with greenhouse gas and aerosol emission scenarios. A wide diversity of GCMs developed by leading climate centres are available for other researchers to evaluate potential impacts of climate change. To ensure that the predictive elements from a GCM are realistic, a statistical downscaling technique should be employed to bridge the local- and synoptic-scale processes. Statistical downscaling uses a correlation between predictands (site measured variables, such as precipitation) and predictors (region-scale variables, such as GCM variables).

Changes in regional temperature and precipitation have important implications for all aspects of the hydrologic cycle. Variations in these parameters determine the amount of water that reaches the surface, evaporates or transpires back to the atmosphere, becomes stored as snow or ice, infiltrates into the groundwater system, runs off the land, and ultimately becomes base flow to streams and rivers.

Hydrological impact assessments of watersheds (and aquifers) require information on changes in evapotranspiration because it is a key component of the water balance. However, climate-change scenarios tend to be expressed in terms of changes in temperature and precipitation. Consequently, the effects of global warming on potential evaporation (or more inclusively, evapotranspiration) are not simple to estimate. Many global scenarios suggest an increase in potential evaporation, but these factors may be outweighed locally or regionally by other factors reducing evaporation. Various models may be used to calculate potential evaporation using data on net radiation, temperature, humidity, and wind speed, and sometimes plant physiological properties. The estimated effect of a change in climate on potential evaporation depends on the characteristics of the site.

Many rivers and streams that are fed by glacier runoff could be significantly impacted as a result of climate change. As glacier retreat accelerates, increased summer runoff could occur. However, when the glaciers have largely melted, the late summer and fall glacial input into streams and rivers may be lost, resulting in a significant reduction in flow in some cases.

Water resource management plans increasingly need to incorporate the affects of global climate change in order to accurately predict future supplies. Numerous studies have documented the sensitivity of streamflow to climatic changes for watersheds all over the world. Most of these studies involve watershed scale hydrologic models, of which validation remains a fundamental challenge. Moreover, outputs from general circulation models (GCM) can be rather uncertain and downscaling their predictions for local hydrologic use can produce inconsistent results. Therefore, the sensitivity of streamflow to climate changes is perhaps best understood by analyzing the historical records.

Building empirical models to link climate and regional hydrological regimes has a long history. In recent years, many researchers have used empirical rainfall–runoff model to study the impacts of climatic change on hydrology. However, applications of these empirical relationships to climate or basin conditions different from those used in the original development of these functions are questionable.

## Impact of Climate Change on Groundwater Resources

Although the most noticeable impacts of climate change could be fluctuations in surface water levels and quality, the greatest concern of water managers and government is the potential decrease and quality of groundwater supplies, as it is the main available potable water supply source for human consumption and irrigation of agriculture produce worldwide. Because groundwater aquifers are recharged mainly by precipitation or through interaction with surface water bodies, the direct influence of climate change on precipitation and surface water ultimately affects groundwater systems.

It is increasingly recognized that groundwater cannot be considered in isolation from the landscape above, the society with which it 'interacts', or from the regional hydrological cycle, but needs to be managed holistically. In understanding the likely consequences of possible future (climate and non-climate) changes on groundwater systems and the regional hydrological cycle, an important (but not exclusive) component to understand is the influence that these factors exert on recharge and runoff.

It is important to consider the potential impacts of climate change on groundwater systems. As part of the hydrologic cycle, it can be anticipated that groundwater systems will be affected by changes in recharge (which encompasses changes in precipitation and evapotranspiration), potentially by changes in the nature of the interactions between the groundwater and surface water systems, and changes in use related to irrigation.

*(a) Soil Moisture*

The amount of water stored in the soil is fundamentally important to agriculture and has an influence on the rate of actual evaporation, groundwater recharge, and generation of runoff. Soil moisture contents are directly simulated by global climate models, albeit over a very coarse spatial resolution, and outputs from these models give an indication of possible directions of change.

The local effects of climate change on soil moisture, however, will vary not only with the degree of climate change but also with soil characteristics. The water-holding capacity of soil will affect possible changes in soil moisture deficits; the lower the capacity, the greater the sensitivity to climate change. Climate change also may affect soil characteristics, perhaps through changes in waterlogging or cracking, which in turn may affect soil moisture storage properties. Infiltration capacity and water-holding capacity of many soils are influenced by the frequency and intensity of freezing.

*(b) Groundwater Recharge and Resources*

Groundwater is the major source of water across much of the world, particularly in rural areas in arid and semi-arid regions, but there has been very little research on the potential effects of climate change.

Aquifers generally are replenished by effective rainfall, rivers, and lakes. This water may reach the aquifer rapidly, through macro-pores or fissures, or more slowly by infiltrating through soils and permeable rocks overlying the aquifer. A change in the amount of effective rainfall will alter recharge, but so will a change in the duration of the recharge season. Increased winter rainfall, as projected under most scenarios for mid-latitudes, generally is likely to result in increased groundwater recharge. However, higher evaporation may mean that soil deficits persist for longer and commence earlier, offsetting an increase in total effective rainfall. Various types of aquifer will be recharged differently. The main types are unconfined and confined aquifers. An unconfined aquifer is recharged directly by local rainfall, rivers, and lakes, and the rate of recharge will be influenced by the permeability of overlying rocks and soils.

Macro-pore and fissure recharge is most common in porous and aggregated forest soils and less common in poorly structured soils. It also occurs where the underlying geology is highly fractured or is characterized by numerous sinkholes. Such recharge can be very important in some semi-arid areas. In principle, "rapid" recharge can occur whenever it rains, so where recharge is dominated by this process it will be affected more by changes in rainfall amount than by the seasonal cycle of soil moisture variability.

Shallow unconfined aquifers along floodplains, which are most common in semi-arid and arid environments, are recharged by seasonal streamflows and can be depleted directly by evaporation. Changes in recharge therefore will be determined by changes in the duration of flow of these streams, which may locally increase or decrease, and the permeability of the overlying beds, but increased evaporative demands would tend to lead to lower groundwater storage. The thick layer of sands substantially reduces the impact of evaporation.

It will be noted from the foregoing that unconfined aquifers are sensitive to local climate change, abstraction, and seawater intrusion. However, quantification of recharge is complicated by the characteristics of the aquifers themselves as well as overlying rocks and soils. A confined aquifer, on the other hand, is characterized by an overlying bed that is impermeable, and local rainfall does not influence the aquifer. It is normally recharged from lakes, rivers, and rainfall that may occur at distances ranging from a few kilometers to thousands of kilometers.

Aside from the influence of climate, recharge to aquifers is very much dependent on the characteristics of the aquifer media and the properties of the overlying soils. Several approaches can be used to estimate recharge based on surface water, unsaturated zone and groundwater data. Among these approaches, numerical modelling is the only tool that can predict recharge. Modelling is also extremely useful for identifying the relative importance of different controls on recharge, provided that the model realistically accounts for all the processes involved. However, the accuracy of recharge estimates depends largely on the availability of high quality hydrogeologic and climatic data. Determining the potential impact of climate change on groundwater resources, in particular, is difficult due to the complexity of the recharge process, and the variation of recharge within and between different climatic zones.

Attempts have been made to calculate the rate of recharge by using carbon-14 isotopes and other modeling techniques. This has been possible for aquifers that are recharged from short distances and after short durations. However, recharge that takes place from long distances and after decades or centuries has been problematic to calculate with accuracy, making estimation of the impacts of climate change difficult. The medium through which recharge takes place often is poorly known and very heterogeneous, again challenging recharge modeling. In general, there is a need to intensify research on modeling techniques, aquifer characteristics, recharge rates, and seawater intrusion, as well as monitoring of groundwater abstractions. This research will provide a sound basis for assessment of the impacts of climate change and sea-level rise on recharge and groundwater resources.

*(c) Coastal Aquifers*

When considering water resources in coastal zones, coastal aquifers are important sources of freshwater. However, salinity intrusion can be a major problem in these zones. Salinity intrusion refers to replacement of freshwater in coastal aquifers by saltwater. It leads to a reduction of available fresh groundwater resources. Changes in climatic variables can significantly alter groundwater recharge rates for major aquifer systems and thus affect the availability of fresh groundwater. Salinization of coastal aquifers is a function of the reduction of groundwater recharge and results in a reduction of fresh groundwater resources.

Sea-level rise will cause saline intrusion into coastal aquifers, with the amount of intrusion depending on local groundwater gradients. Shallow coastal aquifers are at greatest risk. Groundwater in low-lying islands therefore is very sensitive to change. A reduction in precipitation coupled with sea-level rise would not only cause a diminution of the harvestable volume of water; it also would reduce the size of the narrow freshwater lense. For many small island states, such as some Caribbean islands, seawater intrusion into freshwater aquifers has been observed as a result of overpumping of aquifers. Any sea-level rise would worsen the situation.

A link between rising sea level and changes in the water balance is suggested by a general description of the hydraulics of groundwater discharge at the coast. Fresh groundwater rides up over denser, salt water in the aquifer on its way to the sea (Figure 1), and groundwater discharge is focused into a narrow zone that overlaps with the intertidal zone. The width of the zone of groundwater discharge measured perpendicular to the coast, is directly proportional to the discharge rate. The shape of the water table and the depth to the freshwater/saline interface are controlled by the difference in density between freshwater and salt water, the rate of freshwater discharge and the hydraulic properties of the aquifer. The elevation of the water table is controlled by mean sea level through hydrostatic equilibrium at the shore.

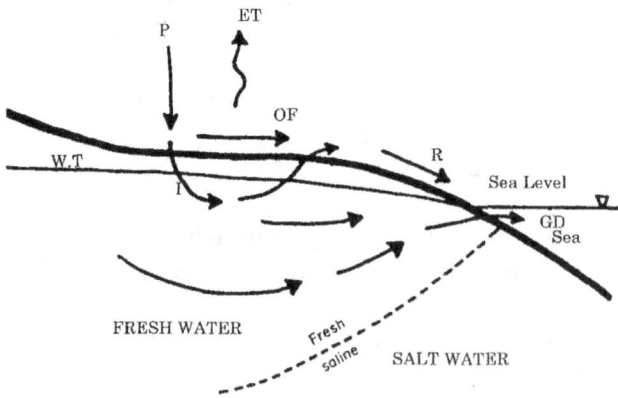

Figure 1: Conceptual Model of the Water Balance in a Coastal Watershed

To assess the impacts of potential climate change on fresh groundwater resources, we should focus on changes in groundwater recharge and sea level rise on the loss of fresh groundwater resources in water resources stressed coastal aquifers.

## Climate Change Scenario for Groundwater in India

Impact of climate change on the ground water regime is expected to be severe. It is to be pointed out that groundwater is the principle source of drinking water in the rural areas. About 85% of the rural water supply in India is dependent on groundwater. India on the whole has a potential of 431 bcm/year of replenishable groundwater (CGWB, 2011), Unfortunately, due to rampant drawing of the subsurface water, the water table in many regions of the country has dropped significantly in the recent years resulting in threat to groundwater sustainability. The overexploited areas are mostly concentrated on three parts of the country. In north western part in Punjab, Haryana, Delhi, Western Uttar Pradesh where through replenishable resources is abundant but there have indiscriminate withdrawals of ground water leading to over-exploitation, in western part of the country particularly in Rajasthan where due to arid climate, ground water recharge itself is less leading to stress on the resource and in peninsular India like Karnataka and Tamil Nadu where due to poor aquifer properties, ground water availability is less.

The most optimistic assumption suggests that an average drop in groundwater level by one metre would increase India's total carbon emissions by over 1%, because the time of withdrawal of the same amount of water will increase fuel consumption. A more realistic assumption reflecting the area projected to be irrigated by groundwater, suggests that the increase in carbon emission could be 4.8% for each metre drop in groundwater levels (Mall et al., 2006). It is recommended to study the aquifer geometry, establish the saline fresh interfaces within few km of the coastal area, the effect of glaciers melting on recharge potential of aquifers in the Ganga basin and its effects on the transboundary aquifer systems, particularly in the arid and semi-arid regions.

Climate change is likely to affect ground water due to changes in precipitation and evapotranspiration. Rising sea levels may lead to increased saline intrusion into coastal and island aquifers, while increased frequency and severity of floods may affect groundwater quality in alluvial aquifers. Sea-level rise leads to intrusion of saline water into the fresh groundwater in coastal aquifers and thus adversely affects groundwater resources. For two small and flat coral islands at the coast of India, the thickness of freshwater lens was computed to decrease from 25 m to 10 m and from 36 m to 28 m, respectively, for a sea level rise of only 0.1 m (Mall et al., 2006).

Increased amount of precipitation in short heavy spells will lead to low infiltration thereby causing low moisture availability for soil. Furthermore, water management systems in the area like number of reservoirs, boreholes etc. would also modify the water availability. Global warming will also affect the water supply by changes in evaporation and ground water recharge. Finally through sea level rise, the global warming may contribute saline intrusion.

Agricultural demand, particularly for irrigation water, which is a major share of total water demand of the country, is considered more sensitive to climate change. A change in field-level climate may alter the need and timing of irrigation. Increased dryness may lead to increased

271

demand, but demand could be reduced if soil moisture content rises at critical times of the year. It is projected that most irrigated areas in India would require more water around 2025 and global net irrigation requirements would increase relative to the situation without climate change by 3.5–5% by 2025 and 6–8% by 2075. In India, roughly 52% of irrigation consumption across the country is extracted from groundwater; therefore, it can be an alarming situation with decline in groundwater and increase in irrigation requirements due to climate change.

Warm air holds more moisture and increase evaporation of surface moisture. With more moisture in the atmosphere, rainfall and snowfall events tend to be more intense, increasing the potential for floods. However, if there is little or no moisture in the soil to evaporate, the incident solar radiation goes into raising the temperature, which could contribute to longer and more severe droughts. Therefore, change in climate will affect the soil moisture, groundwater recharge and frequency of flood or drought episodes and finally groundwater level in different areas. In a number of studies, it is projected that increasing temperature and decline in rainfall may reduce net recharge and affect groundwater levels. However, little work has been done on hydrological impacts of possible climate change for Indian regions/basins.

Existing economic growth scenarios project total power generation capacity in India to increase nine times from 96 GW to 912 GW between 1995-2100. As a result of climate change, it is estimated that approximately 1.5% more power generation capacity will be required. Increased energy demand may arise from a number of sources. For example, increases in average temperature can result in the need for space cooling for buildings, and variability in precipitation can impact irrigation needs and consequent demand for energy from groundwater pumping.

**Status of Research Studies**

The increase of concentration of carbon dioxide and other greenhouse gases in the atmosphere will certainly affect hydrological regimes. Global warming is thus expected to have major implications on water resources management. The observation of long-term trends in climate for many regions of the world has led to considerable research on the impact of greenhouse gases on climate. To this end, several general circulation models (GCMs) have been used to simulate the type of climate that might exist if global concentrations of carbon dioxide (greenhouse gas) were twice their pre-industrial levels. Recent GCM estimates of the projected rise in long-term global average annual surface temperature are between 1 and 4.5 $^{\circ}$C under simulated doubled concentrations of $CO_2$. On the subcontinent scale, there remains considerable uncertainty in the model results and it is not possible to know with confidence the fine details of how the climate will change regionally (Taylor 1997). Consequently, it is customary to use observational data as a baseline and adjust these by the GCM scenarios (Taylor 1997). Because precipitation patterns are significantly influenced by changes in the global-circulation patterns induced by climate change, regional projections for changes in precipitation under doubled $CO_2$ scenarios remain very uncertain.

There have been many studies relating the effect of climate changes on surface water bodies. However, very little research exists on the potential effects of climate change on groundwater, although groundwater is the major source of drinking water across much of the world and plays a vital role in maintaining the ecological value of an area. Available studies show that groundwater recharge and discharge conditions are reflection of the precipitation regime, climatic variables, landscape characteristics and human impacts such as agricultural drainage and flow regulation. Hence, predicting the behavior of recharge and discharge conditions under future climatic and other changes is of great importance for integrated water management.

Studies which consider the indirect effects derived from climate-change-induced alterations in soil, land cover, salt-water intrusion due to rising sea levels and changes in water demand are less common. These studies represent a move away from impact studies (which may be considered to be vertically integrated, in which climate change acts upon an environmental compartment) towards horizontally integrated studies in which environmental compartments interact with each other. However, they remain an incomplete assessment of the pressures facing groundwater resources associated with the direct and indirect effects of future climate and socio-economic change.

Previous studies have typically coupled climate change scenarios with hydrological models, and have generally investigated the impact of climate change on water resources in different areas. The scientific understanding of an aquifer's response to climate change has been studied in several locations within the past decade. These studies link atmospheric models to unsaturated soil models, which, in some cases, were further linked into a groundwater model. The groundwater models used were calibrated to current groundwater conditions and stressed under different predicted climate change scenarios. Some of the recent studies on impact of climate change on groundwater resources have been discussed below.

**Vaccaro (1992)** investigated the degree of variability in climate and its impact on future recharge predictions in a basin in the northwestern United States. In addition to historical records, climate predictions from the synthetic weather generator WGEN (Richardson and Wright, 1984) and three GCMs were considered along with two different land use conditions. The results of the study indicated that the variability in annual recharge was less under the GCM conditions than using the historic data.

**Bouraoui et al. (1999)** presented a general approach to evaluate the effect of potential climate changes on groundwater resources. In the current stage of knowledge, large-scale global climate models are probably the best available tools to provide estimates of the effects of raising greenhouse gases on rainfall and evaporation patterns through a continuous, three dimensional simulation of atmospheric, oceanic and cryospheric processes. However their spatial resolution (generally some thousands of square kilometers) is not compatible with that of watershed hydrologic models. The main purpose of this study is to evaluate the impact of potential climate changes upon groundwater resources. A general methodology is proposed in order to disaggregate outputs of large-scale models and thus to make information directly usable by hydrologic models. As an illustration, this method is applied to a $CO_2$-doubling scenario through the development of a local weather generator, although many uncertainties are not yet assessed about the results of climate models. Two important hydrological variables: rainfall and potential evapotranspiration are thus generated. They are then used by coupling with a physically based hydrological model to estimate the effects of climate changes on groundwater recharge and soil moisture in the root zone.

**Rosenberg et al. (1999)** studied the impact of climate change on the water yield and groundwater recharge of the Ogallala aquifer in the central United States. Three different GCMs were used to predict changes in the future climate due to anticipated changes in temperature and $CO_2$ concentrations. The study found that recharge was reduced under all scenarios, ranging up to 7%, depending on the simulation conditions.

**Sherif and Singh (1999)** investigated the possible effect of climate change on sea water intrusion in coastal aquifers. There is increasing debate these days on climate change and its possible consequences. Much of the debate has focused in the context of surface water systems. In many arid areas of the world, rainfall is scarce and so is surface runoff. These areas rely heavily on groundwater. The consequences of climate change on groundwater are long term and can be far reaching. One of the more apparent consequences is the increased migration of salt water inland in coastal aquifers. Using two coastal aquifers, one in Egypt and the other in India, this study investigated the effect of likely climate change on sea water intrusion. Under conditions of climate change, the sea water levels will rise for several reasons, including variations in atmospheric pressures, expansion of warmer occasions and seas and melting of ice sheets and glaciers. The rise in sea water levels will impose additional saline water heads at the sea side and therefore more sea water intrusion is anticipated. Three realistic scenarios mimicking climate change were considered. Under these scenarios, the Nile Delta aquifer is found to be more vulnerable to climate change and sea level rise. A 50 cm rise in the Mediterranean sea level will cause additional intrusion of 9.0 km in the Nile Delta aquifer. The same rise in water level in the Bay of Bengal will cause an additional intrusion of 0.4 km. Additional pumping will cause serious environmental effects in the case of the Nile Delta aquifer.

**Ghosh Bobba (2002)** analysed the effects of human activities and sea-level changes on the spatial and temporal behaviour of the coupled mechanism of salt-water and freshwater flow through the Godavari Delta of India. The density driven salt-water intrusion process was simulated with the use of a SUTRA (Saturated-Unsaturated TRAnsport) model. Physical parameters, initial heads, and boundary conditions of the delta were defined on the basis of available field data, and an areal, steady-state groundwater model was constructed to calibrate the observed head values corresponding to the initial development phase of the aquifer. Initial and boundary conditions determined from the areal calibration were used to evaluate steady-state, hydraulic heads. Consequently, the initial position of the hydraulic head distribution was calibrated under steady-state conditions. The changes of initial hydraulic distribution, under discharge and recharge conditions, were calculated, and the present-day position of the interface was predicted. The present-day distribution of hydraulic head was estimated via a 20-year simulation. The results indicate that a considerable advance in seawater intrusion can be expected in the coastal aquifer if current rates of groundwater exploitation continue and an important part of the freshwater from the river is channelled from the reservoir for irrigation, industrial and domestic purposes.

Groundwater investigations are presently very active in Andhra Pradesh because of the urgent need for more water to meet the demands for the agricultural, industrial, and domestic purposes of the growing population in the coastal areas. This research has provided numerical simulation of the influence of surface flows, coming from the water management of the projected reservoir,

274

on the regional groundwater behaviour in the delta aquifer. A two-dimensional finite element model, considering open boundary conditions for coasts and a sharp interface between freshwater and salt water, was applied to the aquifer under steady-state conditions for freshwater surplus and deficits at the coastline. When recharges of salt water occur at the coastline, essentially of freshwater deficits, a hypothesis of mixing for the freshwater-salt water transition zone allows the model to calculate the resulting seawater intrusion in the aquifer. Hence, an adequate treatment and interpretation of the hydrogeological data, which are available for the coastal aquifer, were of main concern in satisfactorily applying the proposed numerical model. The results of the steady-state simulations showed reasonable calculations of the water table levels and the freshwater and salt-water thicknesses, as well as the extent of the interface and seawater intrusion into the aquifer for the total discharges or recharges in the delta and along the coastline. As a result of the present hydrogeological conditions, a considerable advance in seawater intrusion would be expected in the coastal aquifer if current rates of groundwater exploitation continue and an important part of the freshwater from the river is channeled from the reservoir for irrigation, industrial and domestic purposes.

**Kirshen (2002)** used the groundwater model MODFLOW to study the impact of global warming on a highly permeable aquifer in the northeastern United States. Groundwater recharge was estimated using a separate model based on precipitation and potential evapotranspiration. Both hypothetical and GCM-predicted changes to the input parameters were used, resulting in higher, no different, and significantly lower recharge rates and groundwater elevations, depending on the climate scenario used.

**Croley and Luukkonen (2003)** investigated the impact of climate change on groundwater levels in Lansing, Michigan. The groundwater recharge rates were based on an empirical streamflow model which was calibrated using the results from two GCMs. The results of the study indicated that the simulated steady-state groundwater levels were generally predicted to increase or decrease due to climate change, depending on the GCM used.

**Eckhardt and Ulbrich (2003)** investigated the impact of climate change on groundwater recharge and streamflow in a small catchment in Germany. The input parameters in their hydrologic model were adjusted based on simulations from five different GCMs. The results of the study indicated that more precipitation will fall as rain in winter due to increased temperatures, resulting in higher recharge and streamflow in January and February. They also found that the increase in recharge from the snowmelt in March disappears, while recharge and streamflow were shown to be potentially reduced in the summer months.

**Loaiciga (2003)** studied a karst aquifer in south-central Texas and considered the impact of climate change not only on streambed recharge, but also on pumping rates (i.e. groundwater use). The impact of climate change on the streambed recharge was estimated using runoff scaling factors based on the ratio of historical and future streamflows predicted from linked general and regional climate models. The study concluded that the rise in groundwater use associated with predicted population growth would pose a higher threat to the aquifer than climate change.

**Allen et al. (2004)** used the Grand Forks aquifer, located in south-central British Columbia, Canada as a case study area for modeling the sensitivity of an aquifer to changes in recharge and river stage consistent with projected climate-change scenarios for the region. Results suggested that variations in recharge to the aquifer under the different climate-change scenarios, modeled under steady-state conditions, have a much smaller impact on the groundwater system than changes in river-stage elevation of the Kettle and Granby Rivers, which flow through the valley. All simulations showed relatively small changes in the overall configuration of the water table and general direction of groundwater flow. High-recharge and low-recharge simulations resulted in approximately a +0.05 m increase and a –0.025 m decrease, respectively, in water table elevations throughout the aquifer. Simulated changes in river-stage elevation, to reflect higher-than-peak flow levels (by 20 and 50%), resulted in average changes in the water-table elevation of 2.72 and 3.45 m, respectively. Simulated changes in river-stage elevation, to reflect lower-than-base flow levels (by 20 and 50%), resulted in average changes in the water-table elevation of -0.48 and -2.10 m, respectively. Current observed water table elevations in the valley are consistent with an average river-stage elevation (between current base flow and peak-flow stages).

**Brouyere et al. (2004)** developed an integrated hydrological model (MOHISE) in order to study the impact of climate change on the hydrological cycle in representative water basins in Belgium. This model considers most hydrological processes in a physically consistent way, more particularly groundwater flows which are modelled using a spatially distributed, finite-element approach. Thanks to this accurate numerical tool, after detailed calibration and validation, quantitative interpretations can be drawn from the groundwater model results. Considering IPCC climate change scenarios, the integrated approach was applied to evaluate the impact of climate change on the water cycle in the Geer basin in Belgium. The groundwater model is described in detail and results are discussed in terms of climate change impact on the evolution of groundwater levels and groundwater reserves. From the modelling application on the Geer basin, it appears that, on a pluri-annual basis, most tested scenarios predict a decrease in groundwater levels and reserves in relation to variations in climatic conditions. However, for this aquifer, the tested scenarios show no enhancement of the seasonal changes in groundwater levels.

**Holman (2006)** described an integrated approach to assess the regional impacts of climate and socio-economic change on groundwater recharge from East Anglia, UK. Many factors affect future groundwater recharge including changed precipitation and temperature regimes, coastal flooding, urbanization, woodland establishment, and changes in cropping and rotations. Important sources of uncertainty and shortcomings in recharge estimation were discussed in the light of the results. The uncertainty in, and importance of, socio-economic scenarios in exploring the consequences of unknown future changes were highlighted. Changes to soil properties are occurring over a range of time scales, such that the soils of the future may not have the same infiltration properties as existing soils. The potential implications involved in assuming unchanging soil properties were described.

**Mall et al. (2006)** examined the potential for sustainable development of surface water and groundwater resources within the constraints imposed by climate change and future research needs in India. In recent times, several studies around the globe show that climatic change is likely to impact significantly upon freshwater resources availability. In India, demand for water has already increased manifold over the years due to urbanization, agriculture expansion, increasing population, rapid industrialization and economic development. At present, changes in cropping pattern and land-use pattern, over-exploitation of water storage and changes in irrigation and drainage are modifying the hydrological cycle in many climate regions and river basins of India. An assessment of the availability of water resources in the context of future national requirements and expected impacts of climate change and its variability is critical for relevant national and regional long-term development strategies and sustainable development.

He concluded that the Indian region is highly sensitive to climate change. The elements/sectors currently at risk are likely to be highly vulnerable to climate change and variability. It is urgently required to intensify in-depth research work with the following objectives:

- Analyse recent experiences in climate variability and extreme events, and their impacts on regional water resources and groundwater availability.
- Study on changing patterns of rainfall, i.e. spatial and temporal variation and its impact on run-off and aquifer recharge pattern.
- Study sea-level rise due to increased run-off as projected due to glacial recession and increased rainfall.
- Sea-water intrusions into costal aquifers.
- Determine vulnerability of regional water resources to climate change and identify key risks and prioritize adaptation responses.
- Evaluate the efficacy of various adaptation strategies or coping mechanisms that may reduce vulnerability of the regional water resources.

It has been the endeavour of this study to summarize some important vulnerability issues associated with the present and potential future hydrological responses due to climate change and highlight those areas where further research is required. The National Environment Policy (2004) also advocated that anthropogenic climate changes have severe adverse impacts on India's precipitation patterns, ecosystems, agricultural potential, forests, water resources, coastal and marine resources. Large-scale planning would be clearly required for adaptation measures for climate change impacts, if catastrophic human misery is to be avoided.

**Ranjan et al. (2006)** evaluated the impacts of climate change on fresh groundwater resources specifically salinity intrusion in water resources stressed coastal aquifers. Their assessment used the Hadley Centre climate model, HadCM3 with high and low emission scenarios (SRES A2 and B2) for years 2000–2099. In both scenarios, the annual fresh groundwater resources losses indicated an increasing long-term trend in all stressed areas, except in the northern Africa/Sahara region. They also found that precipitation and temperature individually did not show good correlations with fresh groundwater loss. However, the relationship between the aridity index and fresh groundwater loss exhibited a strong negative correlation. They also discussed the impacts of loss of fresh groundwater resources on socio-economic activities, mainly population growth and per capita fresh groundwater resources.

**Scibek and Allen (2006)** developed a methodology for linking climate models and groundwater models to investigate future impacts of climate change on groundwater resources. An unconfined aquifer, situated near Grand Forks in south central British Columbia, Canada, was used to test the methodology. Climate change scenarios from the Canadian Global Coupled Model 1 (CGCM1) model runs were downscaled to local conditions using Statistical Downscaling Model

(SDSM), and the change factors were extracted and applied in LARS-WG stochastic weather generator and then input to the recharge model. The recharge model simulated the direct recharge to the aquifer from infiltration of precipitation and consisted of spatially distributed recharge zones, represented in the Hydrologic Evaluation of Landfill Performance (HELP) hydrologic model linked to a geographic information system (GIS). A three-dimensional transient groundwater flow model, implemented in MODFLOW, was then used to simulate four climate scenarios in 1-year runs (1961–1999 present, 2010–2039, 2040–2069, and 2070-2099) and compare groundwater levels to present. The effect of spatial distribution of recharge on groundwater levels, compared to that of a single uniform recharge zone, is much larger than that of temporal variation in recharge, compared to a mean annual recharge representation. The predicted future climate for the Grand Forks area from the downscaled CGCM1 model will result in more recharge to the unconfined aquifer from spring to the summer season. However, the overall effect of recharge on the water balance is small because of dominant river-aquifer interactions and river water recharge.

**Hsu et al. (2007)** adopted a numerical modeling approach to investigate the response of the groundwater system to climate variability to effectively manage the groundwater resources of the Pingtung Plain. The Pingtung Plain is one of the most important groundwater-resource areas in southwestern Taiwan. The overexploitation of groundwater in the last two decades has led to serious deterioration in the quantity and quality of groundwater resources in this area. Furthermore, the manifestation of climate change tends to induce the instability of surface-water resources and strengthen the importance of the groundwater resources. Southwestern Taiwan in particular shows decreasing tendencies in both the annual amount of precipitation and annual precipitation days. A hydrogeological model was constructed based on the information from geology, hydrogeology, and geochemistry. Applying the linear regression model of precipitation to the next two decades, the modeling result shows that the lowering water level in the proximal fan raises an alarm regarding the decrease of available groundwater in the stress of climate change, and the enlargement of the low-groundwater-level area on the coast signals the deterioration of water quantity and quality in the future. Suitable strategies for water-resource management in response to hydrological impacts of future climatic change are imperative.

**Jyrkama and Sykes (2007)** presented a physically based methodology that can be used to characterize both the temporal and spatial effect of climate change on groundwater recharge. The method, based on the hydrologic model HELP3, can be used to estimate potential groundwater recharge at the regional scale with high spatial and temporal resolution. In this study, the method is used to simulate the past conditions, with 40 years of actual weather data, and future changes in the hydrologic cycle of the Grand River watershed. The impact of climate change is modelled by perturbing the model input parameters using predicted changes in the regions climate. The results of the study indicate that the overall rate of groundwater recharge is predicted to increase as a result of climate change. The higher intensity and frequency of precipitation will also contribute significantly to surface runoff, while global warming may result in increased evapotranspiration rates. Warmer winter temperatures will reduce the extent of ground frost and shift the spring melt from spring toward winter, allowing more water to infiltrate into the ground. While many previous climate change impact studies have focused on the temporal changes in groundwater recharge, results of this study suggest that the impacts can also have high spatial variability.

**Toews (2007)** modeled the impacts of future predicted climate change on groundwater recharge resources for the arid to semi-arid south Okanagan region, British Columbia. The hydrostratigraphy of the region consists of Pleistocene-aged glaciolacustrine silt overlain by glaciofluvial sand and gravel. Spatial recharge was modelled using available soil and climate data with the HELP 3.80D hydrology model. Climate change effects on recharge were investigated using stochastically-generated climate from three GCMs. Recharge is estimated to be ~45 mm/year, with minor increases expected with climate change. However, growing season and crop water demands will increase, posing additional stresses on water use in the region. A transient MODFLOW groundwater model simulates increases of water table in future time periods, which is largely driven by irrigation application increases. Spatial recharge was also used in a groundwater model to define capture zones around eight municipal water wells. These capture zones will be used for community planning.

**Woldeamlak et al. (2007)** modeled the effects of climate change on the groundwater systems in the Grote-Nete catchment, Belgium, covering an area of 525 km2, using wet (greenhouse), cold or NATCC (North Atlantic Thermohaline Circulation Change) and dry climate scenarios. Low, central and high estimates of temperature changes were adopted for wet scenarios. Seasonal and annual water balance components including groundwater recharge were simulated using the WetSpass model, while mean annual groundwater elevations and discharge were simulated with a steady-state MODFLOW groundwater model. WetSpass results for the wet scenarios showed that wet winters and drier summers are expected relative to the present situation. MODFLOW results for wet high scenario showed groundwater levels increase by as much as 79 cm, which could affect the distribution and species richness of meadows. Results obtained for cold scenarios depict drier winters and wetter summers relative to the present. The dry scenarios predict dry conditions for the whole year. There is no recharge during the summer, which is mainly attributed to high evapotranspiration rates by forests and low precipitation. Average annual groundwater levels drop by 0.5 m, with maximum of 3.1 m on the eastern part of the Campine Plateau. This could endanger aquatic ecosystem, shrubs, and crop production.

**Carneiro et al. (2008)** applied a density dependent numerical flow model (FEMWATER) to study the climate change impact in an unconfined shallow aquifer in the Mediterranean coast of Morocco. The stresses imposed to the model were derived from the IPCC emission scenarios and included recharge variations, rising sea level and advancing seashore. The simulations show that there will be a significant decline in the renewable freshwater resources and that salinity increases can be quite large but are limited to a restricted area.

**Dragoni and Sukhija (2008)** analysed the main methods for studying the relationships between climate change and groundwater, and presented the main areas in which hydrogeological research should focus in order to mitigate the likely impacts. There is a general consensus that climate change is an ongoing phenomenon. This will inevitably bring about numerous environmental problems, including alterations to the hydrological cycle, which is already heavily influenced by anthropogenic activity. The available climate scenarios indicate areas where rainfall may increase or diminish, but the final outcome with respect to man and environment will, generally, be detrimental. Groundwater will be vital to alleviate some of the worst drought situations.

**Herrera-Pantoja and Hiscock (2008)** outlined a methodology to quantify the effects of climate change on potential groundwater recharge (or hydrological excess water) for three locations in the north and south of Great Britain. Using results from a stochastic weather generator, actual evapotranspiration and potential groundwater recharge time-series for the historic baseline 1961-1990 and for a future 'high' greenhouse gas emissions scenario for the 2020s, 2050s and 2080s time periods were simulated for Coltishall in East Anglia, Gatwick in southeast England and Paisley in west Scotland. Under the 'high' gas emissions scenario, results showed a decrease of 20% in potential groundwater recharge for Coltishall, 40% for Gatwick and 7% for Paisley by the end of this century. The persistence of dry periods is shown to increase for the three sites during the 2050s and 2080s. Gatwick presents the driest conditions, Coltishall the largest variability of wet and dry periods and Paisley little variability. For Paisley, the main effect of climate change is evident during the dry season (April-September), when the potential amount of hydrological excess water decreases by 88% during the 2080s. Overall, it is concluded that future climate may present a decrease in potential groundwater recharge that will increase stress on local and regional groundwater resources that are already under ecosystem and water supply pressures.

**Franssen (2009)** indicated the limitations of the studies that address the impact of climate change on groundwater resources and suggested an improved approach. A general review, both from a groundwater hydrological and a climatological viewpoint, is given, oriented on the impact of climate change on groundwater resources. The impact of climate change on groundwater resources is not the subject of many studies in the scientific literature. Only rarely sophisticated downscaling techniques are applied to downscale estimated global circulation model (GCM) future precipitation series for a point or region of interest. Often it is not taken into account that different climate models calculate considerably different precipitation amounts (conceptual uncertainty). The joint downscaling of the meteorological variables that govern potential evapotranspiration (ET) is never done in the context of a study that assessed the impact of climate change on groundwater resources. It is desirable that actual ET is calculated in (groundwater) hydrological models on a physical basis, i.e. by coupling the energy and water balance at the Earth's surface.

**Holman et al. (2009)** indicated that groundwater resource estimates require the calculation of recharge using a daily time step. Within climate-change impact studies, this inevitably necessitates temporal downscaling of global or regional climate model outputs. This paper compares future estimates of potential groundwater recharge calculated using a daily soil-water balance model and climate-change weather time series derived using change factor (deterministic) and weather generator (stochastic) methods for Coltishall, UK. The uncertainty in the results for a given climate-change scenario arising from the choice of downscaling method is greater than the uncertainty due to the emissions scenario within a 30-year time slice. Robust estimates of the impact of climate change on groundwater resources require stochastic modelling of potential recharge, but this has implications for groundwater model runtimes. It is recommended that stochastic modelling of potential recharge is used in vulnerable or sensitive groundwater systems, and that the multiple recharge time series are sampled according to the distribution of contextually important time series variables, e.g. recharge drought severity and persistence (for water resource management) or high recharge years (for groundwater flooding).

Such an approach will underpin an improved understanding of climate change impacts on sustainable groundwater resource management based on adaptive management and risk-based frameworks.

**Shah (2009)** reviewed the India's opportunities for mitigation and adaptation with reference to climate change and groundwater. For millennia, India used surface storage and gravity flow to water crops. During the last 40 years, however, India has witnessed a decline in gravity-flow irrigation and the rise of a booming 'water-scavenging' irrigation economy through millions of small, private tubewells. For India, groundwater has become at once critical and threatened. Climate change will act as a force multiplier; it will enhance groundwater's criticality for drought-proofing agriculture and simultaneously multiply the threat to the resource. Groundwater pumping with electricity and diesel also accounts for an estimated 16–25 million mt of carbon emissions, 4–6% of India's total. From a climate change point of view, India's groundwater hotspots are western and peninsular India. These are critical for climate change mitigation as well as adaptation. To achieve both, India needs to make a transition from surface storage to 'managed aquifer storage' as the center pin of its water strategy with proactive demand- and supply-side management components. In doing this, India needs to learn intelligently from the experience of countries like Australia and the United States that have long experience in managed aquifer recharge.

**Allen (2010)** examined historical groundwater levels for selected observation wells in the south coastal region of British Columbia, Canada, to gain a better understanding of historical trends. Over a common period (1976-1999), negative trends in groundwater level dominate most records, and appear to be related to longer term negative regional trends in precipitation, although variable trends are evident at the shorter time periods used for this study. To explore potential consequences of varying recharge on groundwater quality, water chemistry data from selected monitoring wells on one island were examined. Chloride concentrations were observed to vary annually in one well by up to 4000 mg/L. Projections for future climate from one global climate model (CGCM1) were used as input to a recharge model to study the sensitivity of recharge to shifts in precipitation and temperature predicted for the region. The recharge model was driven by a stochastic daily weather series, calibrated to historic climate data. Daily weather series represent historic climate, and two future time periods (2020s) and (2050s). Simulated recharge increases progressively in the future using this particular global climate model; however, precipitation projections for this region of British Columbia are highly uncertain. Both positive and negative shifts in annual precipitation were predicted using a range of global climate models.

**Allen et al. (2010)** addressed variations in the prediction of recharge by comparing recharge simulated using climate data generated using a state-of-the-art downscaling method, TreeGen, with a range of global climate models (GCMs). The study site is the transnational Abbotsford-Sumas aquifer in coastal British Columbia, Canada and Washington State, USA, and is representative of a wet coastal climate. Sixty-four recharge zones were defined based on combinations of classed soil permeability, vadose zone permeability, and unsaturated zone depth (or depth to water table) mapped in the study area. One-dimensional recharge simulations were conducted for each recharge zone using the HELP hydrologic model, which simulates percolation through a vertical column. The HELP model is driven by mean daily temperature, daily precipitation, and daily solar radiation. For the historical recharge simulations, the climate data series was generated using the LARS-WG stochastic weather generator. Historical recharge was compared to recharge simulated using climate data series derived from the TreeGen downscaling model for three future time periods: 2020s (2010–2039), 2050s (2040–2069), and 2080s (2070–2099) for each of four GCMs (CGCM3.1, ECHAM5, PCM1, and CM2.1). Recharge results are compared on an annual basis for the entire aquifer area. Both increases and decreases relative to historical recharge are simulated depending on time period and model. By the 2080s, the range of model predictions spans −10.5% to +23.2% relative to historical recharge. This variability in recharge predictions suggests that the seasonal performance of the downscaling tool is important and that a range of GCMs should be considered for water management planning.

**Crosbie et al. (2010)** presented a methodology for assessing the average changes in groundwater recharge under a future climate. The method is applied to the 1,060,000 km$^2$ Murray-Darling Basin (MDB) in Australia. Climate sequences were developed based upon three scenarios for a 2030 climate relative to a 1990 climate from the outputs of 15 global climate models. Dryland diffuse groundwater recharge was modelled in WAVES using these 45 climate scenarios and fitted to a Pearson Type III probability distribution to condense the 45 scenarios down to three: a wet future, a median future and a dry future. The use of a probability distribution allowed the significance of any change in recharge to be assessed. This study found that for the median future, climate recharge is projected to increase on average by 5% across the MDB but this is not spatially uniform. In the wet and dry future scenarios the recharge is projected to increase by 32% and decrease by 12% on average across the MDB, respectively. The differences between the climate sequences generated by the 15 different global climate models makes it difficult to project the direction of the change in recharge for a 2030 climate, let alone the magnitude.

**Dams et al. (2010)** presented a methodology to predict the potential impact of climate change on quantitative groundwater characteristics determining GWDTEs (Groundwater Dependent Terrestrial Ecosystems). The developed methodology includes coupling a distributed hydrological model (WetSpa) with a transient groundwater flow model (MODFLOW) and is tested for the Kleine Nete basin, Belgium. Because the occurrence of phreatophytes is strongly determined by the dynamic properties of the groundwater system, a groundwater flow model with a high temporal and spatial distribution was developed using MODFLOW. The groundwater recharge and river heads are estimated with the WetSpa model using a daily time step to incorporate the impact of changes in rainfall intensity. Potential future hydrological changes are calculated by comparing the hydrological state corresponding to 1960-1991 with future scenarios developed for 2070-2101.

Since the uncertainty in the prediction of the future climate components such as potential evapotranspiration (PET) and precipitation is still high, an ensemble of 28 climate scenarios were chosen from the PRUDENCE database. For each of these scenarios, recharge, river stage, groundwater head and groundwater flow are estimated for 32 years with half monthly time steps. Comparison of the original measured PET with future PET shows that the PET during summer rises in all future scenarios with about 1 mm per day. For winter conditions the scenarios predict little change in PET. Future precipitation shows an increase in precipitation during winter and a decrease during summer. Future groundwater recharge decreases on average with 20 mm per year, the highest decreases are simulated from July until September. Average groundwater heads indicate an average decrease of 7 cm. Groundwater levels in interfluves generally show decreases up to 30 cm. The mean lowest groundwater level decreases on average with 6 cm, while the mean highest groundwater level decreases about 3 cm. On average, the groundwater discharge reduces with 4%, from 5 to 4.8 m³/s. GWDTEs that currently receive a low groundwater discharge, are likely to disappear due to future climate changes.

**McCallum et al. (2010)** used a sensitivity analysis of climate variables using a modified version of WAVES, a soil-vegetation-atmosphere-transfer model (unsaturated zone), to determine the importance of each climate variable in the change in groundwater recharge for three points in Australia. This study found that change in recharge is most sensitive to change in rainfall. Increases in temperature and changes in rainfall intensity also led to significant changes in recharge. Although not as significant as other climate variables, some changes in recharge were observed due to changes in solar radiation and carbon dioxide concentration. When these variables were altered simultaneously, changes in recharge appeared to be closely related to changes in rainfall; however, in nearly all cases, recharge was greater than would have been predicted if only rainfall had been considered. These findings have implications for how recharge is projected to change due to climate change.

**Okkonen et al. (2010)** presented a literature review of the impacts of anticipated climate change on unconfined aquifers, along with a conceptual framework for evaluating the complex responses of surface and subsurface hydrology to climate variables in cold regions. The framework offers a way to conceptualize how changes in one component of the system may impact another by delineating the relationships among climate drivers, hydrological responses, and groundwater responses in a straight-forward manner. The model is elaborated in the context of shallow unconfined aquifers in the boreal environment of Finland. In cold conditions, climate change is expected to reduce snow cover and soil frost and increase winter floods. The annual surface water level maximum will occur earlier in spring, and water levels will decrease in summer due to higher evapotranspiration rates. The maximum recharge and groundwater level are expected to occur earlier in the year. Lower groundwater levels are expected in summer due to higher evapotranspiration rates. The flow regimes between shallow unconfined aquifers and surface water may change, affecting water quantity and quality in the surface and groundwater systems.

**Oude Essink et al. (2010)** focussed on a coastal groundwater system that is already threatened by a relatively high seawater level: the low-lying Dutch Delta. Nearly one third of the Netherlands lies below mean sea level, and the land surface is still subsiding up to 1 m per century. This densely populated delta region, where fresh groundwater resources are used intensively for domestic, agricultural, and industrial purposes, can serve as a laboratory case for other low-lying delta areas throughout the world. Their findings on hydrogeological effects can be scaled up since the problems the Dutch face now will very likely be the problems encountered in other delta areas in the future. They calculated the possible impacts of future sea level rise, land subsidence, changes in recharge, autonomous salinization, and the effects of two mitigation counter-measures with a three-dimensional numerical model for variable density groundwater flow and coupled solute transport (MOCDENS3D). They considered the effects on hydraulic heads, seepage fluxes, salt loads to surface waters, and changes in fresh groundwater resources as a function of time and for seven scenarios.

Their numerical modeling results show that the impact of sea level rise is limited to areas within 10 km of the coastline and main rivers because the increased head in the groundwater system at the coast can easily be produced though the highly permeable Holocene confining layer. Along the southwest coast of the Netherlands, salt loads will double in some parts of the deep and large polders by the year 2100 A.D. due to sea level rise. More inland, ongoing land subsidence will cause hydraulic heads and phreatic water levels to drop, which may result in damage to dikes, infrastructure, and urban areas. In the deep polders more inland, autonomous upconing of deeper and more saline groundwater will be responsible for increasing salt loads. The future increase of salt loads will cause salinization of surface waters and shallow groundwater and put the total volumes of fresh groundwater volumes for drinking water supply, agricultural purposes, industry, and ecosystems under pressure.

**Payne (2010)** observed that sea-level rise and changes in precipitation patterns may contribute to the occurrence and affect the rate of saltwater contamination in the Hilton Head Island, South Carolina area. To address the effects of climate change on saltwater intrusion, a three-dimensional, finite-element, variable-density, solute-transport model (SUTRA 2.1) was developed to simulate different rates of sea-level rise and variation in onshore freshwater recharge. Model simulation showed that the greatest effect on the existing saltwater plume occurred from reducing recharge, suggesting recharge may be a more important consideration in saltwater intrusion management than estimated rates of sea-level rise. Saltwater intrusion management would benefit from improved constraints on recharge rates by using model-independent, local precipitation and evapotranspiration data, and improving estimates of confining unit hydraulic properties.

**Rozell and Wong (2010)** investigated the effects of climate change on Shelter Island, New York State (USA), a small sandy island, using a variable-density transient groundwater flow model (SEAWAT). Predictions for changes in precipitation and sea-level rise over the next century from the Intergovernmental Panel on Climate Change 2007 report were used to create two future climate scenarios. In the scenario most favorable to fresh groundwater retention, consisting of a 15% precipitation increase and 0.18-m sea-level rise, the result was a 23-m seaward movement of the freshwater/ salt-water interface, a 0.27-m water-table rise, and a 3% increase in the fresh-water lens volume. In the scenario supposedly least favorable to groundwater retention, consisting of a 2% precipitation decrease and 0.61-m sea-level rise, the result was a 16-m landward movement of the fresh-water/salt-water interface, a 0.59-m watertable rise, and a 1% increase in lens volume. The unexpected groundwater-volume increase under unfavorable climate change conditions was best explained by a clay layer under the island that restricts the maximum depth of the aquifer and allows for an increase in freshwater lens volume when the water table rises.

**Vandenbohede and Lebbe (2010)** evaluated the effects of sea level rise and future recharge changes on the coastal aquifer of the western Belgian coastal plain with a 3D density dependent groundwater flow model (MOCDENS3D). The area is characterised by a wide dune belt. Sea level rise results in a landward enlargement of the fresh water lens under the dunes and an increase of flow towards the dune-polder transition's drainage system. Recharge increase results also in an enlargement of the dune's fresh water lens and an increase of the amount of water which must be evacuated by the polder's drainage system. Recharge decrease has the reverse effect.

**Zhou et al. (2010)** reported that climate change affects not only water resources but also water demand for irrigation. A large proportion of the world's agriculture depends on groundwater, especially in arid and semi-arid regions. In several regions, aquifer resources face depletion. Groundwater recharge has been viewed as a by-product of irrigation return flow, and with climate change, aquifer storage of such flow will be vital. A general review, for a broadbased audience, is given of work on global warming and groundwater resources, summarizing the methods used to analyze the climate change scenarios and the influence of these predicted changes on groundwater resources around the world (especially the impact on regional groundwater resources and irrigation requirements). Future challenges of adapting to climate change are also discussed. Such challenges include water-resources depletion, increasing irrigation demand, reduced crop yield, and groundwater salinization. The adaptation to and mitigation of these effects is also reported, including useful information for water-resources managers and the development of sustainable groundwater irrigation methods. Rescheduling irrigation according to the season, coordinating the groundwater resources and irrigation demand, developing more accurate and complete modeling prediction methods, and managing the irrigation facilities in different ways would all be considered, based on the particular cases.

**Barthel (2011)** presented an integrated approach for assessing the availability of groundwater under conditions of 'global-change'. The approach is embedded in the DANUBIA system developed by the interdisciplinary GLOWA-Danube Project to simulate the interaction of natural and socio-economic processes within the Upper Danube Catchment (UDC, 77,000 km$^2$ and located in parts of Germany, Austria, Switzerland and Italy). The approach enables the quantitative assessment of groundwater bodies (zones), which are delineated by intersecting surface watersheds, regional aquifers, and geomorphologic regions. The individual hydrogeological and geometrical characteristics of these zones are accounted for by defining characteristic response times and weights to describe the relative significance of changes in variables (recharge, groundwater level, groundwater discharge, river discharge) associated with different states. These changes, in each zone, are converted into indices (*GroundwaterQuantityFlags*). The motivation and particularities of regional-scale groundwater assessment and the background of GLOWA-Danube are described, along with a description of the developed methodology. The approach was applied to the UDC, where several different climate scenarios (2011–2060) were evaluated. A selection of results is presented to demonstrate the potential of the methodology. The approach was inspired by the European Water Framework Directive, yet it has a stronger focus on the evaluation of global-change impacts.

**Yihdego and Webb (2011)** carried out modeling of bore hydrographs to determine the impact of climate and land-use change in a temperate subhumid region of southeastern Australia. To determine the relative impact of climate and human intervention on groundwater elevations in western Victoria, southeast Australia, bore hydrograph fluctuations in three aquifers were modelled using a transfer function noise model (PIRFICT) and an auto-regressive model (HARTT), which give generally comparable results. Most of the groundwater-level fluctuations

(>90%) are explained by climatic variation, particularly rainfall. The overall non-climate-related trend in groundwater level is downward and small but statistically significant (−0.04 to −0.066 m/yr), and is probably due to the widespread replacement of grazing land by wheat and canola cultivation, as these crops use more water than pasture. A large non-climate-related trend (−0.30 m/yr) for bores in an irrigation area is mainly related to groundwater extraction. The response time of the system is rapid (only 4.85 years on average), much faster than previously estimated. Rates of groundwater flow are much slower; groundwater ages are up to ~35,000 years. Response times effectively represent the time for the system to move to a new state of hydrologic equilibrium; this prediction of the time scale of the impacts of land-use change on groundwater resources will allow the development of better strategies for groundwater management.

**Crosbie et al. (2012)** investigated episodic recharge with reference to climate change in the Murray-Darling Basin, Australia. In semi-arid areas, episodic recharge can form a significant part of overall recharge, dependant upon infrequent rainfall events. With climate change projections suggesting changes in future rainfall magnitude and intensity, groundwater recharge in semi-arid areas is likely to be affected disproportionately by climate change. This study sought to investigate projected changes in episodic recharge in arid areas of the Murray-Darling Basin, Australia, using three global warming scenarios from 15 different global climate models (GCMs) for a 2030 climate. Two metrics were used to investigate episodic recharge: at the annual scale the coefficient of variation was used, and at the daily scale the proportion of recharge in the highest 1% of daily recharge. The metrics were proportional to each other but were inconclusive as to whether episodic recharge was to increase or decrease in this environment; this is not a surprising result considering the spread in recharge projections from the 45 scenarios. The results showed that the change in the low probability of exceedance rainfall events was a better predictor of the change in total recharge than the change in total rainfall, which has implications for the selection of GCMs used in impact studies and the way GCM results are downscaled.

**Neukum and Azzam (2012)** investigated the impact of climate change on groundwater recharge in a small catchment in the Black Forest, Germany. Temporal and spatial changes of the hydrological cycle are the consequences of climate variations. In addition to changes in surface runoff with possible floods and droughts, climate variations may affect groundwater through alteration of groundwater recharge with consequences for future water management. This study investigates the impact of climate change, according to the Special Report on Emission Scenarios (SRES) A1B, A2 and B1, on groundwater recharge in the catchment area of a fissured aquifer in the Black Forest, Germany, which has sparse groundwater data. The study uses a water-balance model considering a conceptual approach for groundwater-surface water exchange. River discharge data are used for model calibration and validation. The results show temporal and spatial changes in groundwater recharge. Groundwater recharge is progressively reduced for summer during the twenty-first century. The annual sum of groundwater recharge is affected negatively for scenarios A1B and A2. On average, groundwater recharge during the twenty-first century is reduced mainly for the lower parts of the valley and increased for the upper parts of the valley and the crests. The reduced storage of water as snow during winter due to projected higher air temperatures causes an important relative increase in rainfall and, therefore, higher groundwater recharge and river discharge.

**Raposo et al. (2013)** assessed the impact of future climate change on groundwater recharge in Galicia-Costa, Spain. Climate change can impact the hydrological processes of a watershed and may result in problems with future water supply for large sections of the population. Results from the FP5 PRUDENCE project suggest significant changes in temperature and precipitation over Europe. In this study, the Soil and Water Assessment Tool (SWAT) model was used to assess the potential impacts of climate change on groundwater recharge in the hydrological district of Galicia-Costa, Spain. Climate projections from two general circulation models and eight different regional climate models were used for the assessment and two climate-change scenarios were evaluated. Calibration and validation of the model were performed using a daily time-step in four representative catchments in the district. The effects on modeled mean annual groundwater recharge are small, partly due to the greater stomatal efficiency of plants in response to increased $CO_2$ concentration. However, climate change strongly influences the temporal variability of modeled groundwater recharge. Recharge may concentrate in the winter season and dramatically decrease in the summer–autumn season. As a result, the dry-season duration may be increased on average by almost 30 % for the A2 emission scenario, exacerbating the current problems in water supply.

**Lapworth et al. (2013)** estimated residence times of shallow groundwater in West Africa. Although shallow groundwater (<50 mbgl) sustains the vast majority of improved drinking-water supplies in rural Africa, there is little information on how resilient this resource may be to future changes in climate. This study presents results of a groundwater survey using stable isotopes, CFCs, $SF_6$, and $^3H$ across different climatic zones (annual rainfall 400–2,000 mm/year) in West Africa. The purpose was to quantify the residence times of shallow groundwaters in sedimentary and basement aquifers, and investigate the relationship between groundwater resources and climate. Stable-isotope results indicate that most shallow groundwaters are recharged rapidly following rainfall, showing little evidence of evaporation prior to recharge. Chloride mass-balance results indicate that within the arid areas (<400 mm annual rainfall) there is recharge of up to 20 mm/year. Age tracers show that most groundwaters have mean residence times (MRTs) of 32–65 years, with comparable MRTs in the different climate zones. Similar MRTs measured in both the sedimentary and basement aquifers suggest similar hydraulic diffusivity and significant groundwater storage within the shallow basement. This suggests there is considerable resilience to short-term inter-annual variation in rainfall and recharge, and rural groundwater resources are likely to sustain diffuse, low volume abstraction.

**Mollema and Antonellini (2013)** investigated seasonal variation in natural recharge of coastal aquifers. Many coastal zones around the world have irregular precipitation throughout the year. This results in discontinuous natural recharge of coastal aquifers, which affects the size of freshwater lenses present in sandy deposits. Temperature data for the period 1960–1990 from LocClim (local climate estimator) and those obtained from the Intergovernmental Panel on Climate Change (IPCC) SRES A1b scenario for 2070–2100, have been used to calculate the potential evapotranspiration with the Thornthwaite method. Potential recharge (difference between precipitation and potential evapotranspiration) was defined at 12 locations: Ameland (The Netherlands), Auckland and Wellington (New Zealand); Hong Kong (China); Ravenna (Italy), Mekong (Vietnam), Mumbai (India), New Jersey (USA), Nile Delta (Egypt), Kobe and Tokyo (Japan), and Singapore. The influence of variable/discontinuous recharge on the size of freshwater lenses was simulated with the SEAWAT model. The discrepancy between models with continuous and with discontinuous recharge is relatively small in areas where the total annual recharge is low (258–616 mm/year); but in places with Monsoon-dominated climate (e.g. Mumbai, with recharge up to 1,686 mm/year), the difference in freshwater-lens thickness between the discontinuous and the continuous model is larger (up to 5 m) and thus important to consider in numerical models that estimate freshwater availability.

These studies are still at infancy and more data, in terms of field information are to be generated. This will also facilitate appropriate validation of the simulation for the present scenarios. In summary, climate change is likely to have an impact on future recharge rates and hence on the underlying groundwater resources. The impact may not necessarily be a negative one, as evidenced by some of the investigations. Quantifying the impact is difficult, however, and is subject to uncertainties present in the future climate predictions. Simulations based on general circulation models (GCMs) have yielded mixed and conflicting results, raising questions about their reliability in predicting future hydrologic conditions. However, it is clear that the global warming threat is real and the consequences of climate change phenomena are many and alarming.

**Methodology to Assess the Impact of Climate Change on Groundwater Resources**

The potential impacts of climate change on water resources have long been recognized although there has been comparatively little research relating to groundwater. The principle focus of climate change research with regard to groundwater has been on quantifying the likely direct impacts of changing precipitation and temperature patterns. Such studies have used a range of modelling techniques such as soil water balance models, empirical models, conceptual models and more complex distributed models, but all have derived changes in groundwater recharge assuming parameters other than precipitation and temperature remaining constant.

There are two main parameters that could have a significant impact on groundwater levels: recharge and river stage/discharge. To assess the impact on the groundwater system to changes in these two parameters, it is necessary to have a calibrated flow model and to conduct a sensitivity analysis by varying these two parameters and calculating changes to the water balance (e.g., differences in water levels). The research objectives can be:

- To develop a conceptual model of the hydrogeology of the study region;

- To investigate how regional and local weather events affect recharge;

- To determine potential impacts of climate change on recharge for the study area, and to assess the sensitivity of the results to different global climate models;

- To develop and calibrate a regional-scale three-dimensional groundwater flow model of the region and to use that model to assess the impacts of climate change on groundwater resources; and

> To develop and calibrate a local-scale three-dimensional groundwater flow model, and to undertake a well capture zone analysis for the local community water supply wells for the region.

The methodology consists of three main steps. To begin with, climate scenarios can be formulated for the future years such as 2050 and 2100. This is done by assigning percentage or value changes of climatic variables on a seasonal and/or annual basis only for the future years relative to the present year. Secondly, based on these scenarios and present situation, seasonal and annual recharge, evapotranspiration and runoff are simulated with the WHI UnSat Suite (HELP module for recharge) and/or WetSpass model. Finally, the annual recharge outputs from WHI UnSat Suite or WetSpass model are used to simulate groundwater system conditions using steady-state MODFLOW model setups for the present condition and for the future years.

The main tasks that are involved in such a study are:

1.  Describe hydrogeology of the study area.

2.  Undertake a statistical analysis to separate climate into regional and local events and determine the role of each in contributing to groundwater recharge.

3.  Analyze climate data from weather stations and modelled GCM, and build future predicted climate change datasets with temperature, precipitation and solar radiation variables.

4.  Define methodology for estimating changes to recharge in the model under both current climate conditions and for the range of climate-change scenarios for the study area.

5.  Use of a computer code (such as WHI UnSat Suite or WetSpass) to estimate recharge based on available precipitation and temperature records and anticipated changes to these parameters.

*Recharge estimation by WHI UnSat Suite*

UnSat Suite contains the subprogram, Visual HELP, which contains a more user-friendly interface for the program HELP that is approved by the United States Environmental Protection Agency (US EPA) for designing landfills. Visual HELP enables the modeler to generate estimates of recharge using a weather generator and the properties of the aquifer column.

*Recharge estimation by WetSpass*

WetSpass is a quasi physically distributed seasonal-water balance model, which takes into account detailed soil, land-use, slope, groundwater depth, and hydro-climatological distributed maps with associated parameter tables for estimating groundwater recharge. The model uses seasonal (summer and winter) geographical information systems (GIS) input grids of the mentioned inputs to estimate annual and seasonal groundwater recharge values.

6. Quantify the spatially distributed recharge rates using the climate data and spatial soil survey data.

7. Development and calibration of a three-dimensional regional-scale groundwater flow model (such as Visual MODFLOW). Since one of the inputs required for WetSpass is the groundwater depth data, which is predicted with the MODFLOW model, an interface may be developed in an ArcView GIS platform to couple the two models, facilitating exchange of data between the two models. The coupled WetSpass-MODFLOW model is run for the present situation and for each of the climate change scenarios on an annual basis.

8. Simulate groundwater flow using each recharge data set and evaluate the changes in groundwater flow and levels through time.

9. Undertake sensitivity analysis of the groundwater flow model.

10. Develop a local scale groundwater model for the specific study area and conduct a well capture zone analysis.

A typical flow chart for various aspects of such a study is shown in Figure 2. The figure shows the connection from the climate analysis, to recharge simulation, and finally to a groundwater model. Recharge is applied to a three-dimensional groundwater flow model, which is calibrated to historical water levels. Transient simulations are undertaken to investigate the temporal response of the aquifer system to historic and future climate periods.

Tasks in the upper part of the chart assemble several climate data sets for current and future predicted conditions, which are used to simulate recharge using HELP module of WHI UnSat Suite. The soil layers are parameterized using a pedotransfer function program, which utilizes detailed soil survey measurements. Mapped monthly recharge from HELP is then used in a three-dimensional MODFLOW model to simulate transient saturated groundwater flow.

**Concluding Remarks**

➤ Precipitation is commonly downscaled in climate change impact studies; however, the reliability of the downscaled result is often poor or unreliable, as there is often little correlation between the predictors and the predictands. A poor correlation is often attributed to mesoscale processes occurring at the site-scale that are not represented in regional models due to their representative spatial and temporal sizes in comparison to larger-scale regional precipitation. Mesoscale precipitation processes generally occur in the summer season in the form of convective clouds, which are a result of local-scale evapotranspiration from elevated temperatures and solar radiation magnitudes. As a result, global-scale models may underestimate the summer precipitation measured at a site.

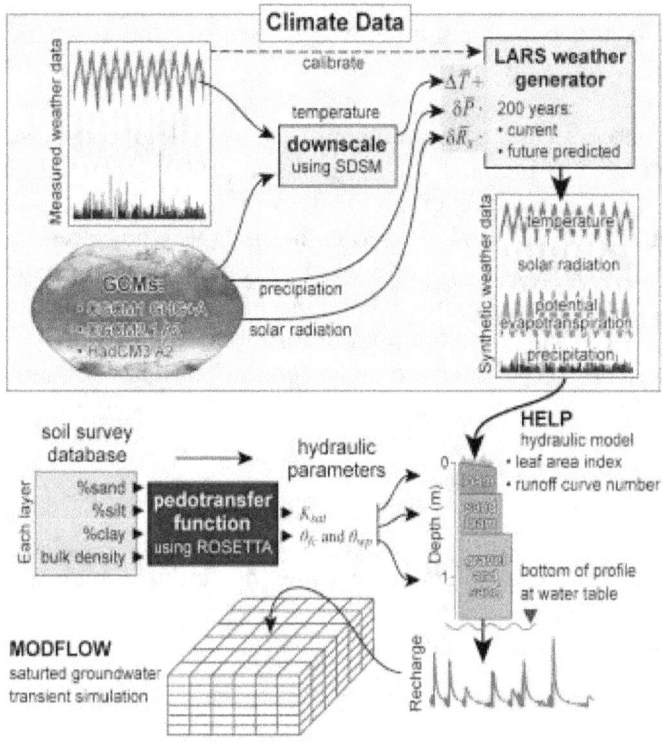

Figure 2: Flow Chart of Tasks (Toews, 2007)

➢ Although climate change has been widely recognized, research on the impacts of climate change on the groundwater system is relatively limited. The reasons may be that long historical data are required to analyze the characteristics of climate change. These data are not always available. Also, the driving forces that cause such changes are yet unclear. The climatic abnormality may occur frequently and last for a period of time. Even if the required data exist, uncertainty is embedded in model parameters, structure and driving force of the hydrological cycle. Predicting the long-term effect of a dynamic system is very difficult because of limitations inherent in the models, and the unpredictability of the forces that drive the earth. A physically based model of a groundwater system under possible climate change based on available data is very important to prevent the deterioration of regional water-resource problems in the future. Although uncertainties are inevitable, new response strategies in water resource management based on the model may be useful.

➢ The investigation of the relationship between climate change and loss of fresh groundwater resources is important for understanding the characteristics of the different regions. The impact of future climatic change may be felt more severely in developing countries such as India, whose economy is largely dependent on agriculture and is already under stress due to current population increase and associated demands for energy, freshwater and food. In spite of the uncertainties about the precise magnitude of climate change and its possible impacts, particularly on regional scales, measures must be taken to anticipate, prevent or minimize the causes of climate change and mitigate its adverse effects.

➤ Groundwater recharge is influenced not only by hydrologic processes, but also by the physical characteristics of the land surface and soil profile. Many climate change studies have focused on modelling the temporal changes in the hydrologic processes and ignored the spatial variability of physical properties across the study area. While knowing the average change in recharge and groundwater levels over time is important, these changes will not occur equally over a regional catchment or watershed. Long-term water resource planning requires both spatial and temporal information on groundwater recharge in order to properly manage not only water use and exploitation, but also land use allocation and development. Studies concerned with climate change should therefore also consider the spatial change in groundwater recharge rates.

➤ If the likely consequences of future changes of groundwater recharge, resulting from both climate and socio-economic change, are to be assessed, hydrogeologists must increasingly work with researchers from other disciplines, such as socio-economists, agricultural modelers and soil scientists.

## References

1.  Allen, Diana M. (2010). Historical trends and future projections of groundwater levels and recharge in coastal British Columbia, Canada. SWIM21 - 21st Salt Water Intrusion Meeting, 21-26 June 2010, Azores, Portugal, pp. 267-270.

2.  Allen, D. M., Mackie, D. C. and Wei, M. (2004). Groundwater and climate change: a sensitivity analysis for the Grand Forks aquifer, southern British Columbia, Canada. Hydrogeology Journal, Vol. 12, pp. 270–290.

3.  Allen, D. M., Cannon, A. J., Toews, M. W. and Scibek, J. (2010). Variability in simulated recharge using different GCMs. Water Resources Research, Vol. 46.

4.  Barthel, Roland (2011). An indicator approach to assessing and predicting the quantitative state of groundwater bodies on the regional scale with a special focus on the impacts of climate change. Hydrogeology Journal, Volume 19, Issue 3, pp. 525-546.

5.  Bouraoui, F., Vachaud, G., Li, L. Z. X., Le Treut, H. and Chen, T. (1999). Evaluation of the impact of climate changes on water storage and groundwater recharge at the watershed scale. Climate Dynamics, Vol. 15, pp. 153-161.

6.  Brouyere, S., Carabin, G. and Dassargues, A. (2004). Climate change impacts on groundwater resources: modelled deficits in a chalky aquifer, Geer basin, Belgium. Hydrogeology Journal, Vol. 12, pp. 123-134.

7.  Carneiro, Júlio F., Boughriba, M., Correia, A., Zarhloule, Y., Rimi, A. and EL Houadi, B. (2008). Climate Change Impact in a Shallow Coastal Mediterranean Aquifer, at

Saïdia, Morocco. 20<sup>th</sup> Salt Water Intrusion Meeting, 23-27 June 2008, Naples, Florida, USA, pp.30-33.

8.  Central Ground Water Board (2002). Master Plan for Artificial Recharge to Groundwater in India. Central Ground Water Board, New Delhi, February 2002, p. 115.

9.  Central Ground Water Board (2011). Dynamic Ground Water Resources of India (as on 31 March 2009). Ministry of Water Resources, Government of India, November 2011, 225p.

10. Chadha, D. K. and Sharma, S. K. (2000). Groundwater management in India: issues and options. Workshop on Past Achievements and Future Strategies, Central Groundwater Authority, New Delhi, 14 January 2000.

11. Climate Change & its Impact on Indian Water Resources: Assessment, Adaptation & Mitigation, Base Paper by National Institute of Hydrology, Roorkee.

12. Croley, T. E. and Luukkonen, C. L. (2003). Potential effects of climate change on ground water in Lansing, Michigan. Journal of the American Water Resources Association, Vol. 39 (1), pp. 149-163.

13. Crosbie, Russell S., McCallum, James L., Walker, Glen R. and Chiew, Francis H. S. (2010). Modelling climate-change impacts on groundwater recharge in the Murray-Darling Basin, Australia. Hydrogeology Journal, Vol. 18, pp. 1639–1656.

14. Crosbie, Russell S., McCallum, James L., Walker, Glen R. and Chiew, Francis H. S. (2012). Episodic recharge and climate change in the Murray-Darling Basin, Australia. Hydrogeology Journal, Volume 20, Issue 2, pp. 245-261.

15. Dams, Jef, Salvadore, Elga and Batelaan, Okke (2010). Predicting Impact of Climate Change on Groundwater Dependent Ecosystems. 2<sup>nd</sup> International Interdisciplinary Conference on Predictions for Hydrology, Ecology and Water Resources Management: Changes and Hazards caused by Direct Human Interventions and Climate Change, 20-23 September 2010, Prague, Czech Republic.

16. Das, P. K. and Radhakrishnan, M. (1991). Proc. Indian Acad. Sci. (Earth Planet. Sci.), Vol. 100, pp. 177–194.

17. Dragoni, W. and Sukhija, B. S. (2008). Climate change and groundwater: a short review. Geological Society, London, Special Publications 2008; v. 288; p. 1-12.

18. Eckhardt, K. and Ulbrich, U. (2003). Potential impacts of climate change on groundwater recharge and streamflow in a central European low mountain range. Journal of Hydrology, Vol. 284, pp. 244-252.

19. Franssen, Harrie-Jan Hendricks (2009). The impact of climate change on groundwater resources. International Journal of Climate Change Strategies and Management, Vol. 1, Issue 3, pp. 241-254.

20. Ghosh Bobba, A. (2002). Numerical modelling of salt-water intrusion due to human activities and sea-level change in the Godavari Delta, India. Hydrological Sciences Journal, Vol. 47(S), August 2002, pp. 67-80.

21. Herrera-Pantoja, M. and Hiscock, K. M. (2008). The effects of climate change on potential groundwater recharge in Great Britain. Hydrological Processes, Vol. 22, Issue 1, January 2008, pp. 73-86.

22. Holman, I. P. (2006). Climate change impacts on groundwater recharge - uncertainty, shortcomings, and the way forward? Hydrogeology Journal, Vol. 14, pp. 637–647.

23. Holman, I. P., Tascone, D. and Hess, T. M. (2009). A comparison of stochastic and deterministic downscaling methods for modelling potential groundwater recharge under climate change in East Anglia, UK: implications for groundwater resource management. Hydrogeology Journal, Vol. 17, pp. 1629–1641.

24. Hsu, Kuo-Chin, Wang, Chung-Ho, Cheu, Kuan-Chih, Chen, Chien-Tai and Ma, Kai-Wei (2007). Climate-induced hydrological impacts on the groundwater system of the Pingtung Plain, Taiwan. Hydrogeology Journal, Vol. 15, Number 5, August 2007, pp. 903-913.

25. IPCC (1995). Report of Working Group I of the Intergovernmental Panel on Climate Change. World Meteorological Organization, United Nations Environment Program, Geneva.

26. IPCC (1997). The regional impact of climate change: an assessment of vulnerability: A Special Report of IPCC working group II. Watson, R. T., Zinyowera, M. C., Moss, R. H., Dokken, D. J. (Eds), http://www.grida.no/climate/ipcc/regional/index.htm, Cited 19 November 2006.

27. IPCC (2000). Emissions Scenarios: A Special Report of Working Group II of the Intergovernmental Panel on Climate Change. In: Nakicenovic, N., Swart, R. (Eds.), Cambridge Univ. Press, Cambridge, UK.

28. IPCC (2001). In: Houghton, J. T., Ding, Y., Griggs, D. J., Noguer, M., van der Linden, P. J., Dai, X., Maskell, K., Johnson, C.A. (Eds.), Climate Change 2001: The Scientific Basis. Contributions of Working Group 1 to the Third Assessment Report of the Intergovernmental Panel on Climate Change, Cambridge University Press, Cambridge, UK.

29.    IPCC (2001). In: James J. McCarthy, Osvaldo F. Canziani, Neil A. Leary, David J. Dokken, Kasey S. White (Eds.), Climate Change 2001: Impacts, Adaptation, and Vulnerability. Contribution of Working Group II to the Third Assessment Report of the Intergovernmental Panel on Climate Change, Cambridge University Press, Cambridge, UK.

30.    IPCC (2007). In: Solomon, S., Qin, D., Manning, M., Chen, Z., Marquis, M., Averyt, K. B., Tignor, M., Miller, H. L. (eds.), 2007. Climate Change 2007: The Physical Science Basis. Contribution of Working Group I to the Fourth Assessment Report of the Intergovernmental Panel on Climate Change, Cambridge University Press, Cambridge, UK and New York, NY, USA, 966 p.

31.    Jyrkama, Mikko I. and Sykes, Jon F. (2007). The impact of climate change on spatially varying groundwater recharge in the grand river watershed (Ontario). Journal of Hydrology, Vol. 338, pp. 237–250.

32.    Kirshen, P. H. (2002). Potential impacts of global warming on groundwater in Eastern Massachusetts. Journal of Water Resources Planning and Management, Vol. 128 (3), pp. 216-226.

33.    Kumar, C. P. and Surjeet Singh (2011). Impact of Climate Change on Water Resources and National Water Mission of India. Proceedings of National Seminar on Global Warming and its Impact on Water Resources, 14 January 2011, Kolkata, pp. 1-9 (Volume 1).

34.    Kumar, C. P. and Surjeet Singh (2012). Impact of Climate Change on Groundwater Resources – Status of Research Studies. Proceedings of 3rd International Conference on Climate Change and Sustainable Management of Natural Resources, 5-7 February 2012, Gwalior.

35.    Kumar, C. P. (2012). Assessing the Impact of Climate Change on Groundwater Resources. India Water Week 2012 - Water, Energy and Food Security: Call for Solutions, 10-14 April 2012, New Delhi.

36.    Kumar C. P. (2012). Assessment of Impact of Climate Change on Groundwater Resources. Water and Energy International, ISSN: 0974-4711, Vol. 69, Number 8, August 2012, pp. 25-31.

37.    Kumar, C. P. (2012). Climate Change and Its Impact on Groundwater Resources. Research Inventy: International Journal of Engineering and Science, ISSN: 2278-4721, Volume 1, Issue 5 October 2012, pp. 43-60.

38.    Lapworth, D. J., MacDonald, A. M., Tijani, M. N., Darling, W. G., Gooddy, D. C., Bonsor, H. C. and Araguás-Araguás, L. J. (2013). Residence times of shallow groundwater in West Africa: implications for hydrogeology and resilience to future changes in climate. Hydrogeology Journal, Volume 21, Issue 3, pp. 673-686.

39.  Loaiciga, H. A. (2003). Climate change and ground water. Annals of the Association of American Geographers, Vol. 93 (1), pp. 30–41.

40.  Mall, R. K., Gupta, Akhilesh, Singh, Ranjeet, Singh, R. S. and Rathore, L. S. (2006). Water resources and climate change: An Indian perspective. Current Science, Vol. 90, No. 12, 25 June 2006.

41.  McCallum, J. L., Crosbie, R. S., Walker, G. R. and Dawes, W. R. (2010). Impacts of climate change on groundwater in Australia: a sensitivity analysis of recharge. Hydrogeology Journal, Vol. 18, pp. 1625–1638.

42.  Mollema, Pauline N. and Antonellini, Marco (2013). Seasonal variation in natural recharge of coastal aquifers. Hydrogeology Journal, Volume 21, Issue 4, pp. 787-797.

43.  National Environment Policy (2004). Ministry of Environment and Forests, Government of India, p. 38.

44.  Neukum, Christoph and Azzam, Rafig (2012). Impact of climate change on groundwater recharge in a small catchment in the Black Forest, Germany. Hydrogeology Journal, Volume 20, Issue 3, pp. 547-560.

45.  Okkonen, Jarkko, Jyrkama, Mikko and Kløve, Bjørn (2010). A conceptual approach for assessing the impact of climate change on groundwater and related surface waters in cold regions (Finland). Hydrogeology Journal, Vol. 18, pp. 429–439.

46.  Oude Essink, G. H. P., van Baaren, E. S. and de Louw, P. G. B. (2010). Effects of climate change on coastal groundwater systems: A modeling study in the Netherlands. Water Resources Research, Vol. 46.

47.  Payne, Dorothy F. (2010). Effects of climate change on saltwater intrusion at Hilton Head Island, SC. U.S.A. SWIM21 - 21st Salt Water Intrusion Meeting, 21-26 June 2010, Azores, Portugal, pp. 293-296.

48.  Ranjan, P., Kazamaa, S. and Sawamoto, M. (2006). Effects of climate change on coastal fresh groundwater resources. Global Environmental Change, Vol. 16, pp. 388–399.

49.  Raposo, Juan Ramón, Dafonte, Jorge and Molinero, Jorge (2013). Assessing the impact of future climate change on groundwater recharge in Galicia-Costa, Spain. Hydrogeology Journal, Volume 21, Issue 2, pp. 459-479.

50.  Ravindranath, N. H., Joshi, N. V., Sukumar, R. and Saxena, A. (2005). Impact of climate change on forests in India. Current Science, Vol. 90, pp. 354–361.

51. Richardson, C. W. and Wright, D. A. (1984). WGEN: A model for generating daily weather variables. ARS-8. Agricultural Research Service, US Department of Agriculture.

52. Rosenberg, N. J., Epstein, D. J., Wang, D., Vail, L., Srinivasan, R. and Arnold, J. G. (1999). Possible impacts of global warming on the hydrology of the Ogallala Aquifer Region. Climatic Change, Vol. 42, pp. 677–692.

53. Rozell, Daniel J. and Wong, Teng-fong (2010). Effects of climate change on groundwater resources at Shelter Island, New York State, USA. Hydrogeology Journal, Vol. 18, pp. 1657-1665.

54. Rupa Kumar, K. (2005). High-resolution climate change scenarios for India for the 21st century. Current Science, Vol. 90, pp. 334–345.

55. Sathaye, Jayant, Shukla, P. R. and Ravindranath, N. H. (2006). Climate change, sustainable development and India: Global and national concerns. Current Science, Vol. 90, Number 3, 10 February 2006, pp. 314-325.

56. Scibek, J. and Allen, D. M. (2006). Modeled impacts of predicted climate change on recharge and groundwater levels. Water Resources Research, Vol. 42, W11405, 18p.

57. Shah, Tushaar (2009). Climate change and groundwater: India's opportunities for mitigation and adaptation. Environmental Research Letter 4, IOP Publishing Ltd, UK.

58. Sherif, Mohsen M. and Singh, Vijay P. (1999). Effect of climate change on sea water intrusion in coastal aquifers. Hydrological Processes, Vol. 13, pp. 1277-1287.

59. Singh, R. D. and C. P. Kumar (2010). Impact of Climate Change on Groundwater Resources. Proceedings of 2nd National Ground Water Congress, 22nd March 2010, New Delhi, pp. 332-350.

60. Taylor B. (1997). Climate change scenarios for British Columbia and Yukon. In: Taylor E, Taylor B. (eds), Responding to global climate change in British Columbia and Yukon, Volume I of the Canada country study: climate impacts and adaptation, Environment Canada and BC Ministry of Environment, Lands and Parks.

61. Toews, Michael W. (2007). Modelling Climate Change Impacts on Groundwater Recharge an a Semi-Arid Region, Southern Okanagan, British Columbia. A thesis submitted in partial fulfillment of the requirements for the degree of Master of Science in the Department of Earth Sciences, Simon Fraser University.

62. UNFCCC (2004). India's Initial National Communications to the United Nations Framework Convention on Climate Change. Ministry of Environment and Forests, New Delhi.

63.     Vandenbohede, Alexander and Lebbe, Luc (2010). Impact of climate change on the phreatic aquifer of the western Belgian coastal plain. SWIM21 – 21$^{st}$ Salt Water Intrusion Meeting, 21-26 June 2010, Azores, Portugal, pp. 297-300.

64.     Vaccaro, J. J. (1992). Sensitivity of groundwater recharge estimates to climate variability and change, Columbia Plateau, Washington. Journal of Geophysical Research, Vol. 97 (D3), pp. 2821-2833.

65.     Woldeamlak, S. T., Batelaan, O. and De Smedt, F. (2007). Effects of climate change on the groundwater system in the Grote-Nete catchment, Belgium. Hydrogeology Journal, Vol. 15, Number 5, August 2007, pp. 891-901.

66.     Yihdego, Yohannes and Webb, John A. (2011). Modeling of bore hydrographs to determine the impact of climate and land-use change in a temperate subhumid region of southeastern Australia. Hydrogeology Journal, Volume 19, Issue 4, pp. 877-887.

67.     Zhou, Yu, Zwahlen, François, Wang, Yanxin and Li, Yilian (2010). Impact of climate change on irrigation requirements in terms of groundwater resources. Hydrogeology Journal, Vol. 18, pp. 1571–1582.

# 22-Mathematical Modelling: Case Studies

## Introduction

Mathematical models provide a scientific and predictive tool for determining appropriate solutions to water allocation, surface water – groundwater interaction, landscape management or impact of new development scenarios. However, if the modelling studies are not well designed from the outset, or the model doesn't adequately represent the natural system being modelled, the modelling effort may be largely wasted, or decisions may be based on flawed model results, and long term adverse consequences may result. This chapter presents two case studies on modeling of soil moisture movement (unsaturated flow) and seawater intrusion (density-dependent groundwater flow).

## Case Study 1 - Modelling of Soil Moisture Movement

The objective of the study was to simulate the movement of soil moisture in Barchi watershed (sub-basin of Kali river in North Kanara district of Karnataka) using the SWIM model. The SWIM (Soil Water Infiltration and Movement) is a software package developed by Division of Soils, CSIRO, Australia (Verburg et al., 1996) for simulating infiltration, evapotranspiration, and redistribution. It has been selected for the present study in view of its simplicity, ease of use, graphical display of intermittent results, and use of input parameters (soil moisture characteristics) which can be directly measured in the field/laboratory.

*Study Area*

The Barchi watershed upstream of Barchi is located in the leeward side of western ghat and is a sub-basin of Kali river. It lies in Haliyala taluk of Karwar (North Kanara) district in Karnataka. The location and drainage system of Barchi watershed is shown in Figure 1.

The Barchinala stream originates from Thavargatti in Belgaum district at an altitude of about 734 m, 20 km north of Dandeli and flows through North Kanara district of Karnataka State. The catchment is relatively short in width and river flows in a southerly direction and joins the main Barchi river near the gauging site. The geographical area covered by Barchi watershed is 21.126 km$^2$. The watershed lies between 74°36' and 74°39' East longitudes, and 15°18' and 15°24' North latitudes.

High land region consists of dissection of high hills and ridges forming part of the foot hills of western ghats. It consists of steep hills and valleys intercepted with thick forest. The slopes of the ghats are covered with dense deciduous forest. Forest cover occupies around 76% of the study area. The watershed is mainly covered with Bamboo, Teak and mixed plantations. The brownish and fine-grained soils are the principal types of soils found in the area. The following land uses were observed in the watershed:

1.      Bamboo plantation    =      04 %

| 2. | Teak plantation | = | 40 % |
| 3. | Mixed forest | = | 32 % |
| 4. | Agricultural land | = | 24 % |

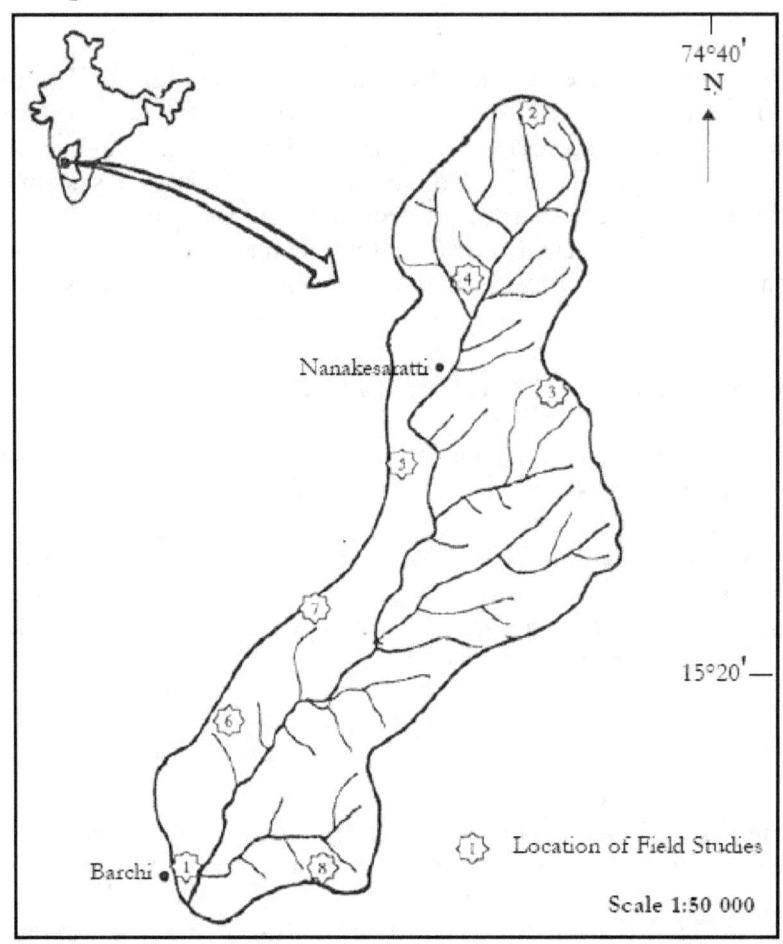

Figure 1:     Drainage system of Barchi watershed

The stream gauging site is located at an elevation of 480 m, where the nala crosses Dandeli-Thavargatti road, about 5 km from Dandeli. The stream is a 4th order stream and joins main Barchi river downstream of the gauging site. A full fledged meteorological station, maintained by Water Resources Development Organisation (WRDO), Karnataka, is located near the gauging site.

The Barchi raingauge station is located at 15°18' N and 74°37' E. Average annual rainfall for the watershed is 1500 mm, majority of which occurs during the south-west monsoon period. Depth to water table varies between 4 to 12 metres during pre- and post-monsoon periods. The yield of borewells in the study area is found to vary between 120 gallons per hour to 1170 gallons per hour.

*Methodology*

The study involves modelling of soil moisture movement in Barchi watershed using the SWIM model. The following steps were undertaken for the study.

(1) Field investigations: Measurement of saturated hydraulic conductivity at 8 locations using Guelph Permeameter and soil sampling.

(2) Laboratory investigations: Determination of saturated moisture content, and soil moisture retention characteristics using the Pressure Plate Apparatus.

(3) Modelling of soil moisture movement using the SWIM model: Daily rainfall and evaporation data of Barchi for the period 1996-97 to 1999-2000 were used for the study. Water balance components like runoff, evapotranspiration and drainage (recharge to groundwater from rainfall) were determined through SWIM.

SWIM is an acronym that stands for <u>S</u>oil <u>W</u>ater <u>I</u>nfiltration and <u>M</u>ovement. It is a software package developed within the CSIRO Division of Soils for simulating infiltration, evapotranspiration, and redistribution. The first version (SWIMv1) was published in 1990 (Ross[2], 1990). Version 2 of the model (identified as SWIMv2.0), which combines water movement with transient solute transport and which accommodates a variety of soil property descriptions and more flexible boundary conditions, was completed in 1992.

SWIMv2 is based on a numerical solution of the Richards' equation (1) and the advection-dispersion equation (2), as given below. The model deals with a one-dimensional soil profile.

$$\frac{\partial \theta}{\partial t} = \frac{\partial}{\partial x} K \left[ \frac{d\psi}{dp} \frac{\partial p}{\partial x} + \frac{dz}{dx} \right] + S \qquad \ldots(1)$$

with

$$-\frac{\psi - \psi_o}{\psi_1} = \sinh p \qquad \psi < \psi_o$$

$$-\frac{\psi - \psi_o}{\psi_1} = p \qquad \psi \geq \psi_o$$

where,

| | | |
|---|---|---|
| $\theta$ | = | volumetric water content ($cm^3/cm^3$); |
| t | = | time (h); |
| x | = | distance into the soil (cm soil); |
| K | = | hydraulic conductivity ($cm^2$ water/cm soil/h); |
| $\Psi$ | = | matric potential (cm); |
| z | = | gravitational potential (cm); |
| S | = | source (or sink, if negative) strength ($cm^3$ water/$cm^3$ soil/h); and |
| $\psi_o$ and $\psi_1$ | = | shifting and scaling parameters, respectively. |

301

Solute movement is based on the following solute transport equation

$$\frac{\partial(\theta c)}{\partial t} + \frac{\partial(\rho s)}{\partial t} = \frac{\partial}{\partial x}\left[\theta D \frac{\partial c}{\partial x}\right] - \frac{\partial(qc)}{\partial x} + \phi \qquad \qquad \text{...(2)}$$

where,

| | | |
|---|---|---|
| c | = | solute concentration in solution ($\mu$mol or $\mu$g solutes/cm$^3$ water); |
| s | = | adsorbed concentration ($\mu$mol/g soil or $\mu$g/g soil); |
| $\rho$ | = | soil bulk density (g/cm$^3$); |
| t | = | time (h); |
| x | = | depth (cm); |
| $\theta$ | = | water content (cm$^3$/cm$^3$); |
| q | = | water flux density (cm/h); |
| D | = | combined dispersion and diffusion coefficient (cm$^2$/h); and |
| $\phi$ | = | source/sink term ($\mu$mol/cm$^3$/h or $\mu$g/cm$^3$/h). |

The SWIM can be used to simulate runoff, infiltration, redistribution, solute transport and redistribution of solutes, plant uptake and transpiration, soil evaporation, deep drainage and leaching. The physical system and the associated flows addressed by the model are shown schematically in Figure 2. Soil water and solute transport properties, initial conditions, and time dependent boundary conditions (e.g., precipitation, evaporative demand, solute input) need to be supplied by the user in order to run the model. The overall purpose of the model is to address issues relating to the soil water and solute balance. As such, it is a research tool that can be integrated in laboratory and field studies concerned with soil water and solute transport.

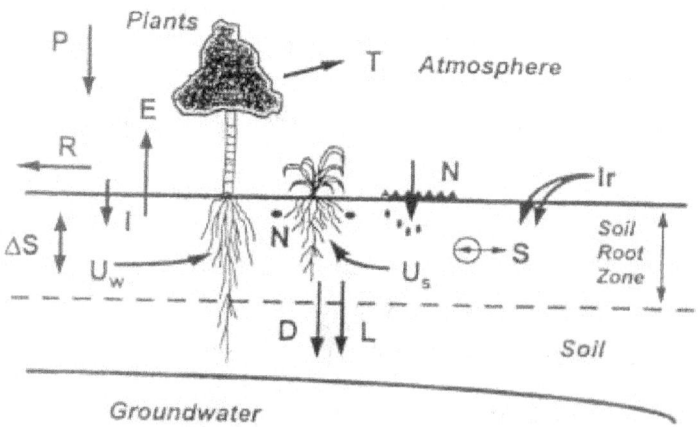

Components of the soil water and solute balances addressed by SWIMv2.1;
P = precipitation, R = runoff, I = infiltration, $U_w$ = water uptake, $U_s$ = solute uptake, T = transpiration, E = evaporation, D = drainage, L = solute leaching, Ir = irrigation/fertigation, N = nutrients/fertiliser, $\Delta$S = storage, S = solute source/sink.

Figure 2: Components of the soil water and solute balances addressed by SWIMv 2.1

To model the retention and movement of water and chemicals in the unsaturated zone, it is necessary to know the relationships between soil water pressure (h), water content ($\theta$) and hydraulic conductivity (K). It is often convenient to represent these functions by means of relatively simple parametric expressions. The problem of characterizing the soil hydraulic properties then reduces to estimating parameters of the appropriate constitutive model.

The measurements of $\theta$(h) from soil cores (obtained through pressure plate apparatus) can be fitted to the desired soil water retention model. Once the retention function is estimated, the hydraulic conductivity relation, K(h), can be evaluated if the saturated hydraulic conductivity, $K_s$, is known. In the present study, parameters of van Genuchten model were derived for soil moisture retention and hydraulic conductivity functions. For the van Genuchten[3] model (1980), the water retention function is given by

$$S_e = (\theta - \theta_r)/(\theta_s - \theta_r) = [\, 1 + (\alpha_v |h|\,)^n\,]^{-m} \qquad \text{for } h < 0$$
$$= 1 \qquad \text{for } h \geq 0$$

$$\ldots (3)$$

and the hydraulic conductivity function is described by

$$K = K_s\, S_e^{1/2}\, [\, 1 - (1 - S_e^{1/m})^m\,]^2 \qquad\qquad \ldots (4)$$

where, $S_e$ is effective saturation; $\theta_r$ is residual water content; $\theta_s$ is saturated water content; $\alpha_v$ and n are van Genuchten model parameters; $m = 1 - 1/n$.

Modelling of soil moisture movement in Barchi watershed was done using SWIM. The model was simulated for 1461 days (1 May 1996 to 30 April 2000). One vegetation type (teak, covered in most parts of the watershed) was considered for the study. Exponential root growth with depth and linear interpolation with time was assumed. The following vegetation parameters were adopted for the simulations:

| | | |
|---|---|---|
| Root radius (rad) | = | 0.5 cm |
| Root conductance (groot) | = | $4.0 * 10^{-7}$ |
| Minimum xylem potential (psimin) | = | -15,000 cm |
| Root depth constant (xc) | = | 150 cm |
| Maximum root length density (rldmax) | = | 4 cm/cm$^3$ |

*Results and Discussion*

Soil moisture retention characteristics were determined in the laboratory using the Pressure Plate Apparatus. The experimental soil moisture retention data were fitted to the van Genuchten[3] model (1980). Residual moisture content ($\theta_r$) was assumed to be equivalent to moisture retained corresponding to 15 bar pressure. The parameters of soil moisture retention function and hydraulic conductivity function were obtained through non-linear regression analysis. Tables 1 and 2 present the van Genuchten parameters $\alpha$ and n (equations 3 and 4) for upper and lower soil layers in Barchi watershed. Average values of these parameters were also determined through non-linear regression analysis and used in modelling of soil moisture movement through SWIM.

Table 1:      van Genuchten Parameters for Upper Soil Layer

| Station | $K_s$ (cm/hour) | $\theta_r$ | $\theta_s$ | van Genuchten Parameters | | Proportion of Variance |
|---------|------------------|------------|------------|------|------|----------------------|
| | | | | $\alpha$ | n | Explained (%) |
| 1 | 0.58 | 0.08 | 0.37 | 0.0073 | 1.434 | 80.78 |
| 2 | 0.57 | 0.14 | 0.37 | 0.0023 | 1.509 | 74.08 |
| 3 | 0.60 | 0.09 | 0.38 | 0.0021 | 1.465 | 79.07 |
| 4 | 0.18 | 0.30 | 0.53 | 0.0067 | 1.523 | 92.00 |
| 5 | 0.20 | 0.28 | 0.53 | 0.0129 | 1.373 | 80.66 |
| 6 | 0.18 | 0.28 | 0.53 | 0.0235 | 1.300 | 64.09 |
| 7 | 0.24 | 0.25 | 0.52 | 0.0020 | 1.580 | 84.07 |
| 8 | 0.16 | 0.30 | 0.54 | 0.0019 | 1.552 | 91.51 |
| **Average** | 0.339 | 0.215 | 0.471 | 0.0047 | 1.4385 | 24.43 |

Table 2:      van Genuchten Parameters for Lower Soil Layer

| Station | $K_s$ (cm/hour) | $\theta_r$ | $\theta_s$ | van Genuchten Parameters | | Proportion of Variance |
|---------|------------------|------------|------------|------|------|----------------------|
| | | | | $\alpha$ | n | Explained (%) |
| 1 | 1.66 | 0.11 | 0.38 | 0.0148 | 1.563 | 97.04 |
| 2 | 0.60 | 0.09 | 0.32 | 0.0045 | 1.760 | 99.52 |
| 3 | 0.007 | 0.06 | 0.43 | 0.0154 | 1.358 | 87.12 |
| 4 | 0.58 | 0.14 | 0.41 | 0.0134 | 1.310 | 81.71 |
| 5 | 0.58 | 0.16 | 0.43 | 0.0070 | 1.444 | 91.68 |
| 6 | 0.18 | 0.28 | 0.53 | 0.0235 | 1.300 | 64.09 |
| 7 | 0.59 | 0.13 | 0.31 | 0.0120 | 1.596 | 95.35 |
| 8 | 0.60 | 0.20 | 0.45 | 0.0123 | 1.688 | 91.97 |
| **Average** | 0.648 | 0.121 | 0.394 | 0.0095 | 1.4212 | 58.31 |

Based upon the available information, two distinct soil layers were identified (0-45 cm and 45-150 cm). Saturated hydraulic conductivity was measured at 8 locations in the study area by using Guelph Permeameter (locations are shown in Figure 1). The average saturated hydraulic conductivity values for the upper layer (0-45 cm) and lower layer (45-150 cm) were found to be 0.339 cm/hour and 0.648 cm/hour respectively.

*Model Conceptualization*

The profile is 150 cm deep with surface at 0 cm and bottom boundary condition applying at 150 cm. Vapour conductivity is not taken into account, nor is the effect of osmotic potential. There are two hydraulic property sets (for upper and lower soil layers) that are applied to 31 depth nodes of the 150 cm deep profile. Hysteresis is not taken into account.

Initially, there is no water ponded on the surface. Runoff is governed by a simple power law function and a surface conductance function. No bypass flow was included. A matric potential gradient of 0, i.e. "unit gradient", has been applied as bottom boundary condition throughout the simulation. Cumulative rainfall and evaporation records (daily) for the period 1996-97 to 1999-2000 were given in the input file for determination of water balance components (runoff, evapotranspiration and drainage).

*Simulation of Water Balance Components*

The model parameters (soil moisture characteristics) were actually measured in the field and laboratory. Therefore, the model does not require any calibration as such. The model was validated by comparing the observed and simulated runoff. However, the observed runoff values were suspected to be erroneous in view of inaccurate positioning of zero of gauge.

Self-recording raingauge data (hourly rainfall values) were not available for the watershed. Therefore, daily rainfall values were used. However, with the available input data and parameters, the model was found to underestimate the runoff values. It happened because daily rainfall data generated low rainfall intensities (distributed over 24 hours) with most of the rainfall infiltrating into the ground and contributing less runoff. Therefore, daily rainfall values were equally distributed to 4 hours for the periods exceeding 20 mm rainfall in a day. This made a better agreement between the observed and simulated runoff and therefore validated the model. The distribution of daily rainfall into 4 hours was decided on the basis of trial simulations by testing varying divisions with part of actual data. The resulting water balance components for the simulation period have been presented in Table 3.

Table 3:     Water Balance Components for the Barchi Watershed

| Year | Rainfall (mm) | Infiltration (mm) | Drainage (mm) | ET (mm) | Runoff (mm) | Runoff Coefficient (%) | Recharge Coefficient (%) |
|------|------|------|------|------|------|------|------|
| 1996-1997 | 1345.85 | 1083.37 | 514.46 | 519.52 | 262.48 | 19.50 | 38.22 |
| 1997-1998 | 1765.25 | 1195.05 | 698.63 | 500.43 | 570.20 | 32.30 | 39.58 |
| 1998-1999 | 1241.30 | 1087.46 | 579.55 | 507.92 | 153.84 | 12.39 | 46.69 |
| 1999-2000 | 1886.80 | 1278.18 | 784.90 | 493.28 | 608.62 | 32.26 | 41.60 |
| **Total** | 6239.20 | 4644.06 | 2577.54 | 2021.15 | 1595.14 | 24.11 | 41.52 |

The yearly rainfall varied between 1241 mm to 1887 mm during the period under study. It can be observed from Table 3 that the drainage (recharge from rainfall) varies from 38% to 47% with the average value being 42%. The runoff coefficient was found to vary between 12% (low rainfall year) to 32% (high rainfall year) with the average value being 24%. Runoff coefficient was found to be lower in low rainfall years (1996-97 and 1998-99). It can be attributed to low rainfall intensities enabling more infiltration and lesser runoff. Antecedent moisture conditions also play an important role in the runoff generation process. Simulation of variable infiltration suggests that it has relatively little effect on evapotranspiration, but considerable effect on point drainage.

*Conclusion*

Application of SWIM model is one of the simplest techniques, which is well suited for unsaturated zone. SWIM is a software package for simulating water infiltration and movement in soils. Water is added as precipitation and removed by runoff, drainage, evaporation from the soil surface and transpiration by vegetation. The simulator assumes that conditions can be treated as horizontally uniform, flow is described by the Richards equation and soil hydraulic properties can be described by simple functions. While this is adequate for many purposes, there are situations where it is not and the simulation results should never be applied uncritically.

Water balance components like runoff, evapotranspiration and drainage were determined through SWIM for the period 1996-97 to 1999-2000. The ground water recharge was found to vary between 38% to 47% of rainfall while the runoff coefficient varied between 12% (low rainfall year) to 32% (high rainfall year) for the study period. Variable infiltration was observed to have relatively little effect on evapotranspiration, but considerable effect on drainage.

The SWIM model demonstrated the possibility of predicting water balance components of the unsaturated zone, but only with careful selection of input parameters. It would appear that when actual observed data is not available, it would be difficult to rely upon numerical models alone.

**Case Study 2 - Modelling of Seawater Intrusion**

Coastal tracts of Goa are rapidly being transformed into settlement areas. The poor water supply facilities have encouraged people to have their own source of water by digging or boring a well. During the last decade, there have been large-scale withdrawals of groundwater by builders, hotels and other tourist establishments. Though the seawater intrusion has not yet assumed serious magnitude, but in the coming years it may turn to be a major problem if corrective measures are not initiated at this stage. It is necessary to understand how fresh and salt water move under various realistic pumping and recharge scenarios. Objectives of this study include simulation of seawater intrusion in a part of the coastal area in Bardez taluk of North Goa, evaluation of the impact on seawater intrusion due to various groundwater pumping scenarios and sensitivity analysis to find the most sensitive parameters affecting the simulation.

*Study Area*

The study area lies in Bardez taluka of North Goa within the watersheds of Baga river and Nerul creek (around 74 km$^2$) and covered by Survey of India toposheets number 48E/10, 48E/14 and 48E/15 on 1:50,000 scale. It is bound by rivers Chapora and Mandovi in north and south directions respectively, besides Arabian sea in the west and encompasses coastal tract from Fort Aguada in the south to Fort Chapora in the north (15 km). The soils are predominantly of lateritic nature. However, the coastal areas are made up of alluvial soils composed of loamy mixed sand and loamy sands. Around 30 km$^2$ area close to the coast (15 km along the coast and 2 km wide) is more prone to seawater intrusion. Layout maps of North Goa and the study area are given in figures 1 and 2 respectively.

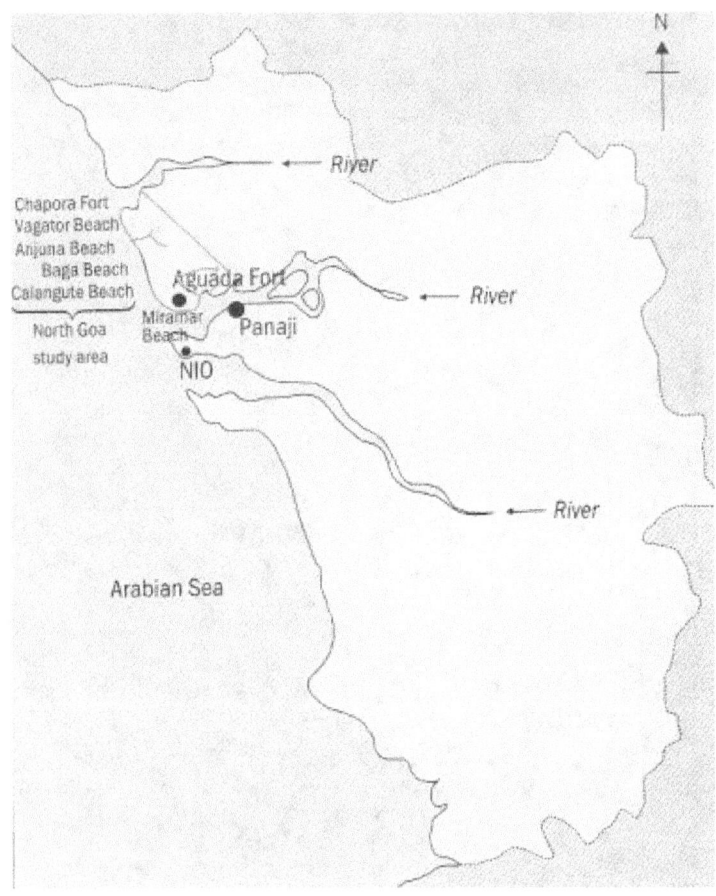

Figure 1:     Location Map of Study Area in Goa

*Laboratory and Field Investigations*

Twenty observation wells were identified in the study area (as indicated in figure 2). Monthly groundwater level data was measured in observation wells (September 2004 to August 2005) and groundwater samples were collected in September, November 2004, January, March, April, May 2005. Salinity for collected groundwater samples was measured in the laboratory. Based upon the bi-monthly measurements of salinity, groundwater quality in all the observation wells was found to be reasonably fresh, both in pre- and post-monsoon periods. It can be attributed to the fact that the transition zone of fresh water-saline water lies below the shallow open wells, as evidenced by vertical electrical soundings.

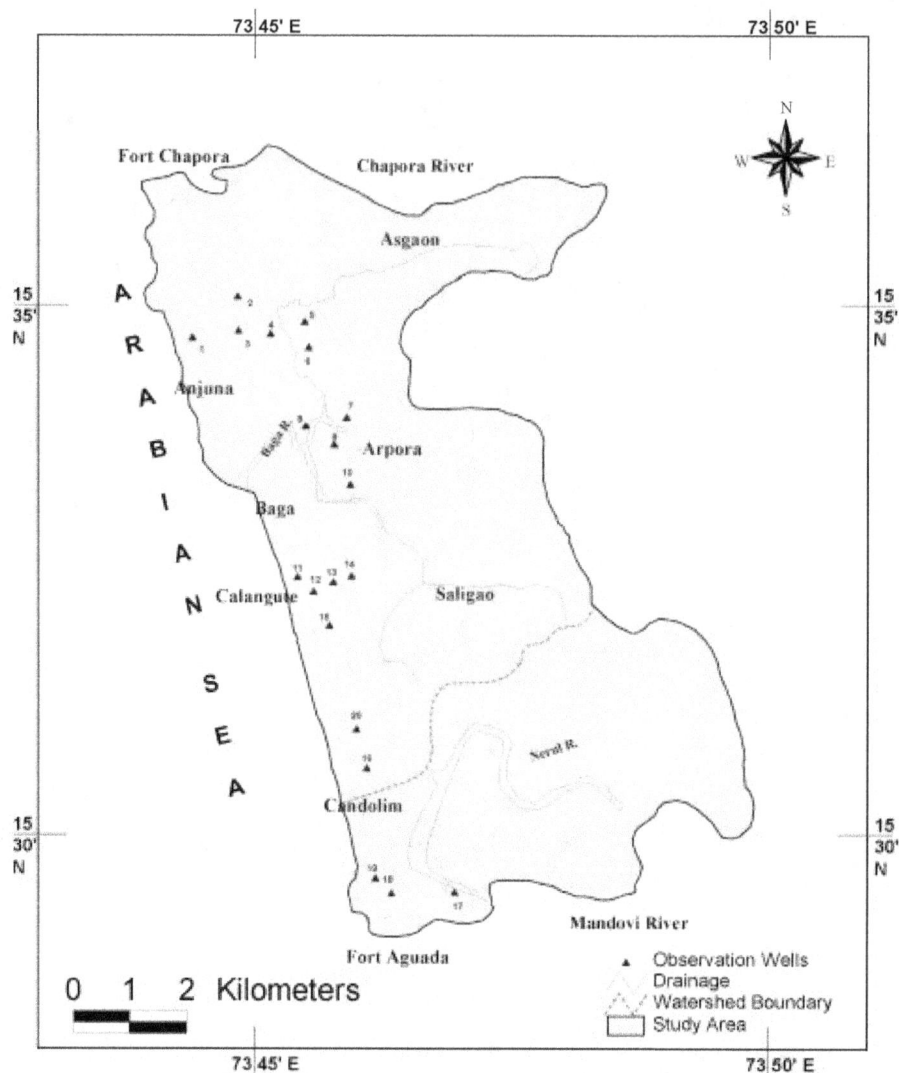

Figure 2:       Layout Map of the Study Area

Apparent electrical resistivity (ohm-m) was measured in four profiles along the Bardez coast (Anjuna, Baga, Calangute, Candolim) at 18 locations upto 525 metres from the coast (Table 1). The inter-electrode separation was kept at 10 meter, that is, the resistivity values measured are at 10 m depth plane.  The seawater mixed zone is witnessed along Anjuna (12 to 45 ohm-m) and Baga beach (4 to 46 ohm-m) sections along the low lying sandy alluvial areas. Very close to the sea, relatively higher apparent resistivity values are due to dry sand dunes. However, along Calangute (75 to 900 ohm-m) and Candolim (142 to 700 ohm-m) beaches, there is no indication of seawater mixing at 10 m depth, as all values are higher.

Seven vertical electrical soundings were carried out at monitoring well sites 1, 3, 6, 7, 8, 15 and 17 (Table 2). These were restricted to a depth of 20 m to know any change in the quality of water vis-à-vis seawater intrusion (3 m to 20 m with 1 m interval). As seen from the apparent resistivity values, well numbers 6, 7 and 8 show low values of resistivity (2 to 33 ohm-m) below about 12 m depth, indicating the presence of seawater or mixed zone below this depth. However, at other sites, there is no indication of the seawater mixing upto 20 m depth. It is noted here that wells located in low lying sandy alluvial areas show seawater mixing than the wells located in laterites at higher altitudes. In both laterite and alluvial soils, the wells are built well above the salt water – fresh water interface and hence no change in water quality was found in summer also.

Table 1:    Apparent Electrical Resistivity Values (ohm-m) in Four Profiles along the Bardez Coast

| S. No. | Distance from Coast (m) | P1 Anjuna | P2 Baga | P3 Calangute | P4 Candolim |
|---|---|---|---|---|---|
| 1 | 15 | 30 | 70 | 150 | 700 |
| 2 | 45 | 40 | 46 | 820 | 555 |
| 3 | 75 | 45 | 35 | 612 | 142 |
| 4 | 105 | 32 | 25 | 360 | 421 |
| 5 | 135 | 26 | 28 | 110 | 281 |
| 6 | 165 | 24 | 22 | 75 | 153 |
| 7 | 195 | 20 | 30 | 125 | 184 |
| 8 | 225 | 15 | 32 | 242 | 255 |
| 9 | 255 | 14 | 24 | 410 | 431 |
| 10 | 285 | 12 | 20 | 623 | 236 |
| 11 | 315 | 13 | 31 | 531 | 165 |
| 12 | 345 | 13 | 32 | 415 | 242 |
| 13 | 375 | 14 | 20 | 324 | 281 |
| 14 | 405 | 16 | 30 | 470 | 641 |
| 15 | 435 | 20 | 20 | 684 | 531 |
| 16 | 465 | 21 | 10 | 650 | 426 |
| 17 | 495 | 24 | 5 | 835 | 186 |
| 18 | 525 | 18 | 4 | 900 | 200 |

Table 2:    Apparent Electrical Resistivity Values (ohm-m) at Observation Well Points in the Study Area

| AB/2 (m) | Apparent Electrical Resistivity Values (ohm-m) at Monitoring Well Numbers | | | | | | |
|---|---|---|---|---|---|---|---|
| | 1 | 3 | 6 | 7 | 8 | 15 | 17 |
| 3 | 410 | 448 | 80 | 102 | 231 | 1988 | 333 |
| 4 | 436 | 446 | 65 | 84 | 277 | 665 | 356 |
| 5 | 467 | 521 | 63 | 72 | 202 | 900 | 373 |
| 6 | 505 | 595 | 68 | 62 | 180 | 256 | 393 |
| 7 | 536 | 613 | 82 | 58 | 138 | 850 | 383 |

| 8 | 541 | 521 | 74 | 51 | 120 | 156 | 381 |
|---|-----|-----|----|----|-----|-----|-----|
| 9 | 533 | 482 | 47 | 45 | 102 | 780 | 381 |
| 10 | 544 | 389 | 62 | 42 | 62 | 194 | 327 |
| 11 | 528 | 339 | 56 | 40 | 61 | 125 | 339 |
| 12 | 539 | 314 | 23 | 25 | 45 | 128 | 314 |
| 13 | 581 | 264 | 24 | 22 | 41 | 158 | 290 |
| 14 | 582 | 245 | 19 | 18 | 31 | 165 | 276 |
| 15 | 563 | 246 | 20 | 16 | 32 | 164 | 246 |
| 16 | 561 | 240 | 21 | 17 | 33 | 215 | 240 |
| 17 | 543 | 226 | 13 | 15 | 25 | 265 | 226 |
| 18 | 609 | 233 | 16 | 12 | 10 | 315 | 203 |
| 19 | 566 | 215 | 12 | 18 | 3 | 452 | 226 |
| 20 | 520 | 214 | 9 | 17 | 2 | 351 | 202 |

*Finite Element Simulation Model*

For this study, a finite-element model (FEFLOW) was selected for model simulations. The FEFLOW is an interactive finite element simulation system (Version 5.1) for three-dimensional (3D) or two-dimensional (2D), i.e. horizontal (aquifer-averaged), vertical or axi-symmetric, transient or steady-state, fluid density- coupled or linear, flow and mass, flow and heat or completely coupled thermohaline transport processes in subsurface water resources (groundwater systems). The package is fully graphics-based and interactive. Pre-, main- and post-processing are integrated. There is a data interface to GIS (Geographic Information System) and a programming interface. The implemented numerical features allow the solution of large problems.

*Model Setup and Simulation Results*

The aquifer domain of the study area (74 km$^2$) was discretized using 6 nodal triangular prism elements with 52,656 mesh elements and 32,053 mesh nodes. The vertical discretization corresponds to 7 slices and 6 layers. The top slice was defined as free and movable (water table). Geological profile in low lying area (ground elevation 0 – 20 m above mean sea level) was assigned as 15 – 30 m deep sandy soil (upto 10 – 15 m below mean sea level) underlain by 1 – 2 m clay layer and then basement rocks (phyllite, graywacke, schist etc). Geological profile in plateau area (ground elevation 20 – 80 m above mean sea level) was assigned as 0 - 75 m deep laterite (upto 0 – 10 m below mean sea level) underlain by 2 – 5 m clay layer and then basement rocks. The boundary conditions for the flow simulation are as follows:

- A coastal head boundary along the coastal zone (western boundary) at the top and bottom slices of the aquifer; FEFLOW uses head (h) instead of pressure with h = (e$_w$ / e$_s$) Z, where e$_w$ and e$_s$ represent ambient and seawater densities respectively and Z is the depth below sea level. The head was calculated in each constant-head boundary node.

- No flow boundaries are specified in the eastern boundary and right part of the northern boundary, where it forms the watershed boundary of Baga and Nerul rivers.

- Southern boundary (Mandovi river) and left part of the northern boundary (Chapora river) are described by third-kind (Cauchy) boundary condition, Transfer. Internal flow boundaries (Baga river and Nerul river) are also described by Transfer boundary condition.

The boundary conditions and initial concentration for the transient state solute transport are dependent on the flow simulation results. For this model, solute transport concentrations are expressed in terms of total dissolved solids (TDS). A concentration of 35,425 mg/l (seawater TDS) is used along the coastal zone where simulated inflow from ocean occurs (mass boundary of $1^{st}$ kind). The initial concentration of the groundwater was set to 0 mg/l.

The aquifer geometry was adopted, as defined in previous studies. Reference zero elevation was assumed at 50 m below mean sea level. Only few measured hydrodynamic data are available and incorporated in the model. Four values of hydraulic conductivity ranging from $0.381 \times 10^{-4}$ to $3.657 \times 10^{-4}$ m/s were measured through pumping tests. Data regionalization for hydraulic conductivity over the study area has been carried out using Akima inter/extrapolation. No measurement of dispersivity has been made, this parameter was therefore estimated by trial and error using prior information from similar cases. Molecular diffusion was assumed as $1.00 \times 10^{-9}$ $m^2$/s. Initial head data have been measured in 20 observation wells. Data regionalization for hydraulic heads over the study area has been carried out using Akima inter/extrapolation.

The transient state simulation of the solute transport was carried out using automatic time step control via predictor-corrector schemes, with initial time step length as 0.001 day and final time as 3650 days (10 years) to reach steady state conditions. Calibration objective for the mass transport was focused mainly at observation wells near Anjuna and Baga beaches and Baga river where resistivity survey has indicated the presence of brackish water.

Mean annual rainfall is estimated to be 2714 mm, based upon daily rainfall data of Panaji for 20 years (1984 to 2003). Rainfall recharge values for laterite and west coast were adopted as 7% and 10% respectively, as recommended by "Groundwater Resource Estimation Methodology - 1997". Annual groundwater draft for the study area was worked out by using the reported density of wells as 25 wells per $km^2$ and average annual groundwater draft per structure as 0.65 ha-m. Porosity for sandy alluvium, laterite and clay were assumed to be 0.32, 0.21 and 0.42 respectively. Specific yield for sandy alluvium, laterite and clay were assumed to be 0.16, 0.025 and 0.03 respectively.

Longitudinal and transverse dispersivity were modified uniformly by trial and error in order to match the measured salinity values from the observation wells. Several runs were carried out to approach the solution. Final calibrated longitudinal and transverse dispersivity are 50 m and 5 m respectively. The calibration process shows that the mass transport model is sensitive to the dispersivity values.

Three-dimensional plot for mass distribution has been presented in figure 3. It indicates 3 peaks where salinity near the coast exceeds 6000 mg/l. Along these three sections, the salinity of groundwater was found to be greater than 500 mg/l upto 300 m inland, the maximum (near the coast) being 9400 mg/l, 9600 mg/l and 6800 mg/l respectively. The computed salinity in the aquifer show a sharp decrease of salinity from the coast towards inland. As an example, for the middle section, the salinity varies from 9,600 mg/l to 500 mg/l from the coastal front to a distance 300 m, as shown in figure 4. The model was not fully calibrated because of uncertainties in the hydrodynamic flow and mass transport data used. However, the results show that the density dependent 3D model is reasonable.

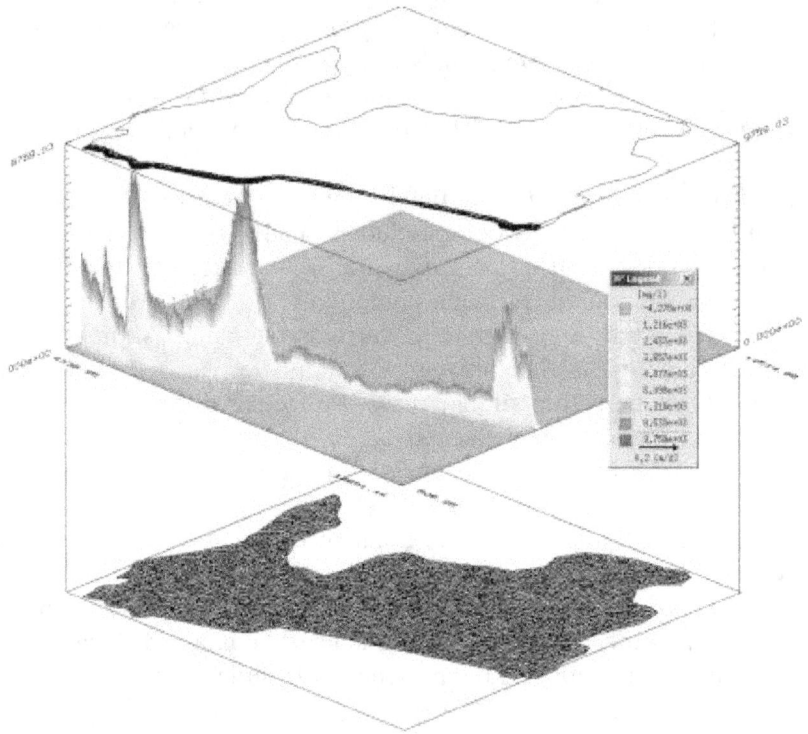

Figure 3:     Three-dimensional Plot for Mass Distribution

*Conclusion*

The above results indicate that presently, seawater intrusion is confined only upto 300 m from the coast under normal rainfall conditions and present draft pattern. It may be slightly more for low rainfall years. However, seawater intrusion may further advance inland if withdrawals of groundwater by builders, hotels and other tourist establishments continue to increase in the coming years. Therefore, corrective measures with proper planning and management of groundwater resources in the area need to be initiated at this stage so that it may not turn to be a major water quality problem in the coming times. This study will guide in making management decisions to monitor and control seawater intrusion and planning of groundwater development in the area.

Mass distribution in [mg/l] along indicated section (Linear plot):

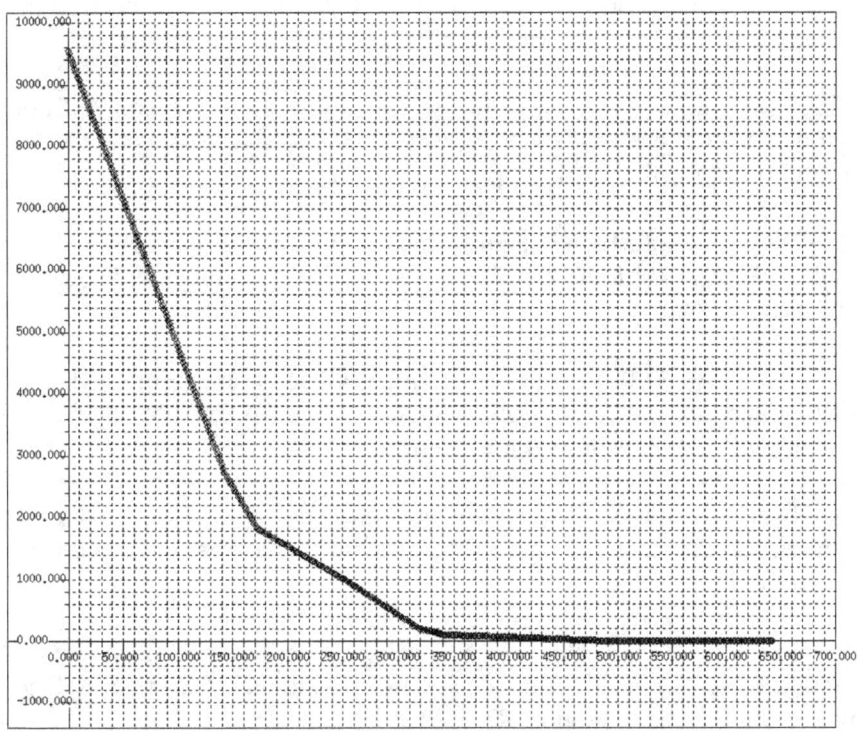

Figure 4:     Mass Distribution along a Section

**References**

1.     Chachadi, A.G. (2000). Strategies for  Water Resources Management in the Coastal Zones of Goa. Coastal  Zone Management, S.D.M.C.E.T, Dharwad & IGCP-367, Special Publication, Volume No. 2, pp. 165-168.

2.     Chachadi, A.G. and C. Joaquim S. John (2000). Geoelectrical Studies  in Ascertaining Fresh-Water Zones in Coastal Goa. Coastal  Zone Management, S.D.M.C.E.T, Dharwad & IGCP-367, Special Publication, Volume No. 2, pp. 169-172.

3.     Chachadi, A. G., J. P. Lobo-Ferreira, Ligia Noronha and B. S. Choudri (2002). Assessing the Impact of Sea-Level Rise on Salt Water Intrusion in Coastal Aquifers using GALDIT Model, Coastin – November 2002, TERI, pp. 27-32.

4.     Cheng, Alexander H.-D. (2003).  Groundwater, Saltwater Intrusion in. Encyclopedia of Water Science, Marcel Dekker, Inc., pp. 404-406.

5.     COASTIN (2001). GIS and Mathematical Modelling for the Assessment of Groundwater Vulnerability to Pollution: Application to an India Case – Study Area in Goa. 2nd Year Report,  April 2001, 69 p.

6.  Diersch, H.-J. G. (2002). FEFLOW Reference Manual, Finite Element Subsurface Flow & Transport Simulation System. WASY Institute for Water Resource Planning and Systems Research Ltd., Berlin, 278 p.

7.  FEFLOW (Finite Element Subsurface Flow & Transport Simulation System) Demonstration Exercise, WASY Institute for Water Resource Planning and Systems Research Ltd., Berlin, 2004, 48 p.

8.  FEFLOW 5.1 (Finite Element Subsurface Flow & Transport Simulation System) User's Manual, WASY Institute for Water Resource Planning and Systems Research Ltd., Berlin, 2004, 168 p.

9.  Goa University (2000). Aquifer Testing, Observation Well Networking, Electrical Profiling and Groundwater Draft Estimation, Baga Watershed, Bardez, Goa. Department of Geology, Goa University, Goa, 46 p.

10. Groundwater Resource Estimation Methodology - 1997. Report of the Groundwater Resource Estimation Committee, Ministry of Water Resources, Government of India, New Delhi, June 1997.

11. Kumar, C. P. and B. K. Purandara (2003). Modelling of Soil Moisture Movement in a Watershed using SWIM. Journal of The Institution of Engineers (India), Agricultural Engineering Division, Volume 84, December 2003, pp. 47-51.

12. Kumar, C. P., A. G. Chachadi, B. K. Purandara, Sudhir Kumar and Raju Juyal (2007). Modelling of Seawater Intrusion in Coastal Area of North Goa. Water Digest, Vol. II, Issue 3, September-October 2007, pp. 80-83.

13. Lobo-Ferreira, J. P., Maria C. Cunha, A. G. Chachadi, Kai Nagel, Catarina Diamantino and Manuel M. Oliveira (2003). Application of Optimization Models for Satisfaction of Water Resource Demand of Tourist Infrastructure. In. Coastal Tourism, Environment, and Sustainable Local Development (Ed. Noronha et al.), Chapter 14, pp. 305-320.

14. Purandara, B. K. and C. P. Kumar (2000). Simulation of Soil Moisture Movement in a Forested Watershed. ICIWRM-2000, Proceedings of International Conference on Integrated Water Resources Management for Sustainable Development, 19-21 December, 2000, New Delhi, India, pp. 833-839.

15. Ross, Peter J. (1990). SWIM - A Simulation Model for Soil Water Infiltration and Movement. Reference Manual to SWIMv1, CSIRO Division of Soils, Australia, 1990.

16. Verburg Kirsten, Peter J. Ross and Keith L. Bristow (1996). SWIMv2.1 User Manual. CSIRO Division of Soils, Australia, Divisional Report No. 130, 1996.

17. van Genuchten, M. Th. (1980). A Closed-form Equation for Predicting the Hydraulic Conductivity of Unsaturated Soils. Soil Sci. Soc. Am. J., Volume 44, 1980, pp. 892-898.

# Online Hydrological Discussion Groups

## Hydrology Forum

This group aims to provide a forum for discussion of scientific research in all aspects of Hydrology including hydrologic design, surface water analysis and modelling, flood studies, drought studies, watershed development, groundwater assessment and modelling, conjunctive use, drainage, mountain hydrology, environmental hydrology, lake hydrology, nuclear hydrology, urban hydrology, forest hydrology, hydrological investigations, remote sensing and GIS applications etc. Exchange of ideas and experiences regarding use and application of hydrological softwares are also welcome.

| | | |
|---|---|---|
| Group home page | : | https://groups.yahoo.com/neo/groups/hydforum/info |
| Post message | : | hydforum@yahoogroups.com |
| Subscribe | : | hydforum-subscribe@yahoogroups.com |
| Unsubscribe | : | hydforum-unsubscribe@yahoogroups.com |
| List owner | : | hydforum-owner@yahoogroups.com |

## Hydrological Modelling

This group aims to provide a forum for discussion of scientific research in modelling of hydrologic systems including

*     Rainfall-Runoff Modelling
*     Overland Flow Modelling
*     Unsaturated Flow Modelling
*     Groundwater Flow Modelling
*     Solute Transport Modelling
*     Sea Water Intrusion Modelling
*     Water Quality Modelling
*     Other relevant aspects of Hydrology and Water Resources Development and Management.

It intends to provide a forum for technical discussions; announcement of new public domain and commercial softwares; calls for abstracts and papers; conference and workshop announcements; and summaries of research results, recent publications, and case studies. Exchange of ideas and experiences regarding use and application of hydrological softwares are also welcome.

| | | |
|---|---|---|
| Group home page | : | https://in.groups.yahoo.com/neo/groups/hydrologymodel/info |
| Post message | : | hydrologymodel@yahoogroups.co.in |
| Subscribe | : | hydrologymodel-subscribe@yahoogroups.co.in |
| Unsubscribe | : | hydrologymodel-unsubscribe@yahoogroups.co.in |
| List owner | : | hydrologymodel-owner@yahoogroups.co.in |

## Ground Water Modelling

An email discussion group related to ground water modelling and analysis. This group is a forum for the communication of all aspects of ground water modelling including technical discussions; announcement of new public domain and commercial softwares; calls for abstracts and papers; conference and workshop announcements; and summaries of research results, recent publications, and case studies.

| | | |
|---|---|---|
| Group home page | : | https://groups.yahoo.com/neo/groups/gwmodel/info |
| Post message | : | gwmodel@yahoogroups.com |
| Subscribe | : | gwmodel-subscribe@yahoogroups.com |
| Unsubscribe | : | gwmodel-unsubscribe@yahoogroups.com |
| List owner | : | gwmodel-owner@yahoogroups.com |

## Groundwater Research and Management

In many regions of the world, the groundwater resource base and the social, economic and environmental systems dependent on it, are under threat from over-abstraction and pollution. Groundwater management, beyond pure development of the resource, is in its incipient phase in many countries and traditionally, management is only weakly hinged on research. When it comes to solving the problems, and putting into place management schemes, the impediments are numerous. Besides the lack of basic knowledge and information of the groundwater resource, knowledge of possible management options is also inadequate.

This group aims to provide an interaction between groundwater researchers and managers to synthesize their knowledge, perceptions and ideas for improved groundwater management and research.

| | | |
|---|---|---|
| Group home page | : | https://in.groups.yahoo.com/neo/groups/gwrm/info |
| Post message | : | gwrm@yahoogroups.co.in |
| Subscribe | : | gwrm-subscribe@yahoogroups.co.in |
| Unsubscribe | : | gwrm-unsubscribe@yahoogroups.co.in |
| List owner | : | gwrm-owner@yahoogroups.co.in |

## Roorkee Hydrology Group

This group aims at interaction amongst professionals, academicians, research scholars and students working in the field of Hydrology and Water Resources. The Roorkee city has got unique distinction of having a number of leading organisations in the water resources sector:

1. Water Resources Development and Management (IIT, Roorkee)
2. Department of Hydrology (IIT, Roorkee)
3. Department of Earth Sciences (IIT, Roorkee)
4. National Institute of Hydrology

5. Irrigation Research Institute
6. Alternate Hydro Energy Centre (IIT, Roorkee)

| | | |
|---|---|---|
| Group home page | : | https://groups.yahoo.com/neo/groups/rhydrology/info |
| Post message | : | rhydrology@yahoogroups.com |
| Subscribe | : | rhydrology-subscribe@yahoogroups.com |
| Unsubscribe | : | rhydrology-unsubscribe@yahoogroups.com |
| List owner | : | rhydrology-owner@yahoogroups.com |

**Isotope Hydrology**

Isotope hydrology is a field of hydrology that uses isotopic dating to estimate the age and origins of water and its movement within the hydrologic cycle. Water molecules carry unique fingerprints, based in part on differing proportions of the oxygen and hydrogen isotopes that constitute all water. Isotopes are forms of the same element that have variable numbers of neutrons in their nuclei.

The isotope techniques have potential to trace the complete hydrological cycle and processes that take place during the exchange and transition of water from one phase to another phase. Eventually, Isotope Hydrology has developed into a multi-disciplinary field. With the development of sophisticated and automated instrumentation for very precise isotopic measurements, new approaches are being developed and new applications and tools are being added to the isotope toolbox. Now isotope techniques can effectively be used in carrying out various hydrological studies.

This group aims to provide a forum for discussion of scientific research and laboratory investigations for application of environmental isotopes (stable and radioactive) in hydrology and management of water resources.

| | | |
|---|---|---|
| Group home page | : | https://groups.yahoo.com/neo/groups/isotope-hydrology/info |
| Post message | : | isotope-hydrology@yahoogroups.com |
| Subscribe | : | isotope-hydrology-subscribe@yahoogroups.com |
| Unsubscribe | : | isotope-hydrology-unsubscribe@yahoogroups.com |
| List owner | : | isotope-hydrology-owner@yahoogroups.com |

**Decision Support Systems in Hydrology**

The decision support system (DSS) includes information systems that perform data acquisition, management and visualisation; and models that perform simulation and optimisation of the system.

This group aims to provide a forum for exchange of ideas and experiences regarding development and use of decision support systems in surface water planning; integrated operation of reservoirs; conjunctive surface water and groundwater planning; drought monitoring, assessment and management; water quality; and other aspects of hydrology and water resources development and management.

| Group home page | : | https://in.groups.yahoo.com/neo/groups/dsshydrology/info |
| --- | --- | --- |
| Post message | : | dsshydrology@yahoogroups.co.in |
| Subscribe | : | dsshydrology-subscribe@yahoogroups.co.in |
| Unsubscribe | : | dsshydrology-unsubscribe@yahoogroups.co.in |
| List owner | : | dsshydrology-owner@yahoogroups.co.in |

## FEFLOW Users Group

FEFLOW (Finite Element subsurface FLOW system) is a sophisticated software package for the modeling of flow and transport processes in porous media under saturated and unsaturated conditions. Integral components are interactive graphics, a GIS interface, data regionalization and visualisation tools and powerful numeric techniques. These components ensure an efficient working process building the finite element mesh, assigning model properties and boundary conditions, running the simulation, and visualizing the results.

This group aims to provide a forum for exchange of ideas and experiences regarding use and application of FEFLOW software.

| Group home page | : | https://in.groups.yahoo.com/neo/groups/feflow/info |
| --- | --- | --- |
| Post message | : | feflow@yahoogroups.co.in |
| Subscribe | : | feflow-subscribe@yahoogroups.co.in |
| Unsubscribe | : | feflow-unsubscribe@yahoogroups.co.in |
| List owner | : | feflow-owner@yahoogroups.co.in |

## Visual MODFLOW Users Group

Visual MODFLOW is a proven standard for professional 3D groundwater flow and contaminant transport modeling using MODFLOW-2000, MODPATH, MT3DMS AND RT3D. Visual MODFLOW seamlessly combines the standard Visual MODFLOW package with Win PEST and the Visual MODFLOW 3D-Explorer to give a complete and powerful graphical modeling environment.

This group aims to provide a forum for exchange of ideas and experiences regarding use and application of Visual MODFLOW software.

| Group home page | : | https://in.groups.yahoo.com/neo/groups/visual-modflow/info |
| --- | --- | --- |
| Post message | : | visual-modflow@yahoogroups.co.in |
| Subscribe | : | visual-modflow-subscribe@yahoogroups.co.in |
| Unsubscribe | : | visual-modflow-unsubscribe@yahoogroups.co.in |
| List owner | : | visual-modflow-owner@yahoogroups.co.in |

## Groundwater Vistas Users Group

Groundwater Vistas (GV) is a sophisticated Windows graphical user interface for 3-D groundwater flow & transport modeling. GV couples a powerful model design system with comprehensive graphical analysis tools. GV is a model-independent graphical design system for MODFLOW MODPATH (both steady-state and transient versions), MT3DMS, MODFLOWT, MODFLOW-SURFACT, MODFLOW2000, GFLOW, RT3D, PATH3D, SEAWAT and PEST, the model-independent calibration software. The combination of PEST and GV's automatic sensitivity analysis make GV a powerful calibration tool.

This group aims to provide a forum for exchange of ideas and experiences regarding use and application of Groundwater Vistas software.

| | | |
|---|---|---|
| Group home page | : | https://in.groups.yahoo.com/neo/groups/groundwater-vistas/info |
| Post message | : | groundwater-vistas@yahoogroups.co.in |
| Subscribe | : | groundwater-vistas-subscribe@yahoogroups.co.in |
| Unsubscribe | : | groundwater-vistas-unsubscribe@yahoogroups.co.in |
| List owner | : | groundwater-vistas-owner@yahoogroups.co.in |

**GMS Users Group**

GMS (Groundwater Modeling System) is an advanced, powerful, and comprehensive groundwater modeling package. GMS provides complete support for the USGS MODFLOW 3D finite difference, MODPATH 3D particle tracking, MT3DMS 3D multi-species contaminant transport, RT3D 3D bioremediation transport, SEAM3D 3D bioremediation transport, SEEP2D 2D finite element, and FEMWATER 3D finite element groundwater models. Tools are provided for site characterization, model conceptualization, mesh and grid generation, geostatistics, telescopic model refinement, automated model calibration, and output post-processing. The program allows common information to be shared among different data types and groundwater models. GMS is completely graphical, and can display a defined groundwater model in either plan view or 3D oblique view, and can be rotated interactively. Tools are provided for site characterization, model conceptualization, mesh and grid generation, geostatistics, and output post-processing.

This group aims to provide a forum for exchange of ideas and experiences regarding use and application of GMS software.

| | | |
|---|---|---|
| Group home page | : | https://in.groups.yahoo.com/neo/groups/gms-users/info |
| Post message | : | gms-users@yahoogroups.co.in |
| Subscribe | : | gms-users-subscribe@yahoogroups.co.in |
| Unsubscribe | : | gms-users-unsubscribe@yahoogroups.co.in |
| List owner | : | gms-users-owner@yahoogroups.co.in |

**SEAWAT Users Group**

The SEAWAT program was developed to simulate three-dimensional, variable-density, transient ground-water flow in porous media. The source code for SEAWAT was developed by combining MODFLOW and MT3DMS into a single program that solves the coupled flow and solute-transport equations. The SEAWAT code follows a modular structure, and thus, new capabilities can be added with only minor modifications to the main program. SEAWAT reads and writes standard MODFLOW and MT3DMS data sets, although some extra input may be required for some SEAWAT simulations. This means that many of the existing pre- and post-processors can be used to create input data sets and analyze simulation results. Users familiar with MODFLOW and MT3DMS should have little difficulty applying SEAWAT to problems of variable-density ground-water flow. SEAWAT-2000 is the latest release of the SEAWAT computer program for simulation of three-dimensional, variable-density, transient ground-water flow in porous media. SEAWAT-2000 was designed by combining a modified version of MODFLOW-2000 and MT3DMS into a single computer program.

This group aims to provide a forum for exchange of ideas and experiences regarding use and application of SEAWAT software.

| Group home page | : | https://in.groups.yahoo.com/neo/groups/seawat/info |
|---|---|---|
| Post message | : | seawat@yahoogroups.co.in |
| Subscribe | : | seawat-subscribe@yahoogroups.co.in |
| Unsubscribe | : | seawat-unsubscribe@yahoogroups.co.in |
| List owner | : | seawat-owner@yahoogroups.co.in |

**archydro - ArcGIS - Geographical Information System**

Geographical Information Systems (GIS) are used to display, manipulate and analyse spatial (map) data. ArcGIS (produced by ESRI) is an integrated collection of software products for building a complete geographic information system (GIS). There are three ArcGIS desktop applications - ArcCatalog, ArcMap, and ArcToolbox. ArcCatalog is the application for managing your spatial data holdings, for managing your database designs, and for recording and viewing metadata. ArcMap is used for all mapping and editing tasks, as well as for map-based analysis. ArcToolbox is used for data conversion and geoprocessing. Using these three applications together, you can perform any GIS task, simple to advanced, including mapping, data management, geographic analysis, data editing, and geoprocessing. There are also server-based ArcGIS products, as well as ArcGIS products for PDAs. Extensions can be purchased separately to increase the functionality of ArcGIS.

GIS is a powerful tool for developing solutions for water resources such as assessing water quality and managing water resources on a local or regional scale. Hydrologists use GIS technology to integrate various data and applications into one, manageable system. ArcGIS with Arc Hydro gives you the flexibility to combine watershed datasets from one map source with stream and river networks. The suite of tools contained in Arc Hydro facilitate the creation, manipulation, and display of hydro features and objects within the ArcGIS environment. Use ArcGIS Spatial Analyst for hydrologic analysis such as calculating flow across an elevation surface, which provides the basis for creating stream networks and watersheds; calculating flow path length; and assigning stream orders.

This group aims to provide a forum for exchange of ideas and experiences regarding application of GIS in Hydrology and Water Resources; and use of ArcGIS software (in general) and Arc Hydro (in particular).

| Group home page | : | https://groups.yahoo.com/neo/groups/archydro/info |
| Post message | : | archydro@yahoogroups.com |
| Subscribe | : | archydro-subscribe@yahoogroups.com |
| Unsubscribe | : | archydro-unsubscribe@yahoogroups.com |
| List owner | : | archydro-owner@yahoogroups.com |

## Hydrology

This group aims to provide a forum for discussion of scientific research in all aspects of Hydrology and modelling of hydrologic processes. Exchange of ideas and experiences regarding use and application of hydrological softwares are also welcome.

| Homepage | : | http://groups.google.com/group/hydrology/ |
| Group email | : | hydrology@googlegroups.com |
| Owner email | : | hydrology-owner@googlegroups.com |

## Groundwater Assessment and Modelling

This group aims to provide a forum for discussion of scientific research in all aspects of ground water assessment and modelling. Exchange of ideas and experiences regarding use and application of groundwater softwares are also welcome.

| Homepage | : | http://groups.google.com/group/groundwater/ |
| Group email | : | groundwater@googlegroups.com |
| Owner email | : | groundwater-owner@googlegroups.com |

## Water Resources

A discussion group on water resources development and management.

| Homepage | : | http://groups.google.com/group/water-resources/ |
| Group email | : | water-resources@googlegroups.com |
| Owner email | : | water-resources-owner@googlegroups.com |

## Indian Hydrology

A forum for discussion of hydrological problems in India.

| Homepage | : | http://groups.google.com/group/indian-hydrology/ |
| Group email | : | indian-hydrology@googlegroups.com |
| Owner email | : | indian-hydrology-owner@googlegroups.com |

## MODFLOW Users Group

A forum for exchange of ideas and experiences regarding use and application of MODFLOW - a modular three-dimensional finite-difference ground-water flow model.

| Homepage | : | http://groups.google.com/group/modflow/ |
|---|---|---|
| Group email | : | modflow@googlegroups.com |
| Owner email | : | modflow-owner@googlegroups.com |

## Modelling of Coastal Aquifers

A forum for discussion of hydrological problems in coastal regions and modelling of seawater intrusion.

| Homepage | : | http://groups.google.com/group/coastal/ |
|---|---|---|
| Group email | : | coastal@googlegroups.com |
| Owner email | : | coastal-owner@googlegroups.com |

# Groundwater Assessment and Modelling

Groundwater development has shown phenomenal progress during past few decades. There has been a vast improvement in the perception, outlook and significance of groundwater resource. The unplanned groundwater development may result in fall of water levels, failure of wells, and salinity ingress in coastal areas. The development and over-exploitation of groundwater resources in many parts of the world have raised the concern and need for judicious and scientific resource management and conservation. This book "Groundwater Assessment and Modelling" is intended to provide a comprehensive treatise on assessment and modelling of groundwater. It presents the related aspects on assessment of groundwater potential, groundwater data requirement and analysis, basic concepts and guidelines for groundwater modelling, groundwater modelling software, modelling of unsaturated flow, modelling of sea water intrusion, and impact of climate change on groundwater resources. The book is expected to be quite useful for undergraduate and postgraduate students (water resources engineering), field engineers and researchers working in the area of assessment, development and management of groundwater resources.

## ABOUT THE AUTHOR

**Mr. C. P. Kumar** post-graduated in Hydraulic Engineering from University of Roorkee in 1985. From 1985, he has been working as a Scientist for National Institute of Hydrology (NIH), Roorkee - 247667 (Uttarakhand), India. His major research areas of interest include assessment of groundwater potential; seawater intrusion in coastal aquifers; numerical modelling of unsaturated flow, groundwater flow and contaminant transport; management of aquifer recharge; and impact of climate change on groundwater. He has authored more than 100 technical papers and reports. You may visit his website at http://www.angelfire.com/nh/cpkumar/

www.ingramcontent.com/pod-product-compliance
Lightning Source LLC
Chambersburg PA
CBHW080758180526
45168CB00006B/2252